Gorda Ridge

Gregory R. McMurray
Editor

Gorda Ridge

A Seafloor Spreading Center in the United States' Exclusive Economic Zone

Proceedings of the Gorda Ridge Symposium
May 11–13, 1987, Portland, Oregon

Sponsored by
The Gorda Ridge Technical Task Force

With 122 Illustrations in 179 Parts

Springer-Verlag New York Berlin Heidelberg
London Paris Tokyo Hong Kong

Dr. Gregory R. McMurray
Oregon Department of Geology and Mineral Industries, 910 State Office Building,
Portland, Oregon 97201;
Dames & Moore, 1750 SW Harbor Way, Portland, OR 97201, USA

On the front cover: Figure 5.7A. Schematic east-west section of SESCA area, p. 86.

Library of Congress Cataloging-in-Publication Data
Gorda Ridge Symposium (1987: Portland, Or.)
Gorda Ridge: a seafloor spreading center in the United States' exclusive economic zone: proceedings of the Gorda Ridge Symposium, May 11–13, 1987, Portland, Oregon/Gregory R. McMurry, editor; sponsored by the Gorda Ridge Technical Task Force.
p. cm.
ISBN-13: 978-1-4612-7937-2 e-ISBN-13: 978-1-4612-3258-2
DOI: 10.1007/978-1-4612-3258-2

1. Sulphides—Gorda Ridge—Congresses. 2. Hydrothermal vent ecology—Gorda Ridge—Congresses. 3. Sea-floor spreading—Gorda Ridge—Congresses. I. McMurray, Gregory R., 1925– . II. Gorda Ridge Technical Task Force. III. Title.
TN495.S9G67 1987
553.6'68—dc20 89-21586

Softcover reprint of the hardcover 1st edition 1990

Typeset by Asco Trade Typesetting Ltd., Hong Kong.

9 8 7 6 5 4 3 2 1

Foreword

This volume is the product of an innovative intergovernmental scientific investigation of Gorda Ridge, an ocean floor spreading center in the northeastern Pacific Ocean which is marked by dynamic geologic and biologic processes. The papers contained herein summarize the results of the investigation as reported at a symposium held in Portland, Oregon, in May 1987. The sections in this volume contain papers that describe the mineral potential of the study area, the technology used in the search for ocean floor minerals, and the benthic ecology of Gorda Ridge and other spreading centers. The interdisciplinary nature of the studies makes this volume especially valuable and comprehensive as a case history of recent study of an active ocean spreading center.

The studies reported in this volume are an outgrowth of the creation by Presidential proclamation of the Exclusive Economic Zone (EEZ) of the United States in March 1983, followed by a proposal to consider leasing of minerals in a broad area encompassing Gorda Ridge offshore of the states of Oregon and California. Gorda Ridge is the only known spreading center wholly within the EEZ of the United States. Important previous scientific discoveries of geologic processes and mineral deposits at other spreading centers suggested that Gorda Ridge might be the locus of potentially important mineral resources.

Public comments regarding the proposed lease sale led to the creation in 1984 of a technical task force appointed by federal and state government officials. The mission of the task force was to investigate the mineral and living resources along Gorda Ridge in order to ascertain the economic and environmental impacts of mineral leasing and possible subsequent development. The task force undertook a series of investigations between 1984 and 1988 using a combination of surface and submersible platforms to locate and sample massive sulfide deposits and associated biologic communities at multiple locations along the 300-km length of the ridge. The research program was designed by the task force during public meetings and involved cooperation among federal agencies, state governments, several universities, and individuals from numerous scientific disciplines.

Federal government agencies from the United States included the U.S. Geological Survey, Minerals Management Service, U.S. Bureau of Mines, National Marine Fisheries Service, National Oceanic and Atmospheric Administration, Environmental Protection Agency, U.S. Fish and Wildlife Service, U.S. Coast Guard, U.S. Army Corps of Engineers, and U.S. Navy. Oregon members represented state government organizations including the Oregon Department of Geology and Mineral Industries, Oregon Department of Land Conservation and Development, and Oregon Coastal Zone Management Association. California state agencies were the Division of Mines and Geology and the Office of the Secretary of Environmental Affairs. Academia was represented by scientists from Oregon State University, Scripps Institution of Oceanography, the University of California at Davis, Woods Hole Oceanographic Institution, the University of Washington, and the University of Hawaii.

The task force is especially grateful to the U.S. Navy, which provided both surface vessels and the deep submersible vehicle *Sea Cliff*. Surface-based oceanographic research was conducted from the *S.P. Lee*, which was made available by the U.S. Geological Survey; the NOAA ships *Surveyor* and *Discoverer*; and the *Wecoma*, which was provided by Oregon State University. The task force's mission could not have been successful without the logistical support of these organizations.

DONALD A. HULL, Cochair
Gorda Ridge Technical Task Force

Preface

The Gorda Ridge Symposium was conceived by the cochairs of the joint federal/state Gorda Ridge Technical Task Force during September 1986. The task force membership voiced support for the symposium idea at a December 1986 meeting in San Francisco, the general concept being that of an exposition of recent Gorda Ridge research results supported by synoptic reviews with which to develop a context for the Gorda papers. Telephone conferences among the task force cochairs, members, and myself resulted in this formal statement of the purpose of the Symposium:

The purpose of the Gorda Ridge Symposium is to present the results of marine polymetallic sulfide mineral resource investigations and biological studies conducted over the past three years. The objectives of the Symposium are to: (1) provide a forum for presentation and discussion of scientific results of recent studies and discoveries; (2) provide for public awareness, information and involvement; (3) provide industry with an awareness of the nature of the mineral discoveries so that they can appraise the leasing potential of the Gorda Ridge and advise the Task Force accordingly; (4) provide for an awareness in the scientific community of the significance of the discoveries and opportunities for future research; and (5) provide a set of recommendations to the Secretary of the Interior and to the Governors of California and Oregon regarding the leasing potential of mineral resources on the Gorda Ridge and the respective roles industry, government and the scientific community can play in further evaluating these resources.

The Symposium is intended as a series of progressive sessions. The first day, May 11, is designed to report the germane research results and set them into the appropriate context. The second day, May 12, is designed to examine the practical aspects of this new information and to review existing and future technology for conducting mineral exploration programs on the Gorda Ridge. The third day, May 13, is designed to elicit responses from the public on the leasing potential of the mineral resources, economic and environmental impacts of leasing, and the respective roles government, industry and the scientific community should play in conducting future programs on the Gorda Ridge. The Gorda Ridge Technical Task Force will convene on Thursday, May 14, in order to discuss the results of the Symposium and to draft a set of recommendations for the Secretary of the Interior and the Governors of the States.

This volume follows the general format of the symposium. The Introduction section is a review paper on the geology and geophysics of Gorda Ridge. The Gorda Ridge Mineral Potential section reports the major results of the task force exploration program. It includes two papers on northern Gorda

Ridge, one on Blanco Fracture Zone, and two on Escanaba Trough. To put the Gorda deposits into some perspective, this section also contains a paper comparing the Gorda Ridge and other known seafloor deposits with the Besshi deposits of Japan, one describing the Atlantis II Deep deposits in the Red Sea and one giving an economic perspective on the Gorda deposits. The Technology section covers developing technologies for undersea mineral exploration and also contains a forward-looking paper on possible mining technologies. The exploration tools include hydrography and geochemistry, passive and active geophysics, and deep-sea drilling. The papers in the Benthic Ecology section address both the preliminary results of work on Gorda Ridge and adjacent nonvent fauna, and summarize vent faunal investigations at other vent and seep areas. An update paper has been added that gives preliminary biological results of 1988 dives which discovered active hydrothermal vents in Escanaba Trough. Lastly, a paper summarizing the symposium results and reporting the recommendations and findings of the five symposium workshops comprises the Conclusions section.

Acknowledgement is due many persons who helped to scope and plan the Gorda Ridge Symposium and consequently this volume. I thank John B. Smith, James F. Davis, and Donald A. Hull, the task force cochairs, for their insightful guidance, patience, and support during the entire project; William R. Normark, Jack Dymond, and James M. Franklin for their assistance in scoping the geology sessions; Fred N. Spiess for helping to plan the technology session; and William G. Pearcy for assistance with the biology sessions. I am grateful to the session chairs for their guidance and editing efforts: TS Ary, Andrew G. Carey, Michael J. Cruickshank, Robert F. Dill, Sylvia A. Earle, James M. Franklin, J. Fred Grassle, John A. Hildebrand, Mark L. Holmes, Ray G. Hunter, Randolph A. Koski, Janet L. Morton, William G. Pearcy, Joseph L. Ritchey, Peter A. Rona, Craig R. Smith, and Gary L. Taghon. The manuscript benefited greatly from review comments supplied by Jo Alexander, Sandra Ridlington, and Joseph S. Cone. I especially thank Kari B. LaPorta for her administrative assistance and support during this project. I thank the National Coastal Resources Institute for administrative support during the final phases of publication. Finally, I acknowledge the support and advice of the editorial and production staff of Springer-Verlag New York, Inc.

Funding for the Gorda Ridge Symposium was provided by the Department of the Interior, Minerals Management Service, Office of Strategic and International Minerals, through Cooperative Agreement No. 14-12-0001-30168 with the State of Oregon, Department of Geology and Mineral Industries.

Gregory R. McMurray

Contents

Contributors

EDWARD T. BAKER
Pacific Marine Environmental Laboratory, NOAA, Seattle, WA 98115-0070, USA

MICHEL BOUDRIAS
Marine Biology Research Division, A-002, Scripps Institution of Oceanography, La Jolla, CA 92093, USA

ANDREW G. CAREY JR.
College of Oceanography, Oregon State University, Corvallis, OR 97331, USA

ALAN D. CHAVE
AT&T Bell Laboratories, Murray Hill, NJ 07974, USA

DAVID A. CLAGUE
Branch of Pacific Marine Geology, U.S. Geological Survcy, MS 999, Menlo Park, CA 94025, USA

ROBERT W. COLLIER
College of Oceanography, Oregon State University, Corvallis, OR 97331, USA

MICHAEL J. CRUICKSHANK
Marine Minerals Technology Center, University of Hawaii, Look Laboratory, Honolulu, HI 96814, USA

DANIEL DESBRUYÈRES
IFREMER, Centre de Brest, 29273 Brest Cedex, France

MARTIN R. FISK
College of Oceanography, Oregon State University, Corvallis, OR 97331, USA

CHRISTOPHER G. FOX
NOAA Marine Resources Research Division, Mark O. Hatfield Marine Science Center, Newport, OR 97365, USA

J. Frederick Grassle
Woods Hole Oceanographic Institution, Woods Hole, MA 02543, USA

Roger Hart
College of Oceanography, Oregon State University, Mark O. Hatfield Marine Science Center, Newport, OR 97365, USA

Benjamin W. Haynes
Division of Environmental Technology, U.S. Bureau of Mines, Washington, DC 20241, USA

Robert R. Hessler
Marine Biology Research Division, A-002, Scripps Institution of Oceanography, La Jolla, CA 92093, USA

John A. Hildebrand
Marine Physical Laboratory, A-005, Scripps Institution of Oceanography, La Jolla, CA 92093, USA

Jochen Hoefs
Geochemisches Institut der Universität Göttingen, D-3400 Göttingen, West Germany

Mark L. Holmes
U.S. Geological Survey, Oceanography WB-10, University of Washington, Seattle, WA 98195, USA

Katherine J. Howard
College of Oceanography, Oregon State University, Corvallis, OR 97331, USA

S. Kim Juniper
Department of Oceanography, University of Quebec at Rimouski, Rimouski, Quebec G5L 3A1, Canada

Randolph A. Koski
Branch of Pacific Marine Geology, U.S. Geological Survey, MS 999, Menlo Park, CA 94025, USA

Gregory R. McMurray
Dames & Moore, Portland, OR 97201, USA

Janet L. Morton
Branch of Pacific Marine Geology, U.S. Geological Survey, MS 999, Menlo Park, CA 94025, USA

William R. Normark
Branch of Pacific Marine Geology, U.S. Geological Survey, MS 999, Menlo Park, CA 94025, USA

Rosemarie F. Petrecca
Woods Hole Oceanographic Institution, Woods Hole, MA 02543, USA

Douglas Pyle
College of Oceanography, Oregon State University, Corvallis, OR 97331, USA

Joseph L. Ritchey
Western Field Operations Center, U.S. Bureau of Mines, Spokane, WA 99202, USA

Peter A. Rona
NOAA Atlantic Oceanographic and Meteorological Laboratories, Miami, FL 33149, USA

Stephanie L. Ross
Branch of Pacific Marine Geology, U.S. Geological Survey, MS 999, Menlo Park, CA 94025, USA

Patrick J.C. Ryall
Department of Geology, Dalhousie University, Halifax, Nova Scotia B3H 3J5, Canada

Scott A. Stebbins
Western Field Operations Center, U.S. Bureau of Mines, Spokane, WA 99202, USA

David L. Stein
College of Oceanography, Oregon State University, Corvallis, OR 97331, USA

Gary L. Taghon
College of Oceanography, Oregon State University, Corvallis, OR 97331, USA

Verena Tunnicliffe
Department of Biology, University of Victoria, Victoria, British Columbia V8W 2Y2, Canada

Cindy Lee Van Dover
Woods Hole Oceanographic Institution, Woods Hole, MA 02543, USA

Nicholas Wetzel
Western Field Operations Center, U.S. Bureau of Mines, Spokane, WA 99202, USA

Robert A. Zierenberg
Branch of Western Mineral Resources, U.S. Geological Survey, MS 901, Menlo Park, CA 94025, USA

Part I Introduction

1—Geology and Geophysics of Gorda Ridge

MARTIN R. FISK and KATHERINE J. HOWARD

Abstract

Gorda Ridge is a relatively short (300 km) segment of volcanic ocean ridge that extends from the Mendocino Fracture Zone (40° 20′ N) to the Blanco Fracture Zone (43° 00′ N) off the coast of Oregon and California. The whole length of the ridge is within 200 nautical miles of the United States and therefore falls within the recently proclaimed U.S. Exclusive Economic Zone. The recent discovery of hydrothermal sulfides at several locations along the ridge may provide impetus for further exploration. This chapter presents the geologic and geophysical background of Gorda Ridge and the surrounding area.

Introduction

The hypotheses of continental drift and sea floor spreading have evolved into the theory of plate tectonics that is now a main unifying theory of geology. Although details and consequences of this theory are still being developed, the main points are that (1) the surface of the earth is made of solid plates that move horizontally over the globe, (2) convection of the upper layers of the earth drives these lithospheric plates, and (3) the boundaries between adjacent plates are either zones of separation (volcanic rifts or ridges), zones of collision (island arcs and mountain belts), or zones of shear motion (transform faults).

The volcanic rifts between plates form a worldwide pattern of which about 65,000 km is below sea level. Surveys of limited areas of these rifts show that they are strings of individual volcanoes averaging 60 km in length. There are about 1,000 of these submarine rift volcanoes compared with a total of 500 active composite and shield volcanoes found above sea level. Each volcano is highest near its center and has two rifts that extend in opposite directions from the center along the ridge axis. The rifts from adjacent volcanoes may be joined end to end, or there may be small offsets where two rifts overlap (overlapping spreading centers). Adjacent volcanoes can grow and shrink in tandem; the growing volcano extending its rift (propagating rift) toward the shrinking volcano (dying rift).

Not all the rift volcanoes are lined up end to end. Large offsets between volcanoes are called fracture zones or transform faults. Transform faults are the boundaries between plates that are sliding past one another. Some of these offsets are only a few kilometers long (as in the overlapping spreading centers), but they can be hundreds of kilometers long. Transform faults are the sites of numerous earthquakes and occasionally of volcanic activity.

The collision of lithospheric plates usually forces one plate under the other. When an oceanic plate and a continental plate collide, the denser plate (oceanic) is driven under the lighter plate. This process is called subduction. An ocean trench (deep) forms as the plate sinks, and a row of volcanoes forms some 100

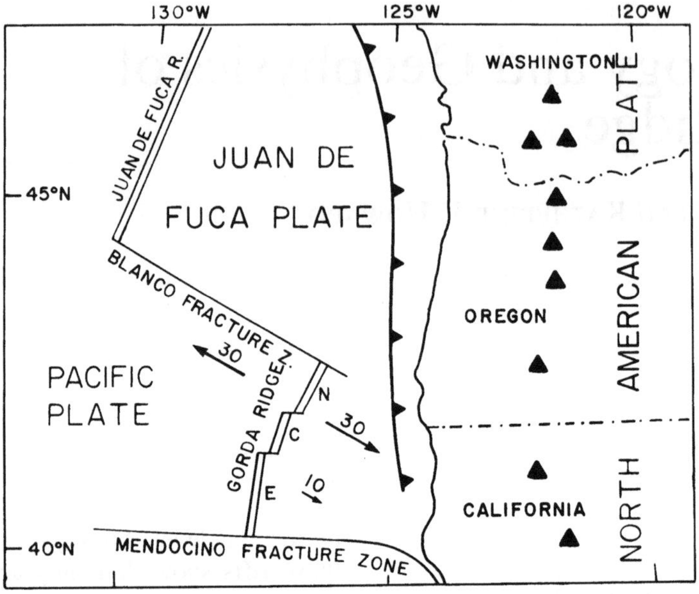

FIGURE 1.1. The lithospheric plates of the northeast Pacific are the North American Plate, the Juan de Fuca Plate, and the Pacific Plate. The boundaries between these plates are shown schematically. The Gorda and Juan de Fuca Ridges are noted by solid double lines, and the fracture zones are indicated with solid straight lines. The major fracture zones are the Mendocino and Blanco. The location of the zone of subduction of the Juan de Fuca Plate beneath the North American Plate is shown with a solid line with triangles. The directions of motion of the plates are shown with arrows, and their speed is given in mm/year. The Cascade volcanoes in California, Oregon, and Washington (triangles) are the result of subduction of the Juan de Fuca Plate beneath North America. Sections of Gorda Ridge labeled N, C, and E are shown in detail in Figures 1.3, 1.4, and 1.5, respectively.

to 300 kms from the trench. These collision zones are the site of frequent earthquakes, and the largest earthquakes occur in these zones.

It has been known since 1977 that the heat from the volcanoes along the submarine rift system produces hot springs (Corliss et al., 1979). These springs are sites of unusual forms of animal life and of mineral deposits that may be of economic value. The water above the springs often has a slightly elevated temperature and contains chemicals and particles that are characteristic of water discharged from the springs. The presence of these indicators can be used to locate hot springs (Baker, this volume).

All of the above features and consequences of plate tectonics are present in the northeastern Pacific Ocean basin. The focus of this volume is one 300-km-long segment of volcanic rift that separates the Pacific Plate from the southern end of the Juan de Fuca Plate. This rift zone, known as Gorda Ridge, is shown in Figure 1.1. Other boundaries between the Pacific Plate and the Juan de Fuca Plates are the Blanco Fracture Zone and the Mendocino Fracture Zone. The other major lithospheric plate in this region is the North American Plate, and the boundary between it and the Juan de Fuca Plate is a subduction zone (Fig. 1.1).

Geography

The major topographic features of the Northeast Pacific region; Gorda Ridge, the Blanco Fracture Zone, and the Mendocino Fracture Zone, have been the topics of investigation for over 20 years (Krause et al., 1964; McManus, 1965, 1967; Atwater and Mudie, 1968). In fact the ocean floor off the northwest coast of North America played a key part in the development of the theory of plate tectonics. The magnetic map of Raff and Mason (1961) was the first demonstration of variations in magnetic intensity in the oceanic crust that could be interpreted as being caused by sea floor spreading at an ocean ridge (Vine and Mathews, 1963). Largely because of this, Gorda Ridge was also one of the first ridges dredged for volcanic rock to test the corollary to the sea-floor spreading hypothesis that ocean ridges should be covered with young volcanic rocks.

The first major compilation of bathymetric and geophysical data for the region between 40° and 46° N off the coasts of Oregon and California was done by Wilde et al. (1978, 1979). The soundings for their topographic maps were collected from ships with imprecise navigation and with single-beam echo sounders, but these early surveys defined the major features of Gorda Ridge. The axial valley of Gorda Ridge resembled that of the Mid-Atlantic Ridge and was unlike the broad flat axis of Juan de Fuca Ridge to the north. Gorda Ridge was seen to be made of three segments, the northern being 100 km long, the middle 70 km long, and the southern 130 km long. Between the segments were offsets of about 10 km. Gorda Ridge was terminated in the north by Blanco Fracture Zone and in the south by the Mendocino Fracture Zone (Fig. 1.1).

The development of multibeam sounding techniques such as SEABEAM and SASS coupled with precise satellite navigation resulted in more accurate and more detailed maps of the ocean floor. Some maps have been published, such as that for the Blanco Fracture Zone (Embley et al., 1987), but many maps remain unpublished (Fox, personal communication), and up to now some survey data have been classified as secret by the U.S. Department of Defense.

High-resolution side scan sonar is another new acoustic technique that is capable of producing bathymetric maps of sea floor, and it also gives information about the type of sea floor being surveyed. It can distinguish between fresh basalt flows and talus and between rock and sediment. This tool was used to survey the northern 70% of the ridge by Clague and Holmes (1987), who revised the existing maps of the area. Their new map of Gorda Ridge showed that it was made of three distinct segments with short right-lateral offsets at 41° 37′ N and 42° 26′ N.

Geophysics

Magnetics

As mentioned above, the first map of the intensity of the earth's magnetic field at sea level over the northeastern Pacific Ocean (Raff and Mason, 1961) was an important step in the development of the theory of plate tectonics. The magnetic pattern, however, is not as regular as would be expected from uniform sea floor spreading. Along both sides of Juan de Fuca Ridge the bands of high- and low-magnetic intensity are parallel to the ridge, but along Gorda Ridge only the bands on the Pacific Plate are parallel to the ridge. On the eastern side of Gorda Ridge the magnetic pattern spreads out on a fan that pivots near the intersection of Gorda Ridge and the Mendocino Fracture Zone (Fig. 1.2).

This unusual pattern of sea floor magnetism indicates that the rate of spreading of the southern end of Gorda Ridge is about 20 mm/year while in the north it is 60 mm/year (Atwater and Mudie, 1973; Riddihough, 1980). This asymmetric spreading along the southern third of Gorda Ridge has continued for the past 2.5 Ma (Wilson, 1986; Riddihough, 1980, 1984). Active faults near the axis of Gorda Ridge just north of the Mendocino Fracture Zone indicate that the ridge is still actively spreading (Clague and Holmes, 1987). A variable spreading rate implies either that there is a transform fault that can accommodate these two different spreading rates (Riddihough, 1980, 1984) or that there is a diffuse zone of deformation in the southern Juan de Fuca Plate (Wilson, 1984; Stoddard, 1987). A transform fault would suggest that the southeast corner of the Juan de Fuca plate is subducting beneath the Mendocino Fracture Zone (Riddihough, 1980), but a diffuse zone of deformation across does not require subduction. Earthquakes from the Juan de Fuca Plate are concentrated along the Blan-

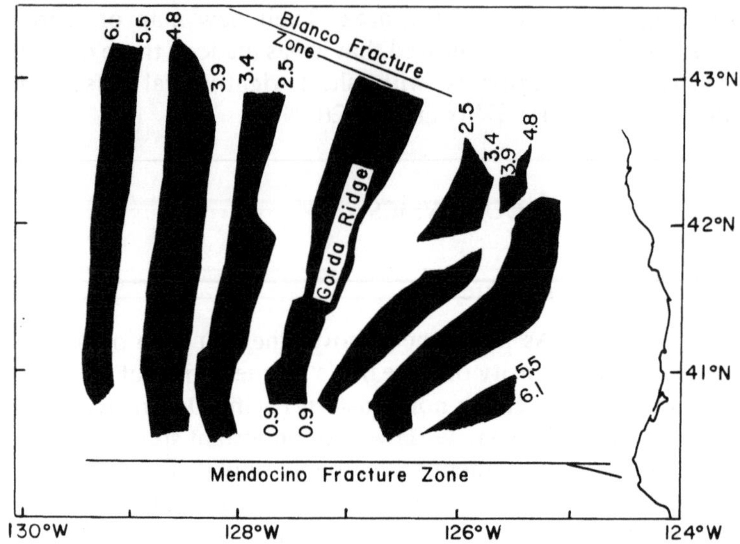

FIGURE 1.2. Map of the magnetic lineations on the Pacific and Juan de Fuca Plates along both sides of the Gorda Ridge. Ages of the lineations are given in millions of years and are derived from Wilson (1986). Fanning of the anomalies in the southeast corner of the Juan de Fuca Plate indicates a variable spreading rate along the southern two thirds of Gorda Ridge.

co and Mendocino Fracture Zones but are also spread over this fan-shaped area of magnetic stripes, which supports the hypothesis that the plate is being deformed (Wilson, 1986).

Heat Flow

Heat generated within the earth by the radioactive decay of potassium, uranium, and thorium results in a worldwide average heat flow out from the surface of the earth of 40 to 80 mW/m^2 (1.0 to 2.0 μcal/cm^2 sec). Within the ocean basins heat flow is highest near oceanic ridges, and about 60% of the earth's internal heat escapes in the vicinity of the ridges (Sclater et al., 1980). Part of this heat (40% of the earth's total) is lost by conduction through the crust (heat flow), and part (20%) is lost by convection of seawater through the crust (hydrothermal activity). Upward convection can be vigorous and localized at hot springs, and the downward convection may be slow and spread over large areas. For this reason, heat flow values at ocean ridge crests are highly variable. Regions of down-welling convection are expected to have anomalously low heat flow, while regions of up-welling convection will have anomalously high heat flow. Most ocean ridges including the Middle and Northern Gorda are free of sediment cover, and water can freely circulate downward through cracks. The Escanaba Trough (southern Gorda), however, is covered with sediment that prevents the rapid circulation of seawater into the oceanic crust.

The first heat flow measurements in the northeastern Pacific were made in the abyssal plain to the west of Gorda Ridge (Foster, 1962). These were highly variable, ranging from 4 to 150 mW/m^2. The 150 mW/m^2 measurement, however, agrees well with a measurement of 172 mW/m^2 made 16 km away by Von Herzen et al. (1964). Other early heat flow measurements made near the southern end of Gorda Ridge range from 160 to 250 mW/m^2 (Von Herzen et al., 1964). Their highest value was to the east of Escanaba Trough at 41° 30′ N, 126° 32′ W in the zone of deformation described by Wilson (1986), but one value within Escanaba Trough (40° 36′N, 127° 25′ W) was 230 mW/m^2. To the south of the Mendocino Fracture Zone near its intersection with Gorda Ridge, the values were much lower (80 mW/m^2).

Additional heat flow measurement was attempted in the southern Gorda Ridge during leg 5 site 35.1 (40° 52.35′ N, 127° 31.21 W) of the Deep Sea Drilling Project (DSDP) where the sediment was 250 to 300 m thick (Burns, 1970). The heat flow through the upper 119 m of sediment was 55 mW/m^2, similar to values for the abyssal plains to the west of Gorda Ridge but much lower than the Escanaba

Trough heat flow value (Von Herzen et al., 1964). Factors that decrease heat flow are the high sedimentation rates (which exists in the Escanaba Trough) and the downward convection of seawater (which could occur at site 35.1).

In 1985, 57 heat flow measurements were made in three transects in Escanaba Trough (Abbott et al., 1986). Two transects crossed the axial valley at 40° 45′ N and at 41° 00′ N; one transect ran along the axis of the valley from 40° 55′ N to 41° 02′ N. Eighteen measurements along the 20-km transect at 40° 45′ N had a background level of 250 mW/m^2 and a maximum of 2,000 mW/m^2 near the center of the valley. Fourteen measurements on the 8-km-long transect at 41° 00′ N had similar background and maximum values with the highest heat flow in the center of the valley. Fifteen measurements along the axis of the ridge show that the high heat flow values are not continuous along the axis of the ridge but are localized near areas where volcanic rock is protruding through the sediment.

The background value of 200 mW/m^2 (Abbott et al., 1986) is close to the value for this area determined by Von Herzen et al. (1964). The new measurements show that the heat flow is higher than the oceanic average, and in places the heat flow is eight times that measured by Von Herzen et al. (1964). The high heat flow appears to be localized in small areas around the small volcanic edifices in the valley. The background value of 200 mW/m^2 for the southern Gorda Ridge (Abbott et al., 1986) is four times that measured in site 35.1 of DSDP. This difference could be due to Site 35.1 being drilled in a region of recharge of the hydrothermal system.

Seismic Activity

Gorda Ridge axis is seismically active (Johnson and Jones, 1978; Solano-Berrego, 1982; Bibee, 1983) and some earthquakes are larger than magnitude six. Ocean bottom seismometers in the middle and northern median valleys recorded as many as 60 microseismic events per hour over short intervals with an average of five to seven events per hour over the duration of their measurements (Solano-Berrego, 1982; Bibee, 1983). For comparison, the Mid-Atlantic Ridge at 23° N had an average of less than one event per hour over the duration of a 10-day experiment (Toomey et al., 1985). The depths of some of these events are as much as 18 km below the ridge axis, but the average depth on the middle Gorda was about 12 km and on the northern Gorda was about 9 km (Solano-Borrego, 1982; Bibee, 1983). Many of the microearthquakes are probably related to fracturing of rock related to magma movement in the axial zone (Solano-Borrego, 1982), which suggests that deep magma storage and brittle crust extends to 18 km beneath the axial valley. An appreciable number of earthquakes occur outside the axis of the ridge, which supports the hypothesis that the Juan de Fuca Plate is deforming.

Gravity and Seismic Reflection

The gravitational field over Gorda Ridge varies slightly primarily because of the shape of the sea floor, but low-density material is required below the axial valley to account for the relatively low gravity field over the axis (Dehlinger et al., 1971). Multichannel seismic reflection of the Escanaba Trough shows that the sediments there are of variable thickness, possibly up to 1 km near the Mendocino Fracture Zone (Clague and Holmes, 1987). From this seismic survey there also appears to have been volcanic activity in the sediment of the Escanaba Trough because the sediment has been lifted and faulted. Faults along the southern end of the Escanaba Trough indicate that it is still spreading. Single-channel seismic reflection in the Escanaba Trough shows that volcanic rocks protrude from the sediment in several locations (Morton et al., 1987).

Geology

The shape of the ridge is similar to that of slowly spreading ridges in other oceans, such as the Mid-Atlantic Ridge, and the average depth of Gorda Ridge is similar to that of the Mid-Atlantic Ridge, but it is deeper than that of the adjacent Juan de Fuca Ridge. The average depth of water over axial valleys of spreading ridges of all oceans is about 2,700 m, and the depth of the nearby Juan de Fuca Ridge is 2,200 to 2,600 m, but Gorda Ridge axial valley is more than 3,000 m deep, and some depressions are as deep as 3,800 m (Clague and Holmes, 1987).

Volcanic Rocks

Through 1984 there were 137 published chemical analyses of whole rock and glass exteriors of basalts from Gorda Ridge, Blanco Fracture Zone, and Mendocino Fracture Zone (Engel and Engel, 1964; Scheidegger, 1973; Kay et al., 1970; Melson et al., 1976, 1977; Wakeham, 1978; Clague et al., 1984). Gorda Ridge basalts are noted for their low iron content relative to lavas from the Juan de Fuca Ridge (Kay et al., 1970; Scheidegger, 1973; Wakeham, 1978). Most of Gorda Ridge basalts have less than 10 wt% FeO, and no reported analyses have more than 12 wt%, while Juan de Fuca basalts rarely have less than 10% FeO and can have more than 14% (Delaney et al., 1981; Melson et al., 1976).

The basalts from Gorda Ridge do not have any remarkable chemical characteristics that set them apart from those from other ocean ridges. Their mineral content and mineral compositions are also typical of ocean ridges. The common minerals are olivine and plagioclase, but pyroxene and chrome spinel are sometimes present. One mineralogical feature of the basalts from Gorda Ridge is the occasional presence of abundant large crystals of plagioclase feldspar that make up 40% of the rock. An oddity of the basalts is the presence of gabbroic xenoliths in a few samples. Xenoliths in general are rare in oceanic rocks, and those found in Gorda Ridge basalts offer one of the few opportunities to look at rock fragments from deep within the ocean crust.

Northern Gorda

The northern segment of Gorda Ridge has been the center of extensive study for the past 9 years. Cruises there in 1980, 1981, 1983, 1985, 1986, and 1988 have lead to the discovery of a new hydrothermal vent field (Rona et al., 1988). During these cruises the ocean floor was mapped with SEABEAM and SeaMARC II sidescan sonar, basalts were dredged from the axial valley of Gorda Ridge and from the Blanco Fracture Zone, sediment cores and water samples were collected, and water temperature was measured. One dive with the submersible *Alvin* was made to the sea floor (Malahoff, 1985), and five additional dives were made in 1988 using the submersible *Sea Cliff* (Rona et al., 1988).

The northern Gorda Ridge appears to be a single rift volcano that has its highest elevation (3,000 m water depth) at 42° 47′ N (Fig. 1.3). At this location the rift valley is narrowest (about 7 km), and the axial valley walls are only about 500 m high (Clague and Holmes, 1987). From 42° 47′ N the valley gets wider and deeper, and the walls of the axial valley are higher, both to the north and to the south. At the intersection of the ridge and the Blanco Fracture Zone the axial valley is more than 15 km wide and has a depth of 3,500 m. South of the saddle the axial valley descends to its greatest depth, 3,800 m (Clague and Holmes, 1987), which may suggest that the magma supply is low in this area. The magnetic anomaly patterns generated along the northern Gorda (Fig. 1.2) suggest that the ridge propagated to the south (Wilson, 1984), and the deep at the southern tip of the ridge is also similar to that found on the Galapagos propagating rift.

Basalts were collected from this segment of the ridge in 1968 and 1976 by Oregon State University, and their chemistry and mineralogy were reported by Kay et al. (1970) and Wakeham (1978). Additional samples were collected in 1983 (Clague et al. 1984) and in 1985 and 1986. These rocks were primarily olivine-plagioclase basalts with minor amounts of chromium spinel, and a few of the samples also contain calcium pyroxene. A few samples close to the shallowest part of the northern ridge contained primary nickel-copper sulfides (Howard, 1989). The most unusual rocks were those from a dredge at 42° 48.5′ N, 126° 41.8′ W, which contained porphyritic basalts with abundant gabboic xenoliths (Davis and Clague, 1988). These xenoliths had four different mineralogies that could reflect chemical zona-

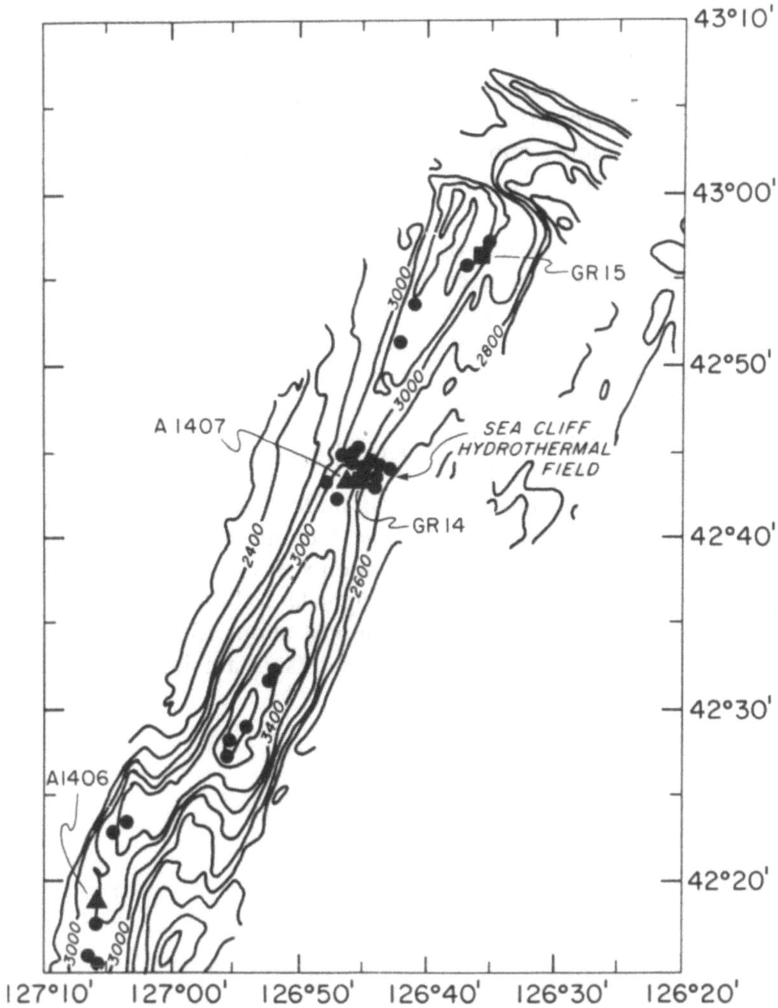

FIGURE 1.3. Northern Gorda Ridge showing axial valley and locations of dredged rocks (circles); *Alvin* dives A1406 and A1407 (triangles); sites of anomalous temperatures, manganese, and particle density, GR14 and GR15 (squares). The *Sea Cliff* hydrothermal field was found at GR14. Additional evidence of hydrothermal activity was found in a dredge at the south end of the northern Gorda Ridge. Two-hundred-meter contour intervals are drawn from data from Fox (personal communication).

tion in the crystalline mush zone of the rift magma chamber or sampling of wall rock during magma ascent.

The first indication of hydrothermal activity on the northern Gorda was the presence of slightly elevated temperatures, high manganese concentrations, and increased particle concentrations in the water within 400 m of the sea floor at 42° 44.7′ N, 126° 43.7′ W, and at 42° 57.1 N, 126° 35.0′ W (Collier et al., 1986; Baker et al., 1987). These were designated site GR14 and site GR15, respectively (Fig. 1.3), and both sites continued to exhibit these anomalous signals later the same year (Collier and Baker, this volume).

Another indication of hydrothermal activity on the northern Gorda was the presence of barite and pyrite on basalts dredged from 3,775 m at the south end of this segment (Clague et al., 1984). Both of these minerals are associated with hydrothermal activity on other oceanic ridges. In 1985 photographs of the axial valley near hydrothermal site GR14 showed that some rocks appeared to be covered with bacterial mats that often grow near areas of hydrothermal activity (Clague and Rona, this volume). Changes in sediment color and texture were also observed in the photographs, which suggested the presence of hydrothermal precipitates. One sediment core from the area also contained high copper content, another indication of hydrothermal action (Nelson, personal communication, 1987).

Thin, black to brightly colored layers of minerals were removed from rocks collected from the axial valley near site GR14 in 1985 and from rocks that had previously been collected from this area (Howard, 1989). These layers proved to contain minerals and chemical indicators that suggested the presence of hydrothermal activity in two areas near GR14. This area was again dredged in 1986, and basalts were recovered with a 1- to 3-mm-thick coating of boehmite.

Boehmite had previously been reported on the Middle Gorda (Clague et al., 1984; Clague and Holmes, 1987), but it was not known whether it was a hydrothermal precipitate or a weathering product of the basalt. Close examination of the sample indicated that the boehmite was a product of weathering of the basalt by reaction with silica-poor water that was 200 to 300°C (Howard and Fisk, 1988). The presence of hydrothermal vents near site GR14 was confirmed in dives with the submersible *Sea Cliff* in 1988 (Rona et al., 1988). The presence of a second hydrothermal system near GR14 and one near GR15 have yet to be confirmed.

Middle Gorda

The middle segment of Gorda Ridge is the most tectonically complex of the three ridge segments. It is bounded to the north and south by right-lateral offsets of about 7 km at 42° 26′ N and of about 15 km at 41° 35′ N (Fig. 1.4). These offsets are interpreted as either ridge-

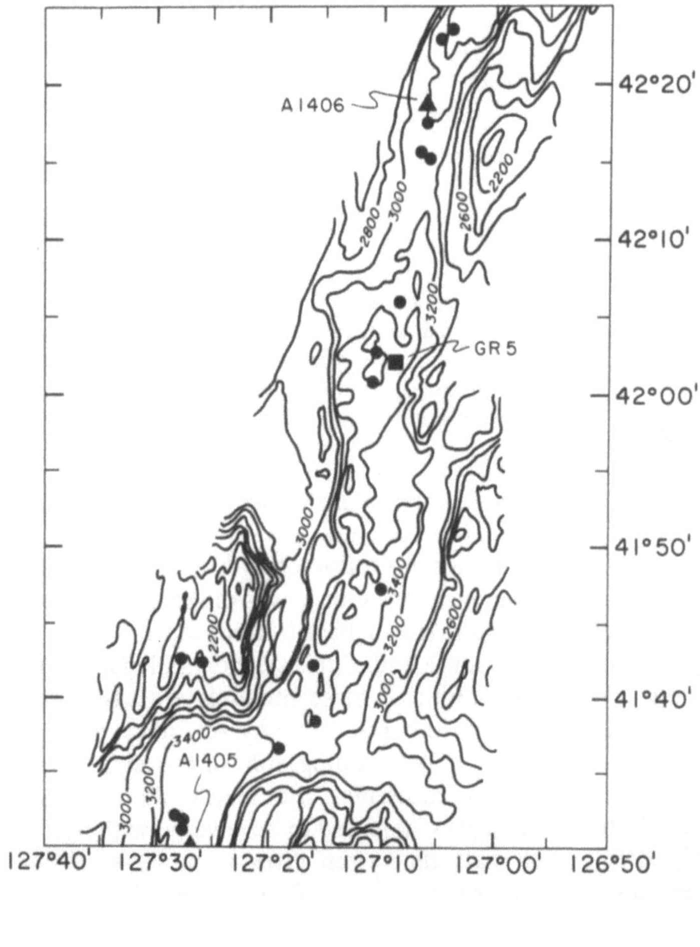

FIGURE 1.4. Central Gorda Ridge showing the location of *Alvin* dives 1405 and 1406; of dredge locations where basalts were recovered; and of site GR5 where water temperature, manganese, and particle densities were found. Symbols are the same as for Figure 1.3. Two-hundred-meter contour intervals are drawn from data from Fox (personal communication).

ridge-transform triple junctions or overlapping spreading centers (Riddihough, 1980; Malahoff, personal communication). At its widest point, seismic reflection records of the middle Gorda show many complex tectonic features such as domes and ridges (Clague and Holmes, 1987), suggesting that this segment of ridge is experiencing major deformation from plate reorientation. This segment of the Gorda is the western boundary of the zone of deformation of the Juan de Fuca Plate (Wilson, 1984), and thus the complexities of this ridge may result from deformation.

The northern third of the middle Gorda has many narrow fault scarps that parallel the ridge axis, but near 42° N high reflectivity in SeaMARC images 20 km east of the axis suggests off-axis volcanism along fractures subparallel to the axial valley trend (Clague and Holmes, 1987). Numerous small volcanoes punctuate the axial valley, and two depressions up to 3,700 m deep (at 42° 05′ N and 41° 50′ N) are second only to the 3,800-m depression in the northern Gorda. At the north end of the middle Gorda, the axial valley is about 5 km wide and overlaps with the southern end of the northern Gorda. Axial valley width increases southward until it is up to about 18 km wide, then at the southern offset the axial valley abruptly narrows again to about 5 km. The flanking ridges are between 1,000 and 1,800 m high.

Recently, samples of basalt from the axial valley were collected in seven USGS dredges (Clague et al., 1984; Davis and Clague, 1987), and other samples were dredged from the axial valley walls in the 1960s and 1970s (Kay et al., 1970; Wakeham, 1978). In addition, one *Alvin* dive (1406) collected basalt from the axial valley in 1984 (Malahoff, 1985). These basalts are chemically and mineralogically similar to those of the northern Gorda segment with one exception (Davis and Clague, 1987). A sample (9D-1) from the northern end of this segment has about 40% plagioclase feldspar megacrysts (1 cm diameter) that are highly anorthitic (up to 91.5% anorthite). These feldspars could not have been in equilibrium with their host magma, and these "xenocrysts" represent plagioclase accumulation. Davis and Clague (1987) report an unusually high strontium isotope signature in this sample.

No strong hydrothermal indicators have been found in this segment of ridge, but a small amount of the hydrothermal tracer, manganese, was measured 300 to 400 m from the bottom in 1985 at 42° 03.6′ N and 127° 08.3′ W (site GR5, Fig. 1.4). The concentration of manganese was only half the value found at sites GR14 and GR15 on the northern Gorda segment, and 6 days later this hydrothermal indicator was absent (Collier et al., 1986). Basalts collected in three dredge hauls near GR-5 have coatings of clay or other minerals that could be interpreted as indicators of hydrothermal activity (Clague et al., 1984). In two dredges near the southern end of the Middle Gorda boehmite was identified on the exteriors of basalts, possibly signifying the presence of 200 to 300°C hydrothermal activity (Howard and Fisk, 1988).

Escanaba Trough

The northern limit of Escanaba Trough is defined by a small right-lateral offset at 41° 30′ N (Fig. 1.5) where the axial valley narrows to 3 to 5 km (Riddihough, 1980). Escanaba Trough is bounded to the south by the Mendocino Fractured Zone (40° 20′ N), where the axial valley is up to 18 km wide. Within Escanaba Trough, the water depth over the axial valley floor averages about 3,200 m, and flanking ridges are 1,200 to 1,700 m high. At the southern end of Escanaba Trough the valley is wide and has upthrown blocks on each side (Atwater and Mudie, 1968; Heinrichs, 1970) while the northern end of the axial valley is narrow with steep sides.

Unlike the northern two segments, Escanaba Trough is filled with sediment (McManus, 1967). A gap between the southern end of the ridge and the Mendocino Fracture Zone creates a channel for turbidites originating on the North American continental margin (Moore, 1970). Much of the sedimentation probably occurred during Pleistocene low sea level stands (Vallier et al., 1973). During the Holocene, however, major continent-derived

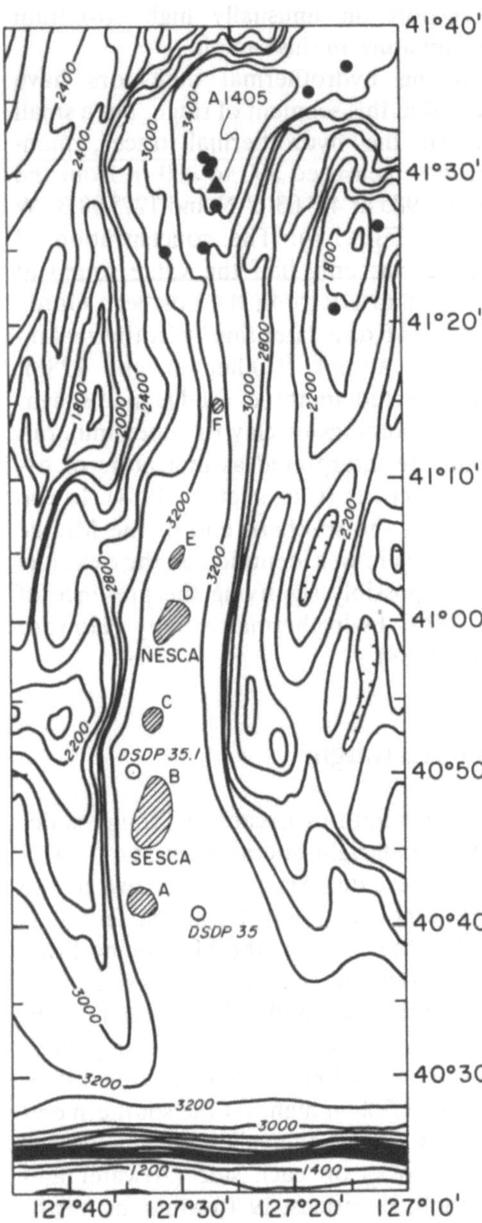

FIGURE 1.5. Escanaba Trough showing the location of dredge locations and *Alvin* dive 1405. Areas where volcanic rock protrudes through sediment are shown with diagonal ruling and are labeled from A to F. Sites of extensive study and dives (Morton et al., this volume; Holmes and Zierenberg, this volume) are labeled NESCA and SESCA. Two-hundred-meter contour intervals are drawn from data from Wilde et al. (1978).

turbidites are not observed, and sedimentation is largely hemipelagic (Karlin and Lyle, 1986).

During Deep Sea Drilling Project leg 5, an attempt was made to drill through this sediment to sample young oceanic crust, but after drilling 390 m of sand, slit, and clay, volcanic basement was not reached (McManus, 1970). The site was spot-cored to 390 m, collecting only 95.3 m of section. Fewer than half of the cores were opened and described, but based on the location of the drill site within the Brunhes magnetic interval, the minimum Pleistocene sedimentation rate is 56 cm/1,000 years (McManus et al., 1970). Layered sediment sections prominent in the seismic reflection records and in the down-hole logs at site 35 are large Pleistocene turbidites derived from the Oregon continental margin. Vallier et al. (1973) have shown that the lowermost large continental turbidites (below 370 m) were derived from the Klamath mountains directly to the east while the more shallow turbidites came from the Columbia River drainage to the north. The number of turbidites apparently decreased upcore, with only 12 major continental turbidites in the upper 250 m of site 35. The sediments are cut by young faults near the margins of the axial valley in the vicinity of the volcanic centers. Deep reflector sequences show offset and folding, suggesting that sediment fill has been subjected to intense tectonism possibly related to left-lateral shear between the southern Gorda Ridge and the Mendocino Fracture Zone (Clague and Holmes, 1987).

The basement surface below the sediment fill is irregular, with as much as 300 m of relief. Disruption of seismic reflectors in the sediments, especially near the center of the axial valley, probably signifies volcanic intrusions. Morton et al. (1987) noted that the absence of extensive drag folding in turbidites around the volcanic centers indicates that the volcanic edifices predate the sedimentary fill or have grown contemporaneously with sedimentation.

Six volcanic centers, located by seismic reflection profiles, are spaced at 15- to 20-km intervals along the axis of Escanaba Trough and dome the otherwise flat-lying sediments up to 100 m high (Morton et al., 1987, this volume). Several of these features had been crossed previously on a site survey for DSDP site 35 (Moore, 1970; Moore and Sharman, 1970), but owing to the lack of adequately spaced seismic reflection lines they were interpreted as one long, continuous basement ridge. These volcanic centers are labeled A through F from south to north (Fig. 1.5), and the two largest volcanic edifices (B and D), located at 40° 45′ N and 41° 00′ N latitude, respectively, were the sites of extensive study and have been named SESCA (B) and NESCA (D). The centers are generally offset to the west of the axis of the valley and can be traced upward through the 300- to 500-m-thick sedimentary fill. Fresh basalts are, in places, exposed on the sea floor, and fresh basalt glass was dredged from edifices B and D.

Heat flow surveys across two of the centers, B and D (Abbott et al., 1986), suggest that both of these edifices are still hot, but heat flow coverage is inadequate to determine if the other centers are active. Based on radiocarbon dating of turbidites, Karlin and Lyle (1989) determined the most recent volcanic activity at domes B and D was 2,400 to 3,000 years ago. This may be an approximate maximum age for chimney deposits that were built up during a postvolcanic quiescent period (Spiess et al., 1980).

The presence of sediment in the axial valley of a ridge provides an insulating blanket that can maintain high temperatures in a hydrothermal system. This can enhance chemical exchange between the hydrothermal fluids and sediments and may have important implications for the chemical composition of hydrothermal fluids and on the longevity of the hydrothermal systems. Evidence of hydrothermal activity at the north end of Escanaba Trough was found in 1983 by Clague et al. (1984). They dredged 100 kg of iron-rich clay and iron-manganese oxides that suggested there had been a long period of production of these hydrothermal minerals by low-temperature (20 to 60°C) springs. One dive (A1405) at this site in 1984 (Malahoff, 1985), however, did not locate these springs.

In 1986 bottom photographs, dredges, and core samples (Morton et al., 1987, this volume) showed that hydrothermal deposits were also located about 50 km south of dive A1405 at volcanic centers B and D (NESCA and SESCA, Fig. 1.5). Samples dredged from the northeastern flank of the volcanic edifice B included Fe-rich, Zn- and Cu-poor massive sulfide fragments (Morton et al., 1987). Some of the sulfides dredged at site D in 1986 contained 4.6 wt% copper and 0.07 wt% silver (Morton et al., this volume). The deposition of sulfides and sulfates in the sediment around the volcanic centers was also documented by Karlin and Lyle (1986, 1989). Dives to sites B and D in 1986 found hydrothermal deposits but no hot springs (Holmes and Zierenberg, this volume). Additional dives in 1988 did locate hot springs with temperatures of 217°C (Zierenberg, 1988).

Fracture Zones

Offsets between the oceanic ridges are zones of strike-slip motion known as transform faults or fracture zones. Fracture zones are seismically active and often have more than 1,000 m of vertical relief across them. Where there is compression across the fracture zone, the transverse ridges may be exceptionally high. The transverse ridges of fracture zones have usually been emplaced by tectonic uplift, but volcanic activity also occurs on the ridges and in the transform valley. Fracture zones are one of the few places that a cross section of the ocean crust can be studied and sampled in place. The rocks recovered from fracture zones include ultramafic mantle material, gabbros, basalts, hydrothermally altered rocks, and sulfides.

Often the direction of motion of the plates on either side of a transform fault is not parallel to the fault. In some cases tension within the fault results in the formation of "pull-apart basins" and volcanism. This type of fracture zone is called a "leaky" transform fault (Thompson and Melson, 1972); the Blanco Fracture Zone is an example of this type.

Blanco Fracture Zone

The Blanco Fracture Zone is a 350-km-long transform fault that connects the northern end of Gorda Ridge with the southern end of the Juan de Fuca Ridge. It is typical of a transform fault with pull-apart basins, and it may be the site of volcanic activity (Selk, 1978; Clague and Holmes, 1987). It is seismically active and is the site of numerous earthquakes of magnitude 5 or larger. The motion along the fault is strike-slip, but there also appears to be some normal faulting (Bolt et al., 1968). The sediments within the transform fault also suggest that part of the basin has dropped 125 m since the Mt. Mazama ash was deposited (6,900 years ago.) The average depth of the sea floor to the south of the fracture zone is about 400 m deeper than that to the north. The Cascadia Channel runs diagonally across the Blanco Fracture Zone between 128° and 129° W and has deposited sediments in the Cascadia Depression (Griggs and Kulm, 1970). This is the main source of sediments to the Tufts Abyssal Plain to the south of the fracture zone.

The strike of the Blanco Fracture Zone is oblique to the direction of motion of the Pacific Plate. This requires a component of spreading within the fracture zone, and between Gorda Ridge and Juan de Fuca Ridge there are five deep basins; the deepest one, the West Blanco Depression, is over 4,800 m. At the eastern end, north of Gorda Ridge, is the Gorda Depression, which is more than 4,400 m deep. DeCharon (1988) has estimated that the rate of spreading in the Cascadia Segment of the Blanco Fracture Zone is 40 to 60 mm/year, similar to the rate of opening of the northern Gorda Ridge.

The Blanco Fracture Zone does not have a continuous transverse ridge, as many fracture zones do, but has discontinuous parallel ridges about 15 km apart. These ridges rise several hundred meters above the adjacent abyssal plain, and between them are six depressions that extend from Gorda Ridge to the Juan de Fuca Ridge (Embley et al., 1987). The depressions are as much as 4,800 m deep and are separated by sills that are about 3,000 m deep. Several of these depressions have evidence of volcanic or hydrothermal activity although no fresh basalts or hydrothermal sulfides have been recovered. Hydrothermal sediments have been found in cores from the East Blanco Depression (Selk, 1978).

Rocks dredged from the north and south walls of the fracture zone are altered and metamorphosed pillow basalts, hydrothermally altered breccias, altered gabbros, and fragments of siliceous vein material (Hart et al., 1986, this volume). Many of the pillow lavas and one of the gabbros have secondary pyrite, and some of the breccias and siliceous material contain copper, nickel, zinc, and lead sulfides. The faulting of the Blanco Fracture Zone has fractured and exposed some of the deeper layers of the oceanic crust, and this crust has been metamorphosed at low pressure and low temperature. This metamorphism was responsible for the movement and concentration of some metals in the altered rock and in the siliceous veins within the fractured rock.

Mendocino Fracture Zone

The Mendocino Fracture Zone is one of the major tectonic features of the northeast Pacific. From its intersection with the San Andreas Fault at the California coast it extends 3,000 km west. At one location it has over 3,000 m of vertical relief, and, next to the major oceanic trenches and the Hawaii-Emperor volcanic chain, it is the most prominent feature of the SEASAT altimeter maps of the free-air gravity field of the Pacific. It is the zone of strike-slip motion between the Gorda and Pacific plates, and it has been the site of numerous earthquakes, some of them magnitude 7. Even though there appears to be some plate reorganization occurring in this region, with the possible migration of the San Andreas Fault northward to meet the Blanco Fracture Zone (Silver, 1971; Riddihough, 1980), which would result in the locking of the Gorda Plate to the Pacific Plate, the Mendocino Fracture Zone remains tectonically active.

The Mendocino Fracture Zone has been surveyed by Krause et al. (1964), Moore (1970), Moore and Sharman (1970), and Clague et al. (1984) as well as by NOAA SEABEAM (unpublished). The major feature of this fracture zone is its high transverse ridge (only 1,100 m below sea level at one point). From this point the ridge drops off to the south 3,300 m to the 4,400-m-deep Pacific Ocean floor, and to the north it drops 2,100 m to Gorda Ridge.

Seismic crossings of the Mendocino Fracture Zone (Moore, 1970; Moore and Sharman, 1970) show that it does not have the same structure as the fracture zones described by Bonatti (1978). There does not appear to be a transform valley, and there is a single transverse ridge rather than two. Moore (1970) suggested that the Mendocino Fracture Zone had spread slightly during the Brunhes, based on the high magnetic intensity of the transverse ridge, and because basalts dredged by Krause et al. (1964) appeared relatively young because of their unaltered glass rims. The age of volcanism of the transverse ridge, however, is not well constrained, as attempts to date the basalts were unsuccessful (Silver, 1971).

The rocks recovered in six dredges by Krause et al. (1964) are the only samples available from the Mendocino Fracture Zone. They are primarily rounded basalt cobbles and pebbles, suggesting that they are from a submerged beach. This indicates at least 1,000 m of subsidence of the Mendocino Ridge. One sample of serpentine was also recovered from a depth of 1,820 to 2,390 m at the eastern end of the fracture zone. Only one chemical analysis of basalt is available from this ridge (Engel and Engel, 1964).

Summary

Gorda Ridge is a 300-km-long volcanic ridge that is bounded by the Mendocino and Blanco Fracture Zones. It is within 200 nautical miles of the coast of Oregon and California and is roughly parallel to the coast. For the past 5 years it has been the site of an extensive, coordinated study by four universities, three federal government agencies, two state agencies, and the U.S. Navy. The purpose of the study was to survey the Gorda Ridge for hydrothermal activity and evaluate the geology, structure, and mineral potential of the ridge and adjacent fracture zones and seamounts. Several methods of hydrothermal prospecting were employed, resulting in the discovery of hot springs on the northern Gorda Ridge and in the Escanaba Trough. Results of this work were presented at the Gorda Ridge Symposium, and detailed reports of the work are included in this volume. This Introduction is meant to give the background geology and geophysics of Gorda Ridge and environs.

References

Abbott DH, Morton JL, Holmes ML (1986) Heat flow measurements on a hydrothermally-active, slow spreading ridge: The Escanaba Trough. Geophys Res Lett 13:678–680.

Atwater TM, Mudie JD (1968) Block faulting on the Gorda Rise. Science 159:729–731.

Atwater T, Mudie JD (1973) Detailed near-bottom geophysical study of the Gorda Rise. Geophys Res 78:8665–8686.

Baker ET, Massoth GJ, Collier RW, et al. (1987) Evidence for high-temperature hydrothermal venting on the Gorda Ridge, northeast Pacific Ocean. Deep-Sea Res 34:1461–1476.

Bibee D (1983) OBS and sonobuoy study of faulting and hydrothermal activity on Gorda Ridge at 42°. Tech Report, Oregon State University, p 14.

Bonatti E (1978) Vertical tectonism in oceanic fracture zones. Earth Planetary Sci Lett 37:369–379.

Bolt BA, Lomnitz C, McKevilly TV (1968) Seismological evidence on the tectonics of central and northern California and the Mendocino Escarpment. Seismol Soc Am Bull 58:1725–1767.

Burns RE (1970) Heat-flow operations at holes 35.0 and 35.1. In: McManus et al. (eds) Initial Reports of the Deep Sea Drilling Project, Vol 5. Washington: U.S. Government Printing Office, pp 551–554.

Clague DA, Friesen W, Quinterno P, et al. (1984) Preliminary geological, geophysical and biological data from Gorda Ridge. USGS Open File Report 84:364, p 41.

Clague DA, Holmes ML (1987) Geology, petrology, and mineral potential of Gorda Ridge. In: Scholl et al. (eds) Geology and Resource Potential of the Continental Margin of Western North America and Adjacent Ocean Basins: Beaufort Sea to Baja California. CircumPacific Council Energy Mineral Resources, Earth Sci Ser 6:563–580.

Collier RW, Holbrook SH, Robbins JM (1986) Studies of trace metals and active hydrothermal venting on Gorda Ridge. Oregon Department of Geology and Mineral Industries, Open-File Report 0-86-13.

Corliss JB, Dymond J, Gordon LI, et al. (1979) Submarine thermal springs on the Galapagos Rift. Science 203:1073–1083.

Davis AS, Clague DA (1987) Geochemistry, mineralogy, and petrogenesis of basalt from Gorda Ridge. J Geophys Res 92:10,467–10,483.

Davis AS, Clague DA (1988) Petrology of gabbroic xenoliths from the northern Gorda Ridge. EOS, Trans Am Geophys Union 69:1493.

DeCharon AV (1988) Structure and tectonics of Cascadia segment, central Blanco Transform Fault Zone. M.S. Thesis, Oregon State University, Corvallis, OR, p 73.

Delaney JR, Johnson HP, Karsten JL (1981) The Juan de Fuca Ridge—hot spot—propagating rift system: New tectonic, geochemical and magnetic data. J Geophys Res 86:11747–11750.

Dehlinger P, Couch RW, McManus DA, Gemperle M (1971) Northeast Pacific structure. In: AE Maxwell (ed) The Sea, Vol 4, Pt 2. New York: John Wiley, pp 133–189.

Embley RA, Kulm LD, Massoth G, Abbott D, Holmes M (1987) Morphology, structure and resource potential of the Blanco Transform Fault Zone. In: Scholl et al. (eds) Geological Resource Potential of Western North America and Adjacent Ocean Basins: Beaufort Sea to Baja California. American Association of Petroleum Geologists Memoir Series pp 549–561.

Engel AEJ, Engel CG (1964) Basalts dredged from the northeastern Pacific Ocean. Science 140: 1321–1324.

Foster TD (1962) Heat-flow measurements in the northeast Pacific and in the Bering Sea. J Geophys Res 67:2991–2993.

Griggs GB, Kulm LD (1970) Physiography of Cascadia deep-sea channel. Northwest Sci 44:82–94.

Hart R, Pyle D, Robbins J (1986) Hydrothermal sulfides, breccias, and greenstones from the Gorda depression. Oregon Department of Geology and Mineral Industries Open-file Report 0-86-16, p 31.

Heinrichs DF (1970) More bathymetric evidence for block faulting on the Gorda Rise. Marine Res 28:330–335.

Howard KJ (1989) Hydrothermal vents of Gorda Ridge, NE Pacific: Mineralogy and chemistry of sulfide chimneys, precipitates and alteration products. M.S. Thesis, Oregon State University, Corvallis, p 245.

Howard KJ, Fisk MR (1988) Hydrothermal alumina-rich clays and boehmite on Gorda Ridge. Geochim Cosmochim Acta 52:2269–2279.

Johnson SH, Jones PR (1978) Microearthquakes located on the Blanco Fracture zone with sonobuoy arrays. J Geophys Res 83:255–261.

Karlin R, Lyle M (1986) Sediment studies on the Gorda Ridge. Oregon Department of Geology and Mineral Industries, Open-File Report 0-86-19.

Karlin R, Lyle M (1989) History of Holocene volcanic and hydrothermal activity in axial valley sediments on Gorda Ridge. Marine Geol (in press).

Kay R, Hubbard NJ, Gast PW (1970) Chemical characteristics and origin of oceanic ridge volcanic rocks. J Geophys Res 75(8):1585–1613.

Krause DC, Menard HW, Smith SM (1964) Topography and lithology of the Mendocino Ridge. J Marine Res 22:236–249.

Malahoff AM (1985) Hydrothermal vents and polymetallic sulfides of the Galapagos and Gorda/Juan de Fuca Ridge systems and of submarine volcanoes. Biol Soc Wash Bull 6:19–41.

McManus DA (1965) Blanco Fracture Zone, northeast Pacific Ocean. Marine Geol 3:429–455.

McManus DA (1967) Physiography of Cobb and Gorda Rises, northern Pacific Ocean. Geol Soc

Am Bull 78:527–546.
McManus DA, et al. (1970) Initial Reports of the Deep Sea Drilling Project, Vol 5. Washington. U.S. Government Printing Office, pp 65–172.
Melson WG, Vallier TL, Wright TL, Byerly G, Nelen J (1976) Chemical diversity of abyssal volcanic glass erupted along Pacific, Atlantic, and Indian Ocean sea-floor spreading centers. In Sutton GH, Manghani MH, Moberly R (eds). The Geophysics of the Pacific Ocean Basin and Its Margin. Washington: American Geophysical Union, Geophysical Monograph 19:351–367.
Melson WG, Byerly GR, Nelen JA, O'Hearn T, Wright TL, Vallier T (1977) A catalog of major element chemistry of abyssal volcanic glasses. In Mason B (ed). Smithsonian Contributions to Earth Sciences. Washington: Smithsonian Institution Press, 19:31–60.
Moore GW (1970) Sea-floor spreading at the junction between the Gorda Rise and the Mendocino Ridge. Geol Soc Am Bull 81:2817–2824.
Moore GW, Sharman GF (1970) Summary of SCAN site 4. In: McManus DA et al. (eds): Initial Reports of the Deep Sea Drilling Project, Vol 5. Washington: U.S. Government Printing Office, pp 761–773.
Morton JL, Holmes ML, Koski RA (1987) Volcanism and massive sulfide formation at a sedimented spreading center, Escanaba Trough, Gorda Ridge. Geophys Res Lett 14:769–772.
Raff AD, Mason RG (1961) Magnetic survey off the west coast of North America, 40 to 52°N latitude. Geol Soc Am Bull 72:1267–1270.
Riddihough RP (1980) Gorda Plate motions from magnetic anomaly analysis. Earth Planetary Sci Lett 51:163–170.
Riddihough RP (1984) Recent movements of the Juan de Fuca plate system. J Geophys Res 89:6980–6994.
Rona PA, Clague DA (1986) Geologic setting of hydrothermal activity at the northern Gorda Ridge, EOS. Trans Am Geophys Union 67:1028.
Rona PA, Denlinger R, Fisk M, et al. (1988) Hydrothermal activity on Gorda Ridge, EOS. Trans Am Geophys Union 69:1588.
Scheidegger KF (1973) Temperatures and compositions of magmas ascending beneath actively spreading mid-ocean ridges. Ph.D. Dissertation, Oregon State University, Corvallis, p 143.
Sclater JG, Jaupart C, Galson D (1980) The heat flow through oceanic and continental crust and the heat loss of the Earth. Rev Geophys Space Physics 18:269–311.
Selk BW (1978) The manganese-enriched sediments of the Blanco Trough: Evidence for hydrothermal activity in a fracture zone. M.S. Thesis, Oregon State University, Corvallis, p 137.
Silver EA (1971) Tectonics of the Mendocino Triple Junction. Geol Soc Am Bull 82:2965–2978.
Solano-Borrego AE (1982) Microseismicity on Gorda Ridge. M.S. Thesis, Oregon State University, Corvallis, p 76.
Spiess FN, Macdonald KC, Atwater T, et al. (1980) East Pacific Rise: Hot springs and geophysical experiments. Science 207:1421–1433.
Stoddard PR (1987) A kinematic model for the evolution of the Gorda Plate. J Geophys Res 92:11524–11532.
Thompson G, Melson WG (1972) The petrology of oceanic crust across fracture zones in the Atlantic Ocean: Evidence for a new kind of sea-floor spreading. J Geol 80:526–538.
Toomey DR, Solomon SC, Purdy GM, Murray MH (1985) Microearthquakes beneath the median valley of the Mid-Atlantic Ridge near 23° N: Hypothesis and focal mechanism. J Geophys Res 90:5443–5448.
Vallier TL, Harold PJ, Girdley WA (1973) Provenances and dispersal patterns of turbidite sand in Escanaba Trough, Northeastern Pacific Ocean. Marine Geol 15:67–87.
Vine FJ, Matthews DH (1963) Magnetic anomalies of oceanic ridges. Nature 199:947–949.
Von Herzen RP (1964) Ocean-floor heat-flow measurements west of the United States and Baja California. Marine Geol 1:225–239.
Wakeham SE (1978) Petrochemical patterns in young pillow basalts dredged from Juan de Fuca and Gorda Ridges. M.S. Thesis, Oregon State University, Corvallis, p 95.
Wilde P, Chase TE, Holmes ML, et al. (1978) Oceanographic data off northern California–southern Oregon, 40° to 43° North including the Gorda Deep Sea Fan. LBL Publ. 251. Berkeley, CA: Lawrence Berkeley Lab.
Wilde P, Chase TE, Holmes ML, et al. (1979) Oceanographic data off Oregon 43° to 46°N including Astoria Deep-Sea Fan. LBL Publ. 253. Berkeley, CA: Lawrence Berkeley Lab.
Wilson DS (1984) Tectonic history of the Juan de Fuca Ridge over the last 40 million years. J Géophys Res 93:11,863–11,876.
Wilson DS (1986) A kinematic model for the Gorda deformation zone as a diffuse southern boundary of the Juan de Fuca Ridge. J Geophys Res 91:10259–10269.
Zierenberg RA (1988) Preliminary results of ALVIN dives on active sediment-hosted massive sulfide deposits in the Escanaba Trough, southern Gorda Ridge. EOS 69:1148.

Part II Gorda Ridge Mineral Potential

2—Hydrography and Geochemistry of Northern Gorda Ridge

ROBERT W. COLLIER and EDWARD T. BAKER

Abstract

The physics and chemistry of the near-bottom water column above Gorda Ridge were studied for indications of hydrothermal venting. Based on initial seafloor surveys, three field programs focused on Northern Gorda Ridge, using state-of-the-art shipboard and in situ determinations of physical, optical, and chemical properties useful in "prospecting" for active venting. The results of these investigations demonstrate the presence of active venting in at least two regions of the northernmost ridge segment—the "Narrowgate" section (GR14), and the east slope of the axial valley near its intersection with Blanco Fracture Zone (GR15). Distinct temperature anomalies (≈30 millidegree C) and suspended particle "plumes" were found at both locations during two separate expeditions in 1985. These plumes had stabilized at depths between 2450 and 2900 m and were separated from the bottom by at least 200 m. Detailed sections through the plume at GR14 demonstrated dimensions of at least 2 km (east-west) and 2 to 4 km (north-south), centered over the east wall of the axial valley. The plumes at GR14 and GR15 were not physically connected along the ridge axis. Chemical anomalies in the plumes included high concentrations of dissolved manganese, helium-3, radon-222, and iron-rich hydrothermal precipitates. The composition and hydrography of the plumes suggest that they originated from high-temperature vents. A brief reoccupation of the GR14 site in 1986 failed to relocate these strong plume features.

Introduction

Hydrothermal circulation has a profound effect on the chemistry of seawater, which can be seen on scales ranging from the global oceans (Clarke et al., 1969; Edmond et al., 1979) to the individual orifices of the dramatic "black smoker" high-temperature vents (Spiess et al., 1980). The input of heat and chemicals to the seafloor may have long-range effects on marine sediments (Dymond et al., 1973; Lyle et al., 1987), on marine biological communities (Grassle, 1986; Cowen et al., 1986), and on the abyssal circulation near the major ridge axes (Stommel, 1982). These hydrothermal "fingerprints" are also powerful prospecting tools for the location of active seafloor vents (see Baker, this volume). This chapter discusses the application of these tools toward the demonstration of active venting on the northern Gorda Ridge.

Gorda Ridge is the only spreading center in the Pacific exhibiting slow-spreading physiography, with an axial valley approximately 700 m deep (see Fisk and Howard, this volume). The tectonics and regional geology of the ridge are discussed by Riddihough (1980), Clague et al. (1984), Malahoff (1985), Fisk and Howard (this volume), Morton et al. (this volume), and Clague and Rona (this volume). While the southern ridge segment, the sediment-filled Escanaba Trough, has been the focus of studies on sediment-hosted sulfide mineralization, the northern ridge has been surveyed primarily for active venting analogous to that recently found on the Mid-Atlantic Ridge (Rona et al., 1986). Previous to the surveys discussed below, the only evidence of hydrothermal activity on the northern ridge segment came from samples of altered basalt and hydrothermal precipitates dredged by the U.S. Geological Survey (Clague et al., 1984).

The ridge lies beneath the main body of the California Current (Hickey, 1986), and the southward-flowing surface waters in the region are largely subarctic in character. A plume of low-salinity water from the Columbia River occasionally extends this far offshore. Very little is known about the abyssal circulation in this region, and current meter measurements are essentially nonexistent. The northern ridge axis is an extremely deep canyon with relatively limited isopycnal exchange routes open to the north and south. Within this environment, currents are most probably tidally dominated and exhibit strong topographic control.

Most of the data discussed in this review were collected during three field expeditions in 1985 and 1986. The first expedition, an 18-day cruise on the National Oceanic and Atmospheric Administration (NOAA) ship *Surveyor* during May 1985, was part of the NOAA VENTS program and was designed to determine if hydrothermal venting was occurring anywhere along the central ridge axis. This program resulted in the first clear evidence of active venting in at least two locations on the northernmost segment of the Gorda Ridge (Baker et al., 1987). These locations were centered on the east wall of the axial valley at the "narrowgate" section (station GR14) and near the ridge's intersection with the Blanco Fracture Zone (station GR15).

A follow-up cruise was organized by the Gorda Ridge Technical Task Force on the R/V *Wecoma* (cruise W8508AA) for 7 days in August 1985. During this expedition, the hydrothermal plumes at GR14 and GR15 were relocated and extensively sampled (Collier et al., 1986). The precise location of the vents producing the plumes was not determined.

In June 1986, the NOAA VENTS program sponsored another expedition to station GR14 using the NOAA Ship *Discoverer*. The hydrographic program was limited and was unable to relocate the active vents. Results of extensive bottom photography suggested at least one specific location of recent venting at fractures along the base of the valley's east wall. However, no active vents were recognized.

The sampling, analytical, and interpretive techniques applied in these specific field programs have been described elsewhere (Collier et al., 1986; Baker et al., 1987). A general discussion of the physical and chemical techniques applied to hydrothermal vent "prospecting" is given by Baker (this volume). The USGS sponsored two cruises on the R/V *S.P. Lee* in July 1985 and in 1986 that collected extensive photographs and samples of the seafloor basalts and sediments on the ridge. These have contributed to our interpretation of the hydrographic results.

Results

Physics

The northern Gorda Ridge survey area is shown in Figure 2.1. Stations GR14 and GR15 were each occupied during the 1985 *Surveyor* cruise with a single CTD (conductivity, temperature, depth) profile. A "SLEUTH" tow (streamlined CTD-transmissometer package; see Baker, this volume) was also run along the base of the east wall at GR15 during the *Surveyor* cruise (dotted line near GR15). Significant temperature and particle anomalies were measured at both sites between the depths of 2,450 and 2,900 m. Similar anomalies were encountered during the follow-up *Wecoma* cruise. A maximum-temperature anomaly of approximately 25 millidegrees (C) was measured several hundred meters north of station GR14 during W8508AA, CTD11 (Figs. 2.1, 2.2). A bathymetric profile (12 kHz) across the ridge axis (between points A and B in Fig. 2.1; profile shown in Fig. 2.3) shows the approximate location of the anomalies with respect to the local topography. The plume has clearly risen more than 200 m above the seafloor, suggesting significant buoyancy (high temperature) at the point of venting (Middleton and Thomson, 1986).

The CTDs collected during W8508AA were towed in a vertical zigzag fashion ("tow-yo") through the plumes. The CTD was repeatedly raised and lowered above the bottom while towing the instrument at a speed of 1 knot. The

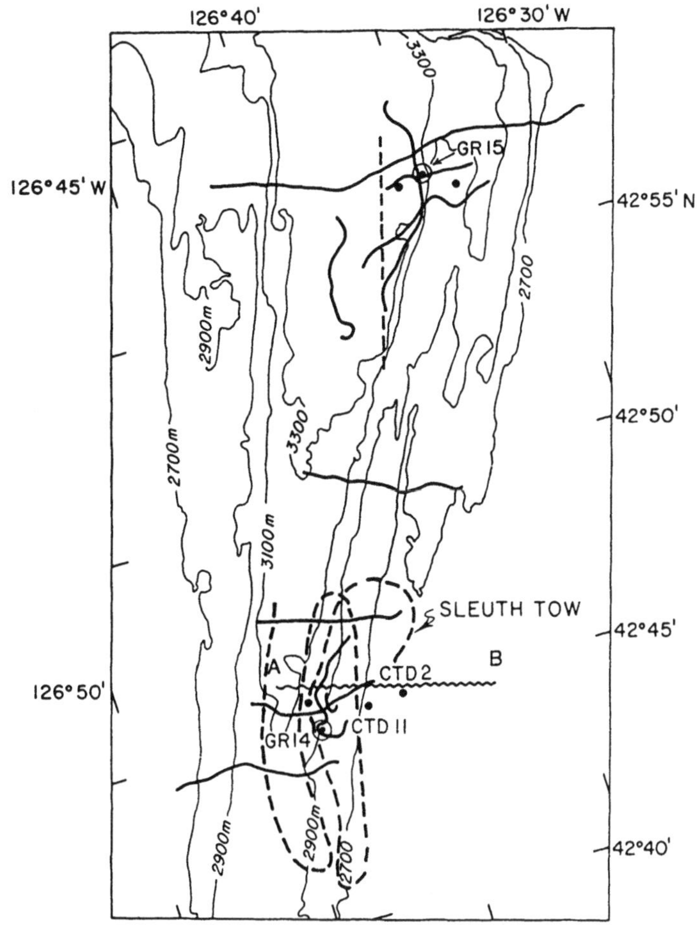

FIGURE 2.1. Map of Northern Gorda Ridge showing sites of the 1985 and 1986 hydrothermal surveys. The primary sites, GR14 and GR15, are circled. Solid track lines represent the path of tow-yo'ed CTD's during W8508AA. Dotted track lines represent the track of SLEUTH tows in 1985 (at GR15) and 1986 (at GR14). Solid dots are the locations of various vertical CTD hydrocasts collected for water analysis. The wavy line (marked AB) is the ship track for the bathymetric profile shown in Figure 2.3.

spatial extent of the plume at GR14 can be seen by comparing the intersecting vertical sections of temperature anomaly developed from CTD2 and CTD11 (Figs. 2.1, 2.4). The plume, when sampled by CTD2 along an east-west survey line (Fig. 2.4a), was located high on the east wall and had two maxima centered at 2,400 and 2,600 m depth. When the site was reoccupied 5 days later with CTD11 (Figs. 2.2, 2.4b), the plume was basically in the same geographic location but was nearly 100 m shallower in depth. The overall dimension of the plume was estimated to be 2 km wide (east-west) and 2 to 4 km long (north-south). The plume at GR15 was not well sampled on the *Wecoma* cruise although chemical anomalies were as large for this northern plume as they were for GR14 during the earlier *Surveyor* cruise. Cross-axis CTD sections (Fig. 2.1) demonstrated that the two hydrothermal plumes were not connected on isopycnals along the east wall of the valley. Based on this observation, there must be at least two separate sources of venting along the northern ridge segment.

As shown in Figure 2.2, the plume carried a particle load that was clearly detectable with the transmissometer. Concentrations of suspended particles in the plume, which are chemical precipitates formed when the hydrothermal effluent mixes with cold, alkaline seawater (Von Damm and Bischoff, 1987), correlate well with the measured temperature anomaly (Fig. 2.5), demonstrating the utility of combining these tracers (Baker, this volume). The intensity of the light absorption relative to temperature in the GR14 plume is somewhat lower than that in plumes encountered on the southern Juan de Fuca and Endeavour Ridges. This may be the result of differences in the

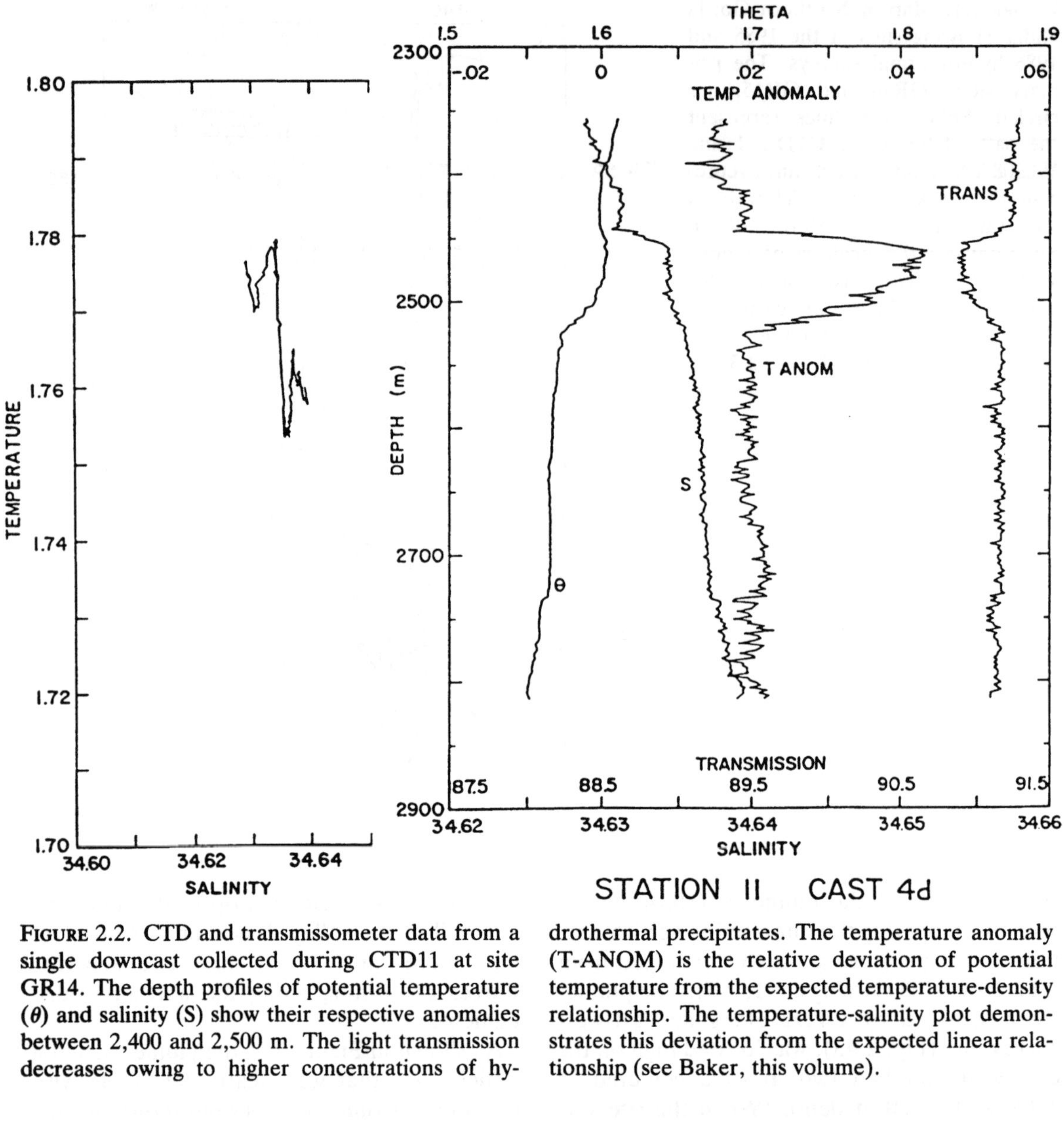

FIGURE 2.2. CTD and transmissometer data from a single downcast collected during CTD11 at site GR14. The depth profiles of potential temperature (θ) and salinity (S) show their respective anomalies between 2,400 and 2,500 m. The light transmission decreases owing to higher concentrations of hydrothermal precipitates. The temperature anomaly (T-ANOM) is the relative deviation of potential temperature from the expected temperature-density relationship. The temperature-salinity plot demonstrates this deviation from the expected linear relationship (see Baker, this volume).

chemistry and temperature of the primary vent fluids at the different sites.

A year later, during 1986, GR14 was revisited by the NOAA Ship *Discoverer*. As part of the reoccupation, a long SLEUTH tow was collected, as well as one CTD cast and one standard hydrocast (Fig. 2.1). Numerous small temperature and particle anomalies were detected, but these were not as well correlated and were definitely smaller than those seen in 1985 using the same equipment. Although this portion of our survey was somewhat limited in extent, we must consider the possibility that the magnitude of venting at GR14 was less in 1986 than in 1985 and may be one of the first indications of temporal variations in venting. In September 1988, high-temperature venting was discovered at this site by scientists using the DSV *Sea Cliff*.

Chemistry

The chemical signatures of hydrothermal venting were also clearly demonstrated in the Gorda Ridge plumes at GR14 and GR15 (Baker et al., 1987; Collier et al., 1986). These included

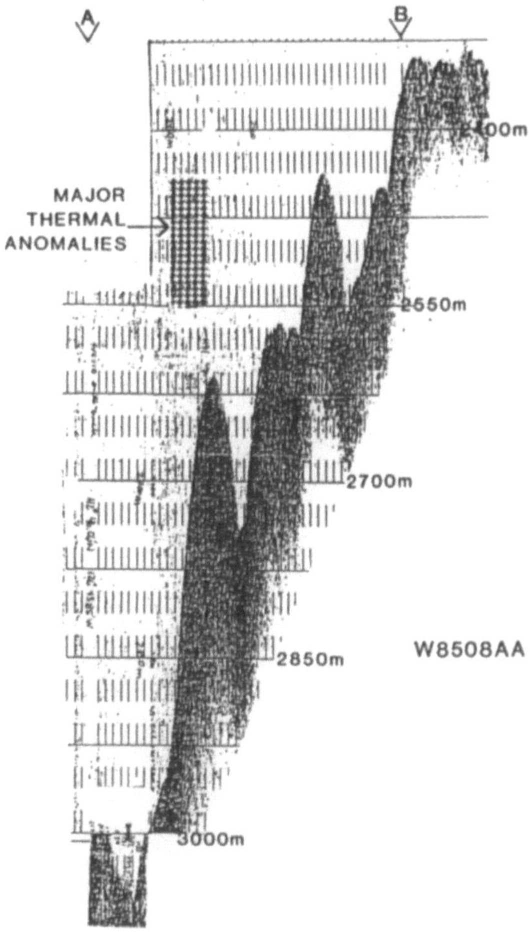

FIGURE 2.3. Bathymetric data (12 kHz) from a transect running west to east out of the axial valley at GR14 (line AB in Fig. 2.1). The approximate location of the hydrothermal temperature and particle anomalies seen in CTD11 and CTD2 (Figs. 2.1, 2.2, 2.4) is shown.

elevated concentrations of helium-3, dissolved manganese, excess radon-222, particulate and dissolved iron, and methane. The presence of the relatively soluble hydrothermal precipitate anhydrite requires an active vent source with high exit temperatures (Baker et al., 1987). These chemical results are completely compatible with the physical plume measurements.

In Figure 2.6, eight vertical profiles of dissolved manganese are shown from the *Surveyor* cruise (Collier et al., 1986). The four stations in Figure 2.6a show background concentrations expected for this environment, with a normal increase toward the seafloor resulting from benthic remobilization of Mn. Stations GR14 and GR15 (Fig. 2.6b) each clearly demonstrate the presence of a hydrothermal plume. Another, smaller, anomalous signal is present at GR3 within the Escanaba Trough (see Fisk and Howard, this volume).

In 1986, the manganese concentrations in the two vertical profiles at GR14 were at background levels (Fig. 2.7). This is consistent with the failure to detect a thermal anomaly during this expedition. It has been shown that even a relatively large plume may not be detected by sampling just "upstream" of it. Because the precise location of the vent source and configuration of the bottom currents are unknown, the two single profiles in 1986 cannot constrain the scale of venting present at that time. However, coupled with the data from the SLEUTH tow, we conclude that the volume of venting in 1986 was relatively small if we missed it with the sample coverage shown in Figure 2.1.

Discussion

The results of these three field expeditions clearly demonstrate the presence of active hydrothermal venting on northern Gorda Ridge. This is occurring in at least two locations, known as GR14 and GR15 (Fig. 2.1), but may

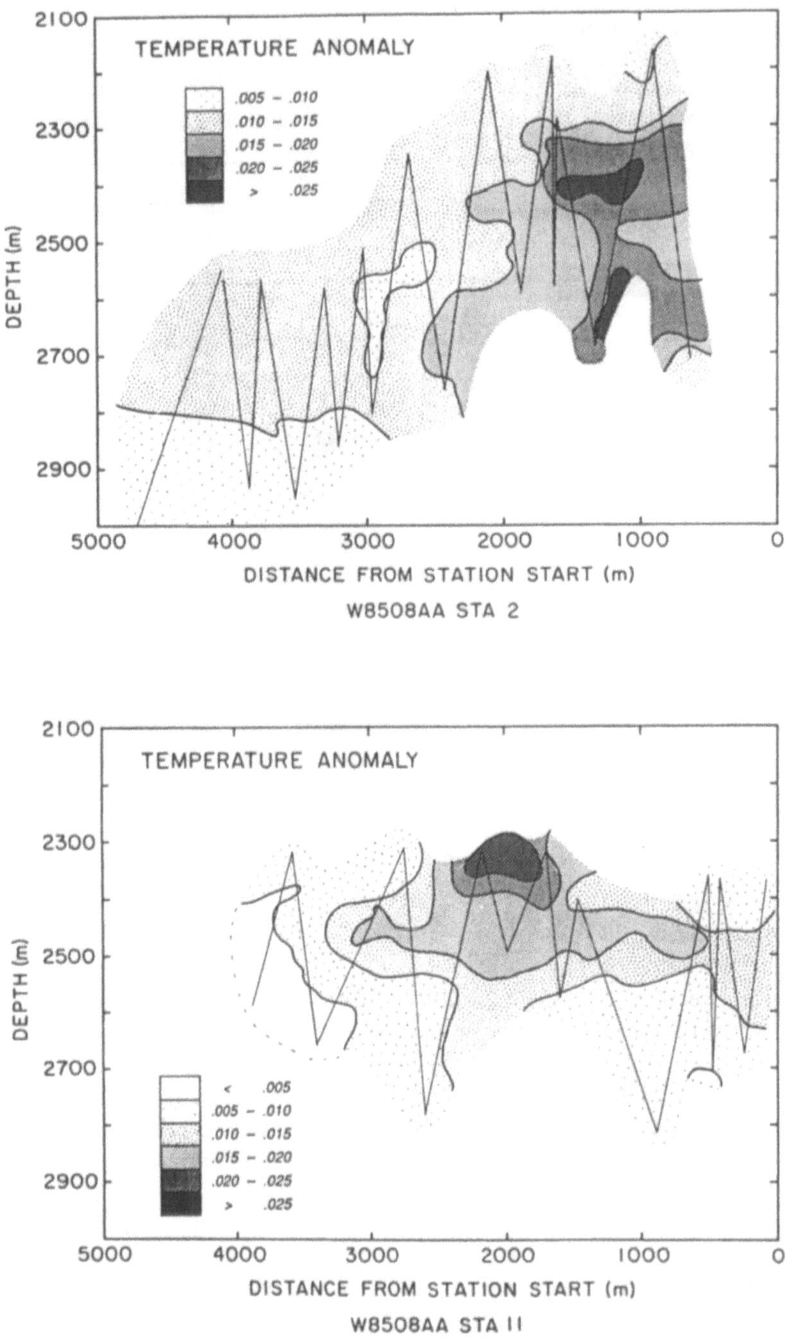

FIGURE 2.4. Contour plots of temperature anomaly detected during CTD tow-yo's (W8508AA). (a) Section made by CTD2—the view is looking toward the north (up the valley). (b) Section made by CTD11—the view is looking toward the west. The sections are roughly normal to each other (see Fig.2.1) and were collected approximately 4 days apart. The path of the CTD instrument is shown by thin zigzag lines.

also occur elsewhere along the ridge (e.g., GR3, 4, 5). The physical and chemical "fingerprints" present in these plumes are directly analogous to those detected in the water column above all other midocean ridge vent systems. Furthermore, the elevation and chemistry of the plumes observed in 1985 suggest that these were "high-temperature" vents (>100°C).

A remarkable feature of the surveys at GR14 was the apparent disappearance of the plume between the 1985 and 1986 surveys. This could be related to temporal variation in venting (see Baker, this volume), or it could dem-

FIGURE 2.5. Correlation of transmissometer (light absorbance) and temperature anomaly data from CTD11. High particle concentrations (high light absorbance, low transmission) correspond with high-tempeature anomalies. These data, along with the elevation of the plume above, suggest that high-temperature venting is occuring.

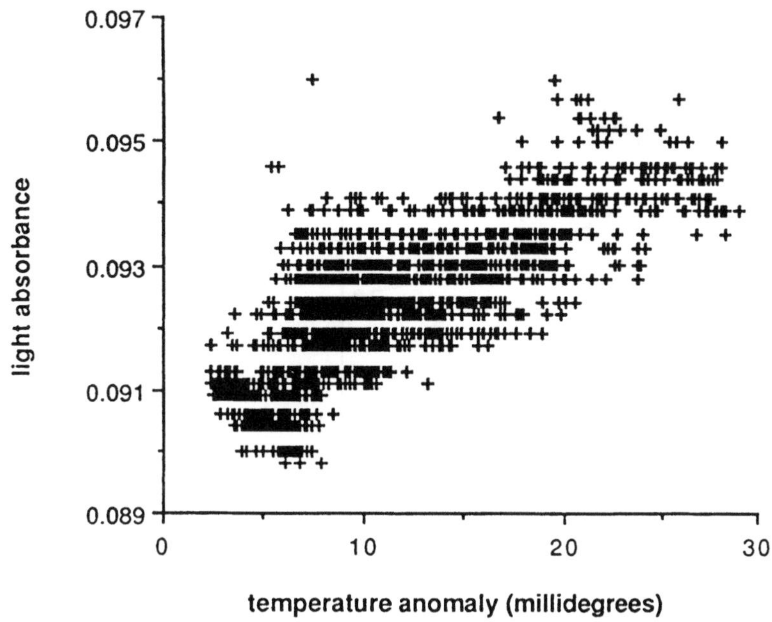

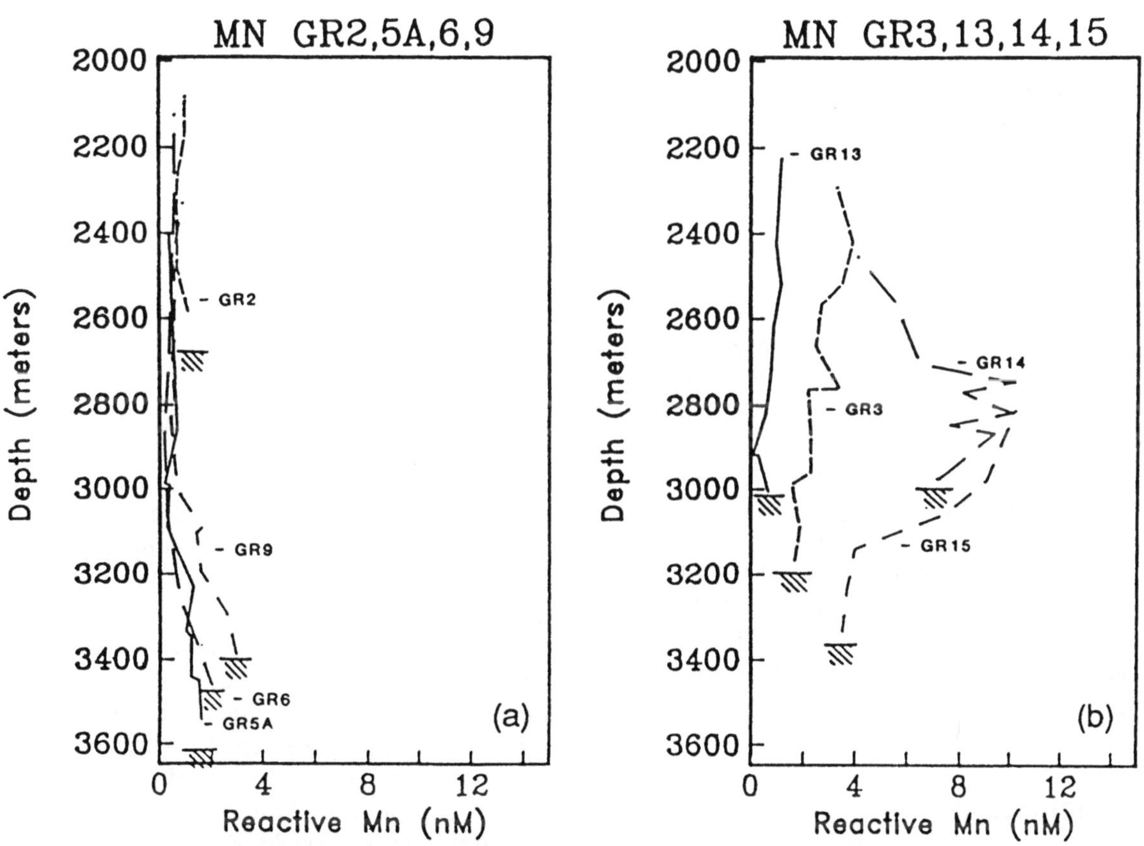

FIGURE 2.6. Vertical profiles of total reactive manganese (TRMn) from Gorda Ridge (Baker et al., 1987; Collier et al., 1986). (a) Overlays of 4 "background" stations. (b) Overlays of GR14, GR15 hydrothermal stations. GR3 shows possible hydrothermal influence.

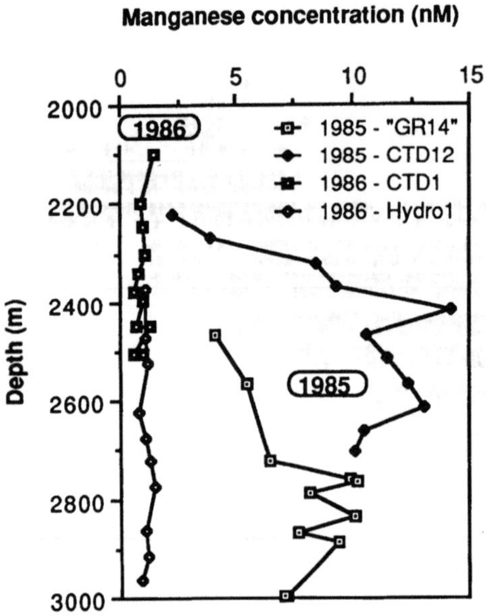

FIGURE 2.7. Vertical profiles of TRMn at station GR14. The 1985 data show that active venting was sampled; the 1986 data show no evidence of a hydrothermal component above background (see Figure 2.6).

onstrate that the areal extent of the plume was quite limited in 1986. It is not surprising to find variations in intensity of venting because hydrothermal vents are short-lived geophysical events. However, the relative constancy of most of the other systems sampled over similar or longer time scales (Galapagos, 21° N, Juan de Fuca) would make this observation unique. Currently, the heat and chemical output of the hydrothermal system at GR14 appears to be small in comparison to systems like the Endeavour Ridge (Kadko et al., 1986).

Although the hydrothermal anomalies sampled at GR15 were as strong as any measured at GR14, our coverage of the site is extremely poor. GR15 lies on the edge of a large, relatively flat bench, which is relatively unsampled. Although the specific area has been the focus of many small seismic events recorded on coastal networks (Bibee, personal communication), it is not known if or how these seismic signals are related to hydrothermal activity.

Conclusions

Active hydrothermal venting has been demonstrated in at least two separate locations on the northern Gorda Ridge by the use of physical and chemical tracers. No other known oceanographic processes can account for the observed phenomena. In September 1988, a group of scientists cooperating through the Gorda Ridge Technical Task Force discovered high-temperature venting at site GR14 using the DSV *Sea Cliff*. The existence of these hydrothermal systems, which lie within a 12-hour radius of several capable oceanographic ports on the West Coast of the United States, will offer a unique research opportunity to upcoming geochemical, biological, and geophysical studies.

Acknowledgments. The work reviewed is the product of the cooperative efforts of the many principal investigators cited in the text. The two NOAA field programs (NOAA Ships *Surveyor* and *Discoverer*) were supported by the NOAA VENTS program, and the *Wecoma* expedition (W8508AA) was supported by the U.S. Department of the Interior, Minerals Management Service, through the Oregon Department of Geology and Mineral Industries. We thank all the scientists who have cooperated under the umbrella of the Gorda Ridge Technical Task Force, which has made much of this work possible. R. Collier was supported by DOGAMI (contracts 63-630-8506 and 63-630-8602), and E. Baker was supported by the NOAA VENTS program.

References

Baker ET, Massoth GJ, Collier RW, et al. (1987) Evidence for high-temperature hydrothermal venting on the Gorda Ridge, northeast Pacific Ocean. Deep-Sea Res 34:1461–1476.

Clague D, Friesen W, Quinterno P, et al. (1984) Preliminary geological, geophysical, and biological data from the Gorda Ridge. USGS Open-File Report 84-364, 49 pp.

Clarke WB, Beg MA, Craig H (1969) Excess helium-3 in the sea: Evidence for terrestrial primordial helium. Earth Planetary Sci Lett 6:213–220.

Collier RW, Holbrook SH, Robbins JM (1986) Studies of trace metals and active hydrothermal venting on the Gorda Ridge. State of Oregon, Department of Geology and Mineral Industries Open-File Report 0-86-13, 36 pp.

Cowen JP, Massoth GJ, Baker ET (1986) Bacterial scavenging of Mn and Fe in a mid- to far-field hydrothermal plume. Nature 322:169–171.

Dymond J, Corliss JB, Heath GR, Field CW, Dasch EJ, Veeh HH (1973) Origin of metalliferous sediments from the Pacific Ocean. Geol Soc Am Bull 84:3355–3372.

Edmond JM, Measures C, McDuff RE et al. (1979) Ridge crest hydrothermal activity and the balances of the major and minor elements in the ocean: The Galapagos Data. Earth Planetary Sci Lett 46:1–18.

Grassle JF (1986) The ecology of deep-sea hydrothermal vent commonities. Adv Marine Biol 23:301–362.

Hickey B (1986) The California Current System–hypotheses and facts. Prog Oceanogr 8:191–279.

Kadko D, Lupton JE, Collier R, Lilley MD, Pak H (1986) Endeavour Ridge hydrothermal plume study. EOS 67:974 (abstract).

Lyle M, Leinen M, Owen RM, Rea DK (1987) Late tertiary history of hydrothermal deposition at the East Pacific Rise, 19°S': Correlation to volcano-tectonic events. Geophys Res Lett 14:595–598.

Malahoff A (1985) Hydrothermal vents and polymetalic sulfides of the Galapagos and Gorda/Juan de Fuca Ridge systems and of submarine volcanos. Biol Soci Wash Bull 6:19–41.

Middleton JM, Thompson RE (1986) Modeling the rise of hydrothermal plumes. Can Tech Rep Hydrogr Ocean Sci 69, 18 pp.

Riddihough RP (1980) Gorda plate motions from magnetic anomaly analysis. Earth Planetary Sci Lett 51:163–170.

Rona PA, Klinkhammer G, Nelsen TA, Trefry JH, Elderfield H (1986) Black smokers, massive sulfides and vent biota at the Mid-Atlantic Ridge. Nature 321:33–37.

Speiss FN, Macdonald KC, Atwater T, et al. (1980) East Pacific Rise: Hot springs and geophysical experiments. Science 207:1421–1433.

Stommel H (1982) Is the South Pacific helium-3 plume dynamically active? Earth Planetary Sci Lett 61:63–67.

Von Damm KL, Bischoff JL (1987) Chemistry of hydrothermal solutions from the southern Juan de Fuca Ridge. J Geophys Res 92:11,334–11,346.

3—Geology of the GR14 Site on Northern Gorda Ridge

DAVID A. CLAGUE and PETER A. RONA

Abstract

The northern Gorda Ridge is spreading at an intermediate rate but exhibits a deep, well-defined rift valley and petrologic diversity more like the slow-spreading Mid-Atlantic Ridge than other oceanic ridges with intermediate rates of spreading. A detailed investigation of a 10 km by 10 km area of the floor and eastern wall of the rift valley was conducted in order to define the geologic setting of active high-temperature hydrothermal vents regionally located using water column studies. High-temperature vents are apparently located on the east wall of the rift valley at the southern end of a linear ridge at a location where ridge-parallel and anomalous ridge-oblique tectonic trends intersect. Additional hydrothermal deposits were discovered along the axis of the valley floor near the shoal point of the axial valley of the northern Gorda Ridge, adjacent to an area of sheet flows. These deposits are of two types—a massive material, probably nontronite, and white coatings on talus that consist of mixed-layer clay and boehmite. These axial deposits do not appear to be actively forming. This study suggests that past and present hydrothermal activity occurs near the shallowest and volcanically most productive segment of the ridge, and the locations of the vents are controlled by the intersection of tectonic trends.

Introduction

At a time when hydrothermal activity is being identified at sites on all oceanic ridges, such activity on the Gorda Ridge is of particular interest. The Gorda Ridge is a 300-km-long spreading center located 200 to 300 km west of northern California and southern Oregon; it is the only oceanic spreading center within the proclaimed U.S. Exclusive Economic Zone of the conterminous United States (Fig. 3.1).

An early hydrographic reconnaissance along the northern 230 km of the Gorda Ridge discovered hydrothermal discharge at two stations in the northern half of the northern section (Fig. 3.1, stations GR14 and GR15; Baker et al., 1987). We focused our work on the southern of these water column signals. Subsequent detailed water column surveys revealed that the indicators of ongoing hydrothermal discharge converge over a linear ridge on the east wall of the axial valley (Collier and Baker, this volume). The depth at which the hydrothermal plume achieves neutral bouyancy and spreads out suggested that the source vents are located at depths of 2,600 to 2,700 m, well above the depth to the floor of the axial valley in this region (Collier et al., 1986; Kadko, 1986; Baker et al., 1987). On the basis of this interpretation, we surveyed the eastern wall of the axial valley and a linear ridge surmounted by hills defined by the 2,600- and 2,720-m isobaths in an effort to locate the hydrothermal vents responsible for the water column anomalies. In this paper we describe the geologic and structural setting of off-axis and on-axis hydrothermal activity in the area of the GR14 station at about 42° 45′ N, 126° 44′ W. The descriptions included here complement our prior work on the geologic controls of hydrothermal discharge in this area (Rona and Clague, 1986, 1989).

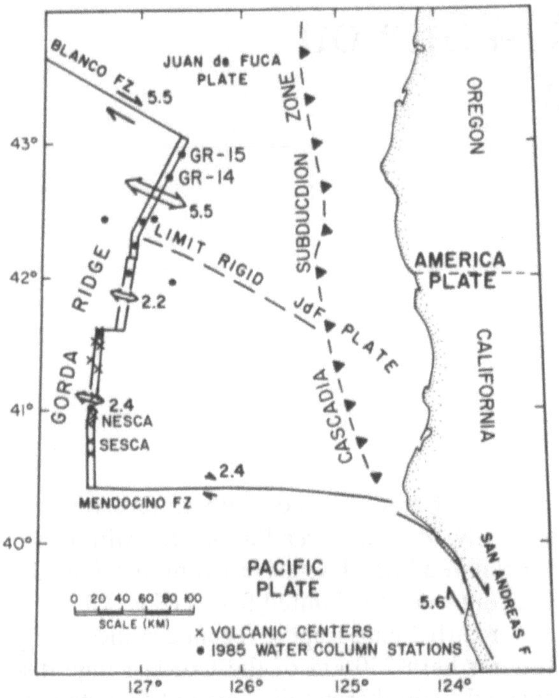

FIGURE 3.1. Regional map of the Gorda Ridge showing the axial zone with positions of volcanic centers (Morton et al., 1987) and water column stations (Baker et al., 1987), full-rates of sea floor spreading, and surrounding plates and their boundaries. (Modified from Riddihough, 1980; Clague and Holmes, 1987.)

The data used in this interpretation were mainly collected during a series of coordinated U.S. Geological Survey and NOAA cruises between 1983 and 1986. Additional work in the area included water column studies and core sampling in 1985 coordinated between NOAA and Oregon State University. The data set includes water column surveys in 1985 and 1986 (Collier et al., 1986; Kadko, 1986; Baker et al., 1987; Collier and Baker, this volume), GLORIA side-scan imagery collected in 1984 by the U.S. Geological Survey and the Institute of Oceanographic Sciences, England, during cruises F3-84 and F4-84 (EEZ-Scan, 1986), multibeam bathymetry collected by NOAA in 1981 and 1982 (Malahoff, 1985), narrow-beam bathymetry, SeaMARC II side-scan sonar imagery collected by the USGS on cruise KK1-83-NC (Clague and Holmes, 1987), SeaMARC IA side-scan sonar imagery collected by NOAA in 1986, deep-towed camera and video collected during USGS cruise L5-85-NC and a 1986 NOAA cruise, and bottom sampling by dredging and coring during USGS cruises KK2-83-NC, L5-85-NC, and L3-86-NC. The USGS cruise L3-86-NC was conducted in support of a diving program using the Navy's *Seacliff* submersible, but poor weather and rough sea conditions precluded dives and limited other operations.

Regional Geology of the Gorda Ridge

The Gorda Ridge is the only midoceanic ridge in the Pacific Ocean with slow-spreading characteristics. On the north it is separated from the intermediate-spreading Juan de Fuca Ridge (full-rate 6 cm/year) by the Blanco Fracture Zone (at about 43° 00′ N). It terminates to the south at the Mendocino Fracture Zone (at about 40° 24′ N). Fisk and Howard (this volume) present a more detailed regional perspective of the Gorda Ridge. The Gorda Ridge has been interpreted as the western boundary of an area extending eastward to the Cascadia Subduction Zone that is undergoing deformation which accommodates differential motion between the Juan de Fuca plate to the north and the Pacific plate to the west and south (Fig. 3.1; Wilson, 1986; Stoddard, 1987). This area, including the Gorda Ridge, is seismically active with a complex pattern of microearthquakes extending to subcrustal depths (Tobin and Sykes, 1968; Bolt et al., 1968; Seeber et al., 1970; Chandra, 1974). These microearthquakes indicate that brittle fracture extends into the upper mantle (Harper, 1985).

The Gorda Ridge is divided into three sections by two right-lateral structural discontinuities: a 20-km offset near 42° 25′ N, 126° 50′ W, and an 18-km offset near 41° 37′ N, 127° 20′ W (Fig. 3.1; Clague and Holmes, 1987). Each section has a well-defined rift valley with relief up to 1.5 km between the axial valley floor (depth 3,050 to 3,900 m) and the crests of flanking rift mountains (depth 2,000 to 2,500 m). The northern section is spreading at an intermediate full rate of 5.5 cm/year (Riddihough, 1980), although its morphologic (Malahoff, 1985) and petrologic (Davis and Clague, 1987) characteristics are more like those of the Mid-Atlantic Ridge than other oceanic ridges in the Pacific. The central and southern sections of the Gorda Ridge are slow-spreading with full rates of 2.2 to 2.4 cm/year (Atwater and Mudie, 1973; Riddihough, 1980). The rift valley is sediment starved north of latitude 41° 15′ N, whereas the southern 100 km is filled with terrigenous sediment up to 650 m thick (Clague and Holmes, 1987). A series of volcanic centers consisting of irregular basaltic domes that project through the sediment are spaced 15 to 20 km apart along the southern section, which is referred to as the Escanaba Trough (Morton et al., 1987). Massive sulfide deposits occur in faulted sediment overlying the volcanic edifices (Morton et al., 1987, this volume; Koski et al., 1988).

The northern section of the Gorda Ridge is 70 km in length with maximum axial depths of about 3,750 m at the northern and southern ends and an axial trend of 022°. A minimum axial depth of about 3,050 m occurs midway between the two ends. The minimum axial depth coincides with a narrowing of the rift valley floor between the 3,000-m isobaths to 2.8 km at this area called Narrowgate (42° 44′ N) from widths of 14.3 and 9.8 km at the northern and southern ends, respectively.

Basaltic lavas from this northern section are petrologically complex, with compositions that indicate variable percentages of partial melting and variable major and trace element compositions in flows that are spatially and temporally closely associated (Davis and Clague, 1987). Sediment accumulation rates, determined by radiocarbon dating of cores W6 and W9 from this area, range from 6 to 15 cm/1,000 years (Karlin and Lyle, 1986).

GR14 Area (44° 44.7′ N, 126° 43.7′ W)

The detailed studies in this area, coordinated between the USGS, NOAA, and Oregon State University, were designed to start with regional studies and to focus the study area to a successively smaller area using finer scale tools. We began with regional coverage bathymetry, side-scan sonar imagery, and reconnaissance water sampling and progressed to detailed sampling and bottom-towed video and camera work.

Water Column Surveys

Water column surveys of the GR14 area revealed the presence of ongoing hydrothermal discharge (Collier et al., 1986; Kadko, 1986; Baker et al., 1987; Collier and Baker, this volume). The water column surveys included profiles of hydrographic properties and suspended particle concentrations made in vertical and towed mode with a CTD (conductivity, temperature, depth/transmissiometer instrument) and discrete water samples analyzed for dissolved and particulate trace metals, ^{222}Rn, methane, ^{3}He, and particle mineralogy; these samples were collected by rosette- and wire-mounted Niskin bottles (Baker et al., 1987). The signals indicative of hydrothermal discharge were strongest over the east wall of the axial valley in an area characterized by a linear ridge.

Suspended particulate matter filtered from the water samples contained abundant particles of the hydrothermal minerals anhydrite (Baker et al., 1987) and atacamite (Nelsen et al., 1987). Anhydrite precipitates from hydrothermal solutions at temperatures of >130°C at 30 bars pressure (Janecky and Seyfried, 1983). Anhydrite particles of the size recovered (<10 μm) dissolve in seawater within a few days (Feely et al., 1987). Precipitation of Cu-bearing phases that weather to atacamite suggests high-temperature discharge. The presence of these minerals as suspended particulate matter in solutions with anomalous concentrations of ^{3}He, dissolved Mn, particulate Fe, and methane suggests ongoing high-temperature venting in the immediate area.

Bathymetry

A roughly 10 by 10 km area of the rift valley around station GR14 extending northward

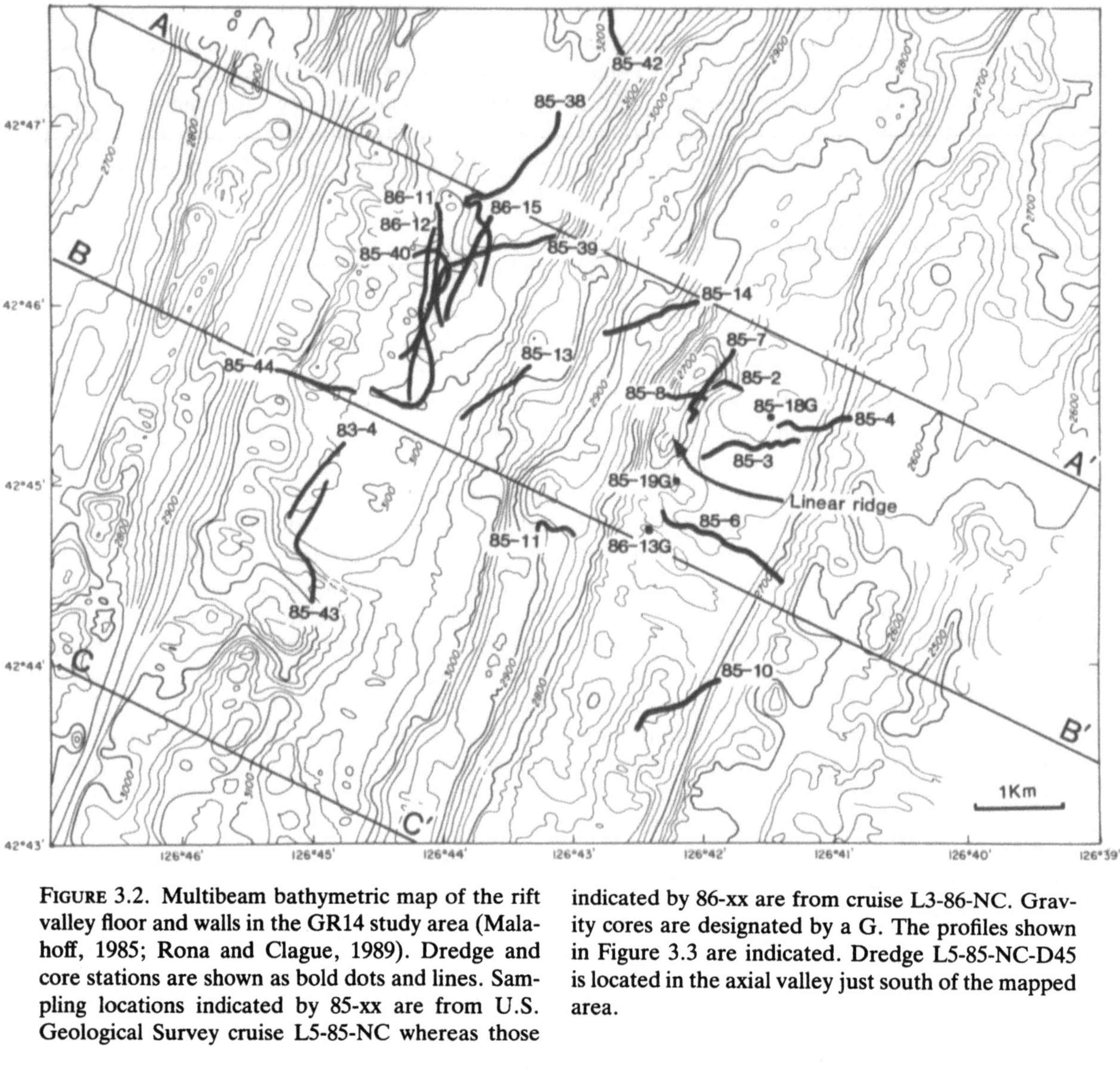

FIGURE 3.2. Multibeam bathymetric map of the rift valley floor and walls in the GR14 study area (Malahoff, 1985; Rona and Clague, 1989). Dredge and core stations are shown as bold dots and lines. Sampling locations indicated by 85-xx are from U.S. Geological Survey cruise L5-85-NC whereas those indicated by 86-xx are from cruise L3-86-NC. Gravity cores are designated by a G. The profiles shown in Figure 3.3 are indicated. Dredge L5-85-NC-D45 is located in the axial valley just south of the mapped area.

from Narrowgate was investigated to relate the water column anomalies to hydrothermal vents, associated mineralization, and tectonic and volcanic features (Fig. 3.2). The bathymetry is based on NOAA multibeam surveys (Malahoff, 1985; Rona and Clague, 1989).

Bathymetric profiles across the rift valley in the GR14 area exhibit a series of four steplike terraces cumulatively rising about 800 m between the floor of the rift valley and the rift mountain crests on either side (Fig. 3.3). The terraces are inferred to be the surfaces of underlying fault blocks (Atwater and Mudie, 1973). A ridge about 200 m high and <1 km wide is crossed by the northern bathymetric profile between about 2,600 and 2,720 m at the level of terrace 2 at middepth on the east wall of the rift valley (Fig. 3.3A). The bathymetric map (Fig. 3.2) shows this feature to be a linear ridge about 0.5 km wide and 2 km long; the ridge is parallel to the axis of the rift valley at the 2,720-m isobath. The linear ridge is surmounted by several small hills attaining a minimum depth of 2,620 m.

The rift valley has a central volcano that rises from the regional depth of the axis of about

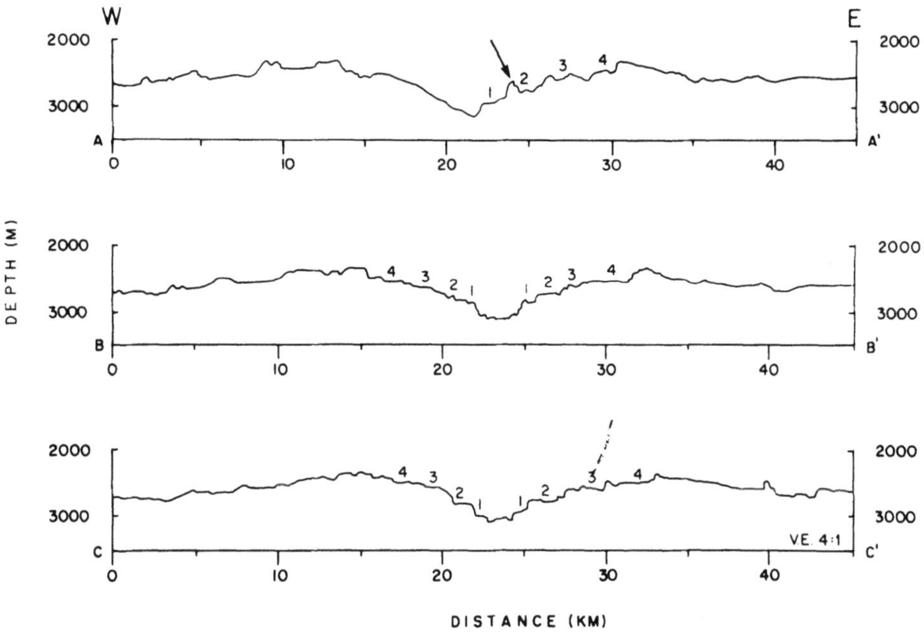

FIGURE 3.3 Narrow-beam bathymetric profiles across the rift valley in the GR14 area along track lines partially shown in Figure 3.2. The linear ridge is crossed at the arrow on profile A. Numbers 1–4 refer to terraces on the tops of inferred fault blocks.

3,100 m to a depth of 2,875 m at about 42° 44′ N, and a series of progressively deeper basins extends to the north and south. The central volcano has a deep moat on the southeast side but has partially filled the valley toward the west wall.

Side-Scan Imagery

A regional mosaic of side-scan sonar images with a SeaMARC II system (10 km swath width, 50 m resolution) is presented in Clague and Holmes (1987), and a detailed analysis of the images in the GR14 area (Rona and Clague, 1989) shows linear reflectors with two consistent trends in the GR14 area. The SeaMARC II imagery and an interpretive line drawing of the identified trends are shown as Figure 3.4. The predominant trend is subparallel to the 022° axial trend of the rift valley and is inferred to be reflections from inward-facing fault scarps spaced hundreds of meters apart on the terraced walls of the rift valley. Similarly oriented reflectors on the floor of the rift valley have wider spacings. A subsidiary 350° linear trend of reflectors, also interpreted as fault scarps, crosses the central portion of the area and intersects the prominent 022° trend on the floor and walls of the rift valley. The deep-towed camera also imaged narrow fissures up to several meters wide trending 350° on the rift valley floor coincident with the subsidiary 350° trend on the side-scan sonar records. This same trend can be seen as larger scale lineations in the GLORIA mosaics south of Narrowgate along the ridge axis.

A side-scan image of the central portion of the GR14 area made with the deep-towed SeaMARC IA system (5 km swath width, 10 m resolution) resolves volcanic features in addition to the tectonic trends already described (Rona and Clague, 1989). The SeaMARC 1A imagery and interpretive line drawings are shown as Figure 3.5. The linear ridge at middepth on the east wall of the rift valley appears as a broad zone of predominantly high reflectivity that shadows the area to the east. A deep-towed camera crossing of the southern end of the linear ridge shows fractured pillow lava flows cut by inward-facing fault scarps and surrounded by sediment. The predominant 022° trend of fault scarps on the east wall is in-

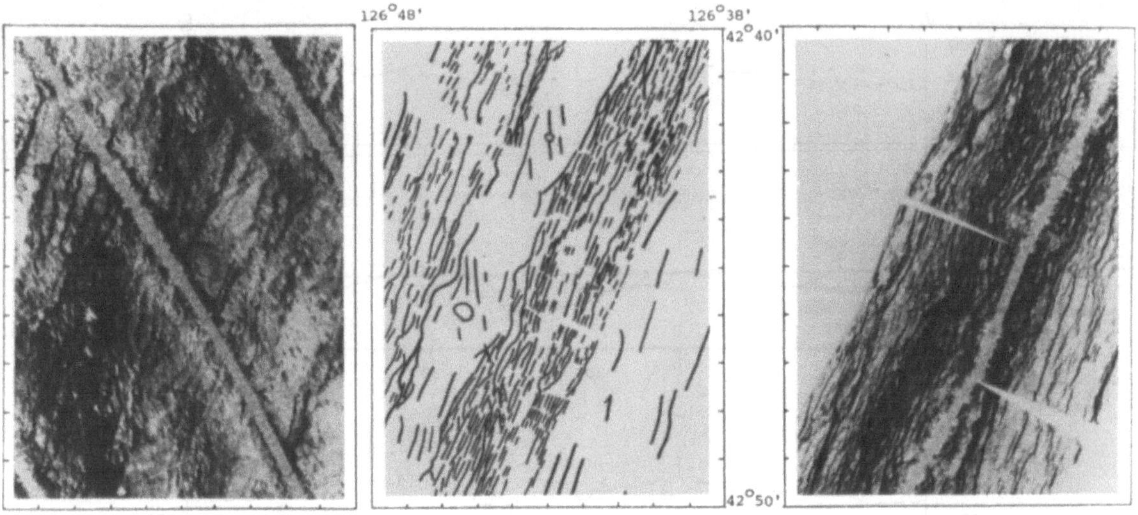

FIGURE 3.4. Mosaics of SeaMARC II side-scan sonar imagery (10-km swaths) and an interpretive line drawing. The mosaics of the SeaMARC II data are from Rona and Clague (1989). Reflections are dark in the SeaMARC II imagery. The outer box is the GR14 study area, and the inner box is the area of the SeaMARC 1A image shown in Figure 3.5. The dominant trend is oriented 022° and is parallel to the ridge. A secondary oblique trend oriented about 350° also is evident.

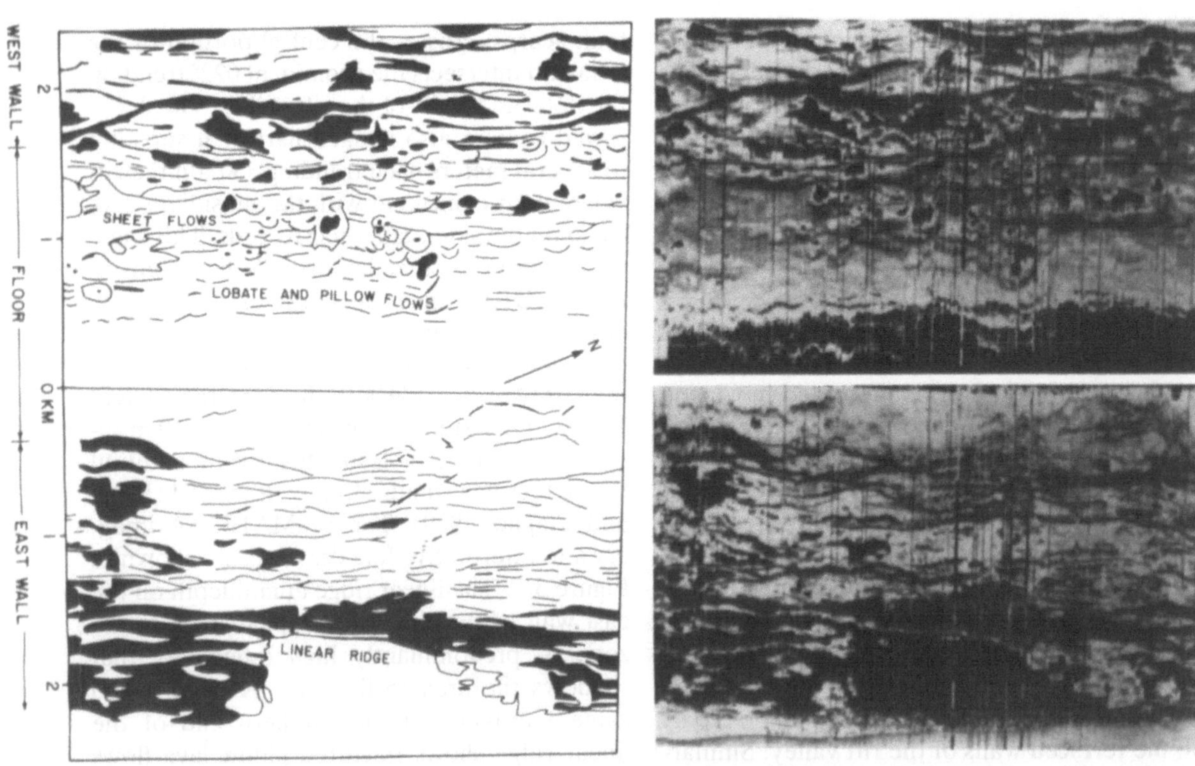

FIGURE 3.5. SeaMARC 1A side-scan sonar imagery (5-km swath) and an interpretive line drawing. The SeaMARC 1A images are from Rona and Clague (1989). Reflections are dark in the SeaMARC 1A imagery. Prominent features are labeled.

tersected by a few linear features with the subsidiary 350° trend that strike toward the southern end of the linear ridge.

Reflections from the rift valley floor have long linear forms with the predominant 022° axial trend of fault scarps, as well as arcuate and circular forms with diameters of tens of meters surrounding dark patches. The arcuate and circular forms are inferred to be mainly pillow lava flows associated with the small volcanic cones and stacks represented by the circular forms surrounding both shadowed and reflective centers. Lobate and pillow flows are the predominant lava flow forms imaged by camera tows over the rift valley floor. Fresh, glassy, nearly sediment-free surfaces of lava flows in bottom photographs and on samples recovered from the axial valley attest to the recency of volcanic activity in the area. An area of uniform gray reflectivity in the southwestern portion of the imaged rift valley floor is inferred to be composed of sheet flows. A small area of sheet flows (about 1 km by 1 km) was imaged by the camera tows in that portion of the rift valley floor. Another small area of sheet flows occurs about 1 km to the north in the axial valley. A third small area of sheet flows occurs about 5.5 km south of the first area. The occurrence of sheet flows indicates higher effusion rates within these small areas. In other regions, a significant statistical correlation exists between sheet flows, fracturing, and hydrothermal activity (Fox et al., 1988). The inferred hydrothermal vents in the GR14 area are close to, but not directly associated with, areas of sheet flows.

Bottom Photography

Bottom video and 35-mm stills were shot during nine stations in 1985 using the U.S. Geological Survey's deep-towed camera sled (Chezar and Lee, 1985). This sled records 4.5 hours of color video and shoots color stills at closely spaced intervals, commonly every 8 seconds. The first six camera deployments were on the upper east wall in the area identified using the water column anomalies as the most likely source of hydrothermal discharge. The last three stations were along the center of the axial valley. This work was followed by an additional four camera stations using NOAA's towed camera system in 1986. These camera stations focused on the axial valley and the lower part of the east wall in a continuing effort to pinpoint the site of hydrothermal discharge. All camera stations were transponder navigated, and the tracks shown in Figure 3.6 are of the relay positioned on the wire above the camera sled.

The camera tows reveal that the upper terrace and walls are nearly completely sediment covered: only rare steep deposits of talus and even rarer in place pillow lavas are seen through the thick sediment cover. The photointerpretation of camera tows in this upper area is shown in Figure 3.7, and selected photographs, keyed to the camera tracks, are shown in Figure 3.8. The sediment is generally light in color and heavily bioturbated (Fig. 3.8A); many animal tracks are seen in the surface sediment. It is difficult to determine the orientation of the talus slopes from the photography, but we infer that the heavily sedimented talus (Fig. 3.8B) has accumulated along scarps produced by ridge-parallel faulting. The larger faults are easily identified in the bathymetry and the SeaMARC 1A and II imagery, but the camera imaged talus at many more, presumably smaller, faults as well. Occasionally, broken pillows can be seen through the thick sediment cover. Many of the hard-substrate outcrops are characterized by abundant sessile organisms.

The photointerpretation of the faulted inner wall of the axial valley and the terrace at the top of the first major fault are shown in Figure 3.9, and a selected photograph is shown in Figure 3.10. The percentage of bottom that is not buried by sediment is much higher than on the terrace to the east. Most of the rock seen is talus, much of it on steep slopes. Most of the talus is at least dusted with sediment (Fig. 3.10A), and some talus deposits are barely visible above the sediment cover. Pillow lavas are commonly seen in this area, but nearly all are fractured and broken; many are rounded, and little of the delicate ornamentation seen on fresh pillows is present. The sediment at the south end of the linear ridge at the top of the

FIGURE 3.6. Multibeam bathymetry with the tracks of the deep-towed camera stations shown. The boxes indicate the boundaries of the photogeologic maps shown as Figures 3.7, 3.9, 3.11, and 3.12. The two areas where the oblique fissure trend transects the axis are shown. The camera stations identified as 85-xx are from the U.S. Geological Survey cruise L5-85-NC whereas those identified N86-x are from the 1986 NOAA cruise.

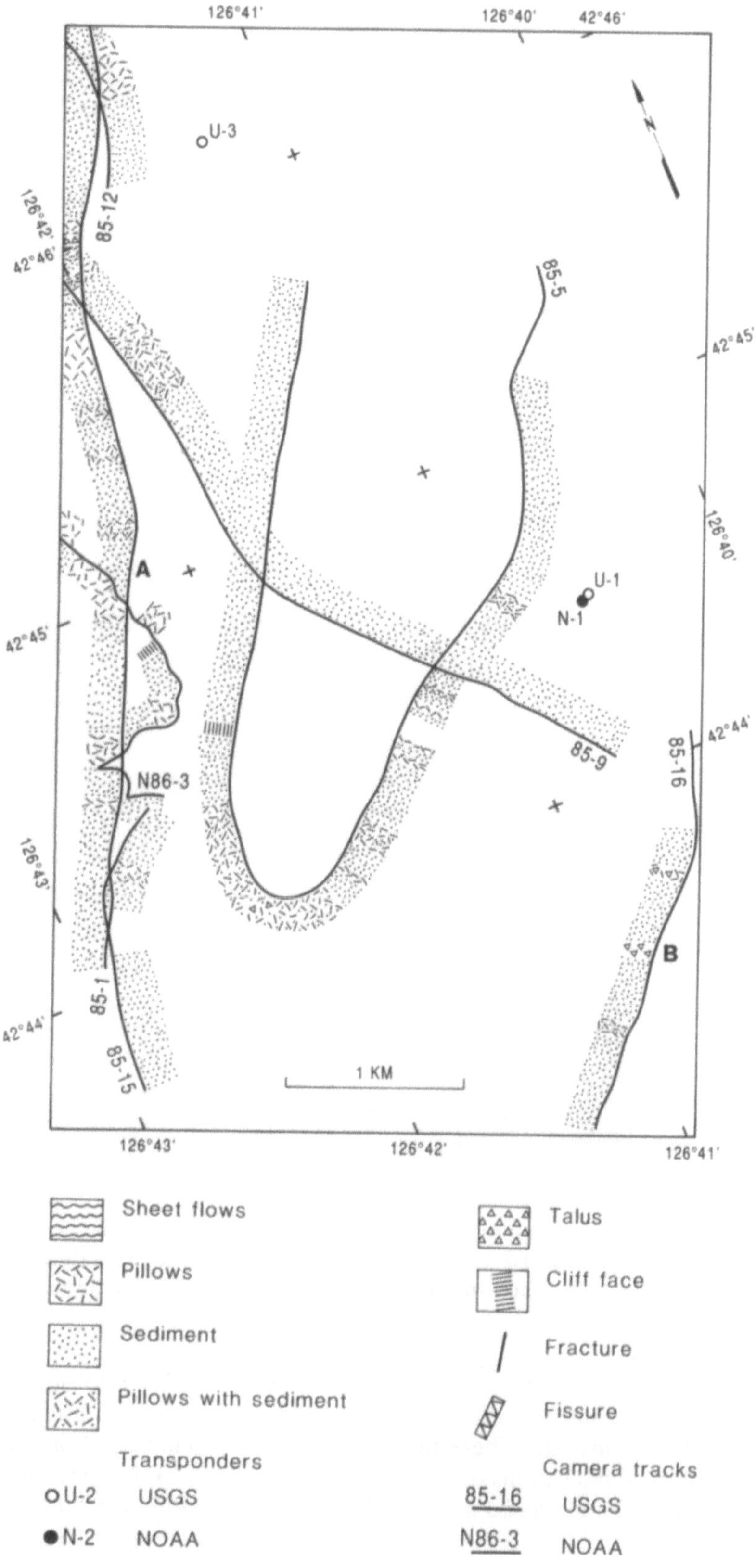

FIGURE 3.7. Photogeology of the upper terrace. The letters indicate the locations of bottom photographs shown in Figure 3.8. The gaps in the geologic symbols indicate data gaps where the camera sled was too high or too close to the bottom. Nearly the entire upper terrace is characterized by thick, bioturbated sediment cover. Only rarely do talus or broken pillow lavas project through the sediment cover.

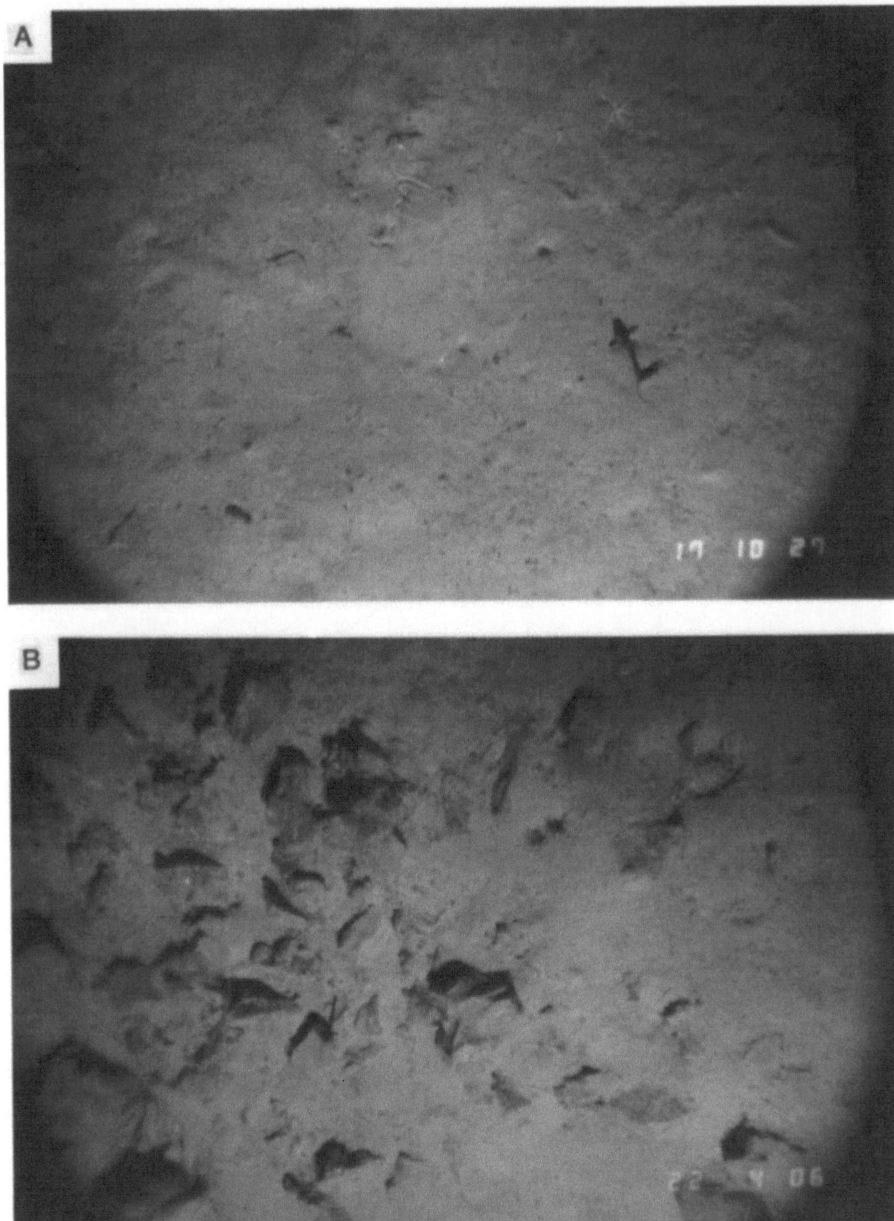

FIGURE 3.8. Photographs located on Figure 3.7. The bottom photographs have variable unknown dimensions but generally image an area 2 to 5 m across. (A) Bottom photograph of heavily bioturbated sediment. (B) Bottom photograph of talus with thick sediment cover.

inner fault scarp is darker in color and coarser grained, and many puffy white "snowflakes" drift in the currents. These features are interpreted to be indications of hydrothermal activity and will be discussed in more detail in a later section.

The photointerpretation of the axial valley is shown in Figures 3.11 and 3.12, and some selected photographs are presented in Figure 3.13. Much of the axial valley floor consists of in-place pillow lavas with delicate surface ornamentation (Fig. 3.13A) and rare sheet flows

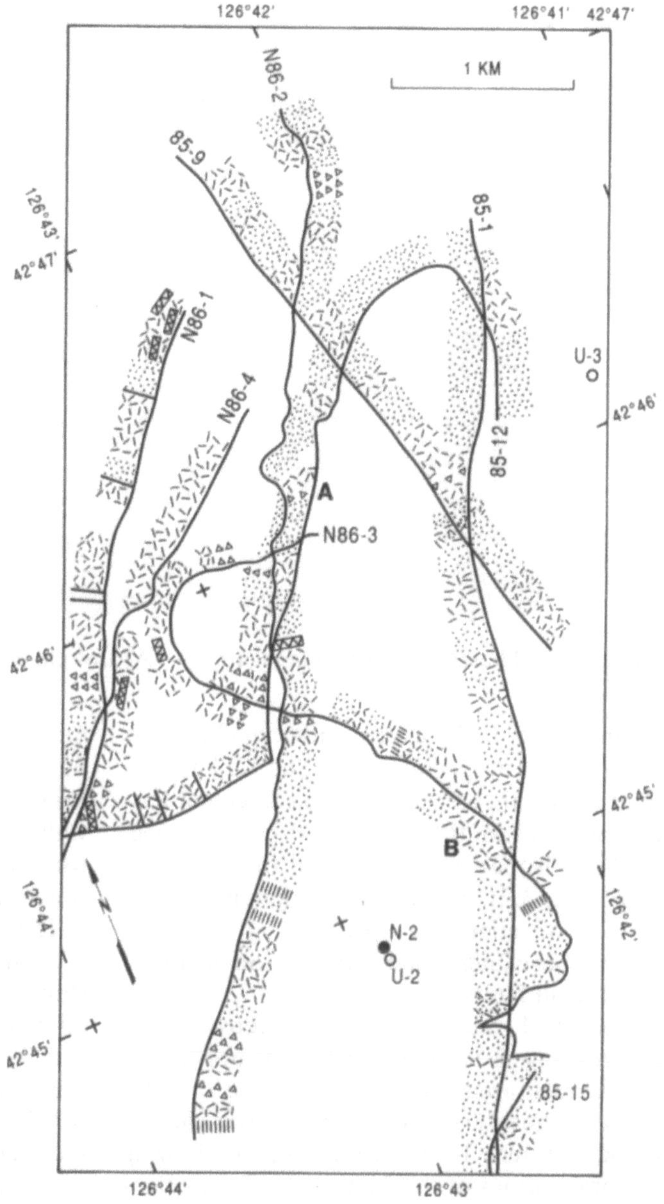

FIGURE 3.9. Photogeology of the wall of the axial valley. (A) Location of the photograph shown in Figure 3.10. (B) Location of the darker and coarser-grained sediment at the southern end of the linear ridge. This sediment is inferred to be of hydrothermal origin. The other symbols are explained in Figure 3.7. Much of the wall area is characterized by talus with much thinner sediment cover than that seen on the upper terrace. Pillow lavas, commonly broken and fractured, are also seen.

(Fig. 3.13B). Most of these lavas have only thin to moderate sediment cover, so the lava forms are easily identified. Talus is also frequently encountered in the axial valley, both near the base of the inner wall as extensive steep-sloped deposits and as broken pillows exposed along small ridge-perpendicular faults. A third tectonic trend oriented about 292°, perpendicular to the predominant 022° axial trend, is inferred from successive crossings of fault scarps by camera tows along the axis of the rift valley floor and from an alternation of topographic highs and lows at spacings of between 0.6 and 3 km along the axis coincident with left-lateral offsets of isobaths transverse to the axis (Fig. 3.2).

The orientation of the side-scan sonar traverse to the rift valley axis precluded imaging these transverse scarps on the SeaMARC 1A records. These transverse scarps are apparently too small to be seen in the Sea MARC II or GLORIA images. From north to

FIGURE 3.10. Photograph located in Figure 3.9. (A) Bottom photograph of talus with thin sediment cover.

south, three fault scarps with angular talus were imaged at spacings of 2.9 and 1.7 km. The fault scarps are associated with coincident left-lateral offsets of 0.5, 0.2, and 0.7 km, respectively, of isobaths at the base of the rift valley walls (Fig. 3.2). These faults commonly have 5 to 15 m of relief and appear to step up to the north; the camera sled hung up on many of these small faults.

The axial valley also has numerous open fissures (Fig. 3.13C) not seen in the upper wall and terrace areas. We measured the orientations of these fissures relative to the camera track; Figure 3.14 is a rose diagram showing the orientations of the 46 fissures whose orientations we could determine. In measuring the orientation of the fissures, we assumed that the camera sled towed parallel to the ship track. Despite the errors undoubtedly introduced by the methodology, the fissures appear to be of three main orientations: ridge-parallel, ridge-perpendicular, and an oblique trend of about 350° orientation. The ridge-parallel fissures are only seen in some, presumably older, sections of the axial valley. The fissures vary in width from roughly 0.3 m to widths beyond the range of the photograph. The ridge-perpendicular fissures are rare, quite narrow, and located near talus deposits. The oblique-trending fissures occur in two areas as fissure swarms (Fig. 3.2); one of these projects up the east wall to the southern end of the linear ridge, whereas the other is just north of the small central volcano in the axis at 42° 44.2′ N.

The abundance of open fissures, talus, broken pillows (Fig. 3.13D), and fractures (Fig. 3.13E) indicates that much of the valley floor has experienced considerable tectonic activity since eruption of many of the flows in the valley. Some flows, however, are unfractured and unfissured, have thin to nearly no sediment cover, and have delicate ornamentation, such as small buds and striated original glass surfaces, on the pillows. These features indicate relatively young posttectonic volcanic activity in the area. Palagonite thicknesses on the basalt samples recovered from the axial valley indicate that the lavas are older than a few hundred years. Nearly all the sediment-free hard substrate is populated by sessile organisms (Fig. 3.13F).

Dredged Lavas

Eighteen dredges from the U.S. Geological Survey's cruises in 1985 and 1986 (Fig. 3.2) and

FIGURE 3.11. Photogeology of the northern part of the axial valley. Letters indicate the locations of photographs shown in Figure 3.13, and the symbols are explained in Figure 3.7. Station L5-85-46C is not fully depicted to avoid confusion where it is in close proximity to stations L5-85-41C and L5-85-37C.

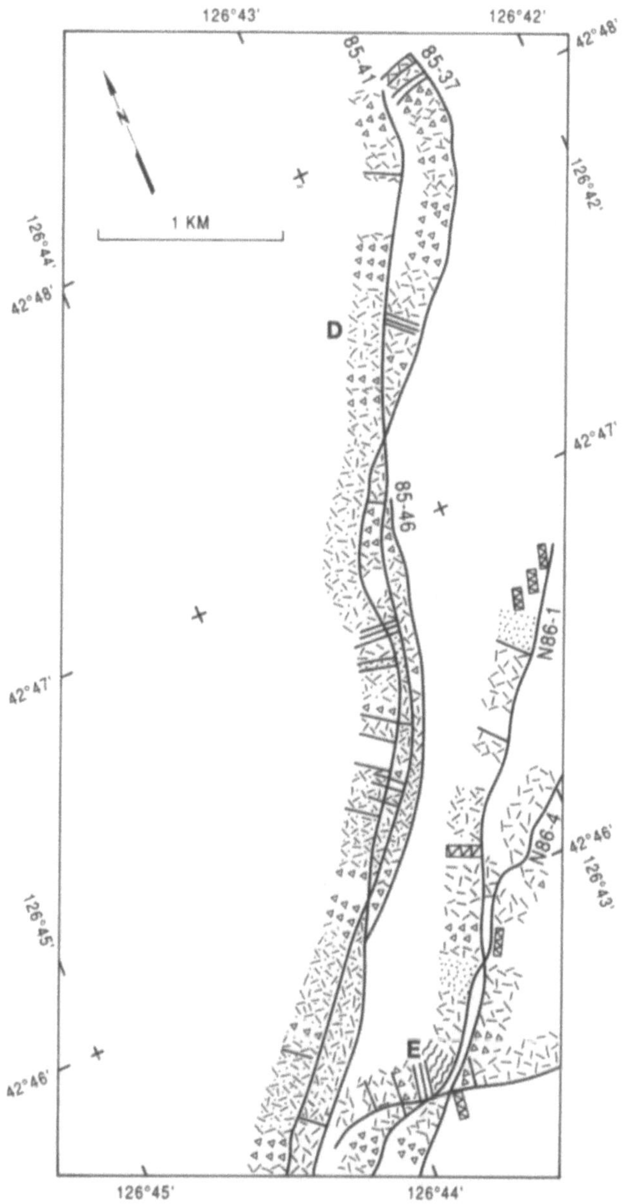

three camera tows (L5-85-12C, L5-85-41C, and NOAA 86-1) recovered samples of midocean ridge basalt from the floor, lower west wall, and the east wall of the axial valley. The L5-85-NC dredge tracks shown are the transponder navigated positions of the ship from the time the dredge was on bottom to the time it was off bottom. The L3-86-NC dredge tracks shown are the GPS-navigated positions of the ship from the time the dredge was on bottom to the time it was off bottom. The depths and start and end positions of the dredges are listed in Table 3.1.

The lavas recovered are as diverse as those previously described from the entire Gorda Ridge (Davis and Clague, 1987). The lavas range from ferrobasalt with Mg numbers of 56 to near-primary lavas having Mg numbers of 69. The lavas from the walls are more homogeneous than those from the axial valley, but evolved and near-primary lavas occur in the axial valley and on the walls. Microprobe

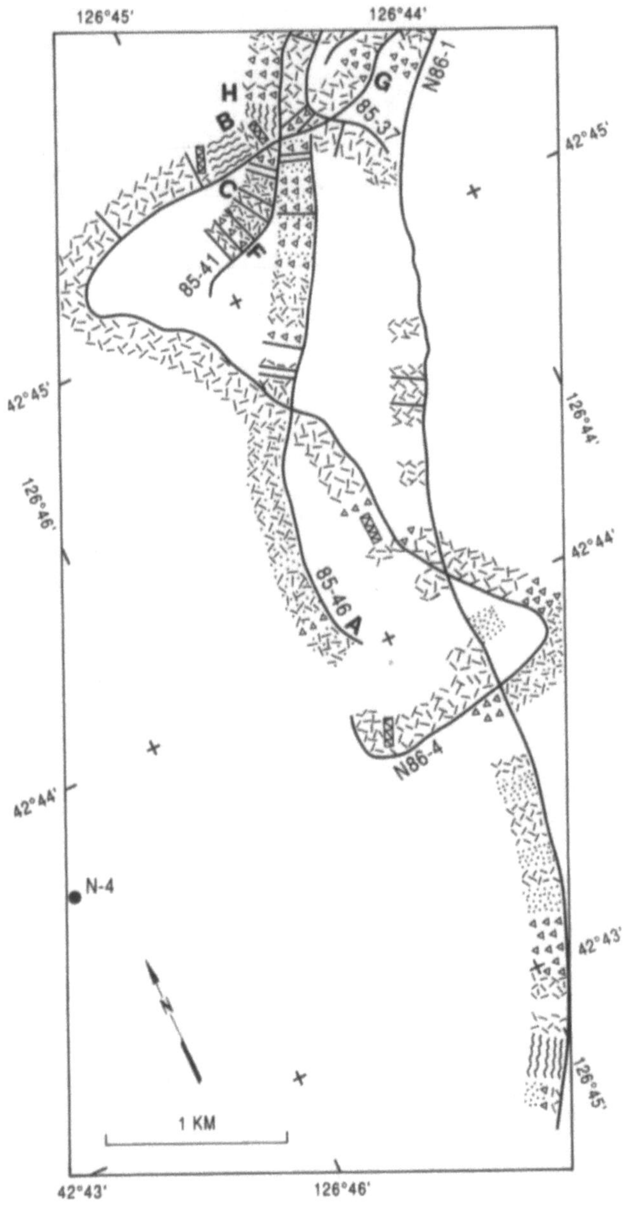

FIGURE 3.12. Photogeology of the southern part of the axial valley. Letters indicate the locations of photographs shown in Figure 3.13. The symbols are explained in Figure 3.7.

analyses of glasses from the lavas from the GR14 areas are presented by Davis and Clague (1989). None of the areas of sheet flows were sampled. One dredge on the floor of the axial valley slightly north of the linear ridge recovered a large suite of gabbroic xenoliths inferred to be derived from the roof and walls of a magma chamber; descriptions and analyses of these xenoliths are also in Davis and Clague (1989). There is no obvious correlation of lava chemistry and proximity to the oblique tectonic trend. The petrologic diversity suggests that magma chambers in the area may be small and discontinuous (David and Clague, 1987) and possibly at greater depth than observed along the East Pacific Rise (Detrick et al., 1987). However, recent geophysical modeling of ridge crests indicates that petrologic diversity may also be produced from large, long-lived magma chambers (Wilson et al., 1988).

Dredges L3-86-11 and L3-86-15 (Fig. 3.2) recovered lavas coated with white deposits of boehmite of hydrothermal origin (Howard and Fisk, 1988); these samples will be discussed in more detail in a later section. The 1986 dredges were specifically located to sample a white precipitate photographed during one of the 1986 NOAA camera tows. None of the dredges recovered samples of stratiform hydrothermal deposits.

Results from Coring

Several short gravity cores were taken on the U.S. Geological Survey cruises in 1985 and 1986. The locations of these cores are shown in Figure 3.2. Results from these and other cores are presented by Nelsen et al. (1987). Additional cores from this area collected by Oregon State University are described by Karlin and Lyle (1986).

Hydrothermal Mineralization

There is evidence for hydrothermal vents at two areas within the GR14 region: at the southern end of the linear ridge on the east wall of the axial valley, and directly downslope from the first site but in the axial valley. A deep-towed camera profile across the southern end of the linear ridge shows an abrupt change in sea floor sediment color and texture from nearly ubiquitous light tan lutite to dark yellow gray, coarser-grained material indicative of metalliferous sediment (location B in Fig. 3.9). An anomalously high Cu content was measured in sediment core L3-86-13G from this area (Nelsen et al., 1987). White fragmentary material scattered on the surface of the dark sediment appears to be either bacterial mats and/or sulfate deposits. The number of fishes imaged near the seafloor in this area increased about 10-fold, from 1/km (background levels) to 10/km. The concurrence of water column and sea floor evidence suggests that a hydrothermal venting zone is present within hundreds of meters of these camera observations at the southern end of the linear ridge (Rona and Clague, 1986, 1989).

Camera tows over the floor of the rift valley in the GR14 area imaged two types of hydrothermal deposits in a small area (about 1 km by 1 km) near the axis. The first deposit is greenish-gray massive material with fine-scale irregularities on its surface (Fig. 3.13G from location G on Fig. 3.12). The appearance of the material suggests that it is nontronite, but altered sulfide is also possible. The hydrothermal deposit of nontronite(?) occurs near the center of the axial valley in one of the small deeps to the north of the central volcano. The second type of deposit is a white coating on dark-colored angular talus fragments (Fig. 3.13H from location H on Fig. 3.12). This deposit was photographed at the base of one of the transverse fault scarps and was subsequently sampled by dredges L3-86-11 and L3-86-15. The talus is composed of basalt pillow fragments with fresh glass on one surface. The white coating occurs on several different faces of each fragment including the glassy rims but never coats the entire fragment. Analysis of the white material indicates that it is an alteration crust up to 1 mm thick composed of an inner layer of Al-rich interlayered smectite/chlorite and an outer layer composed of the aluminum hydroxide mineral boehmite (Howard and Fisk, 1988). The mixed smectite/chlorite layer is inferred to have formed by high-temperature hydrothermal alteration at about 300°C, whereas the boehmite layer apparently formed by high-temperature leaching of the underlying clay layer. It is unclear whether the alteration processes occurred by flow of hydrothermal solutions through fractured basalt pillows along a fault zone prior to accumulation of the talus or by flow through the talus pile after it accumulated at the base of the fault scarp.

Discussion

The GR14 area is located close to the narrowest and shallowest portion of the northern section of the northern Gorda Ridge. Francheteau and Ballard (1983) have proposed that such locations mark centralized regions of magmatic sources and that hydrothermal discharge zones are concentrated in these sites. The location of hydrothermal discharge zones

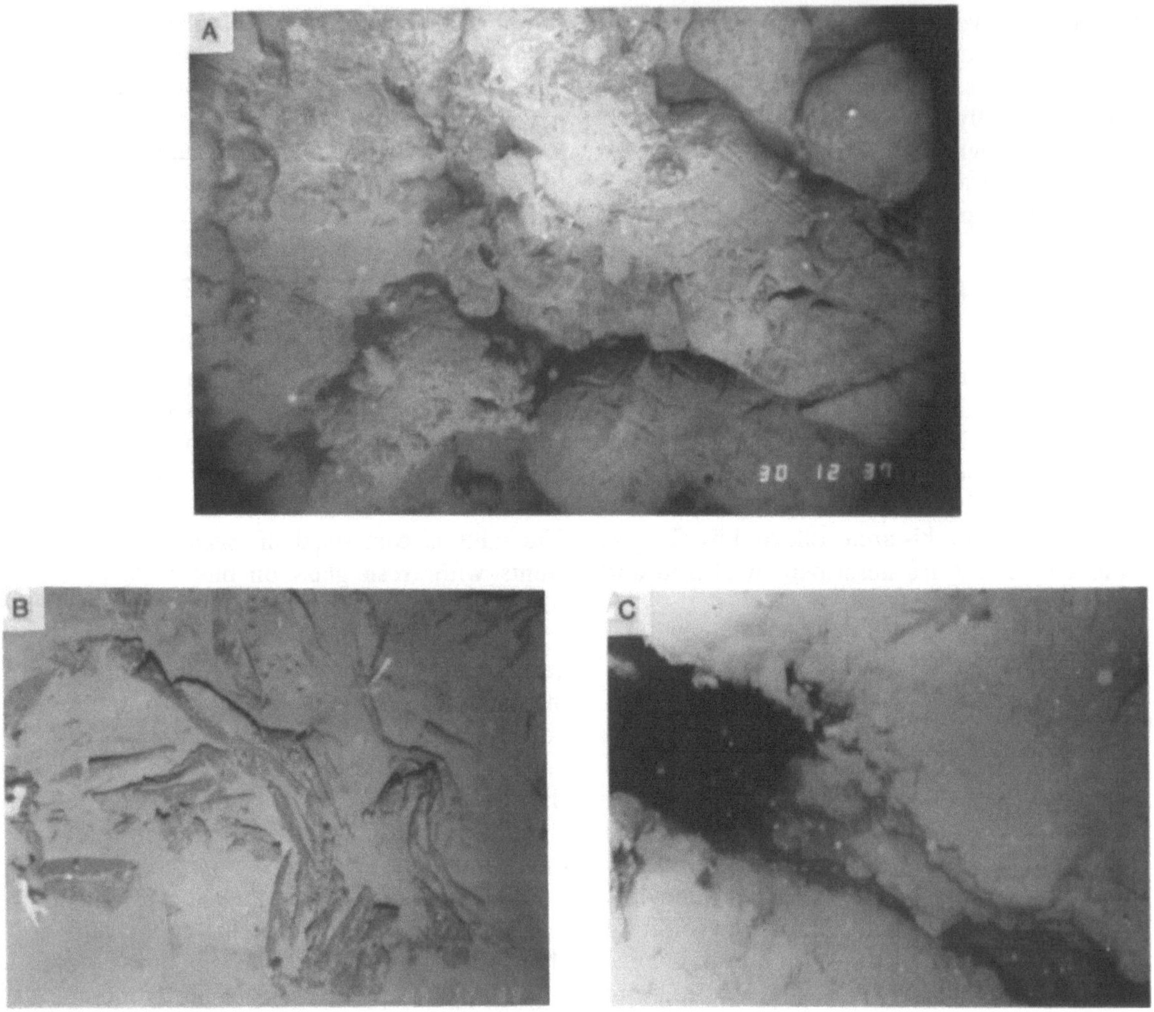

Figure 3.13. Photographs located in Figures 3.11 and 3.12. (A) Bottom photograph of young pillow lava with small buds and knobs. (B) Rumpled sheet flows from the southern area of sheet flows. (C) Portion of a 350° trending fissure about 1 m wide in pillow lava. (D) Broken pillow lava. (E) Fractured pillow lava. (F) Abundant sessile organisms on talus. (G) A stratiform hydrothermal precipitate tentatively identified as nontronite. (H) Talus of pillow basalt fragments with a white coating identified as hydrothermal boehmite. (From Howard and Fisk, 1988.)

in the GR14 area of Gorda Ridge supports their contention, although more thorough exploration may find other sites of discharge along the axis, such as at the GR15 area north of the axial high (Fig. 3.1).

The GR14 area has two anomalous features: the subsidiary oblique 350° trending faults and fissures, and the presence of the linear ridge on the east wall of the axial valley. These features may well be related, as the linear ridge is bounded to the south by the projection of a cluster of 350° trending fissures located in the axial valley. The oblique tectonic trend is probably related to the deformation distributed in the Gorda area to the southeast that accommodates convergent motion of surrounding plates (Wilson, 1986, 1989; Stoddard, 1987). The model presented by Wilson (1986, 1989) suggests that the sense of motion along this trend (his model predicts a more northwest orientation for this trend) is mainly right-lateral strike-slip motion near the ridge axis. The linear ridge

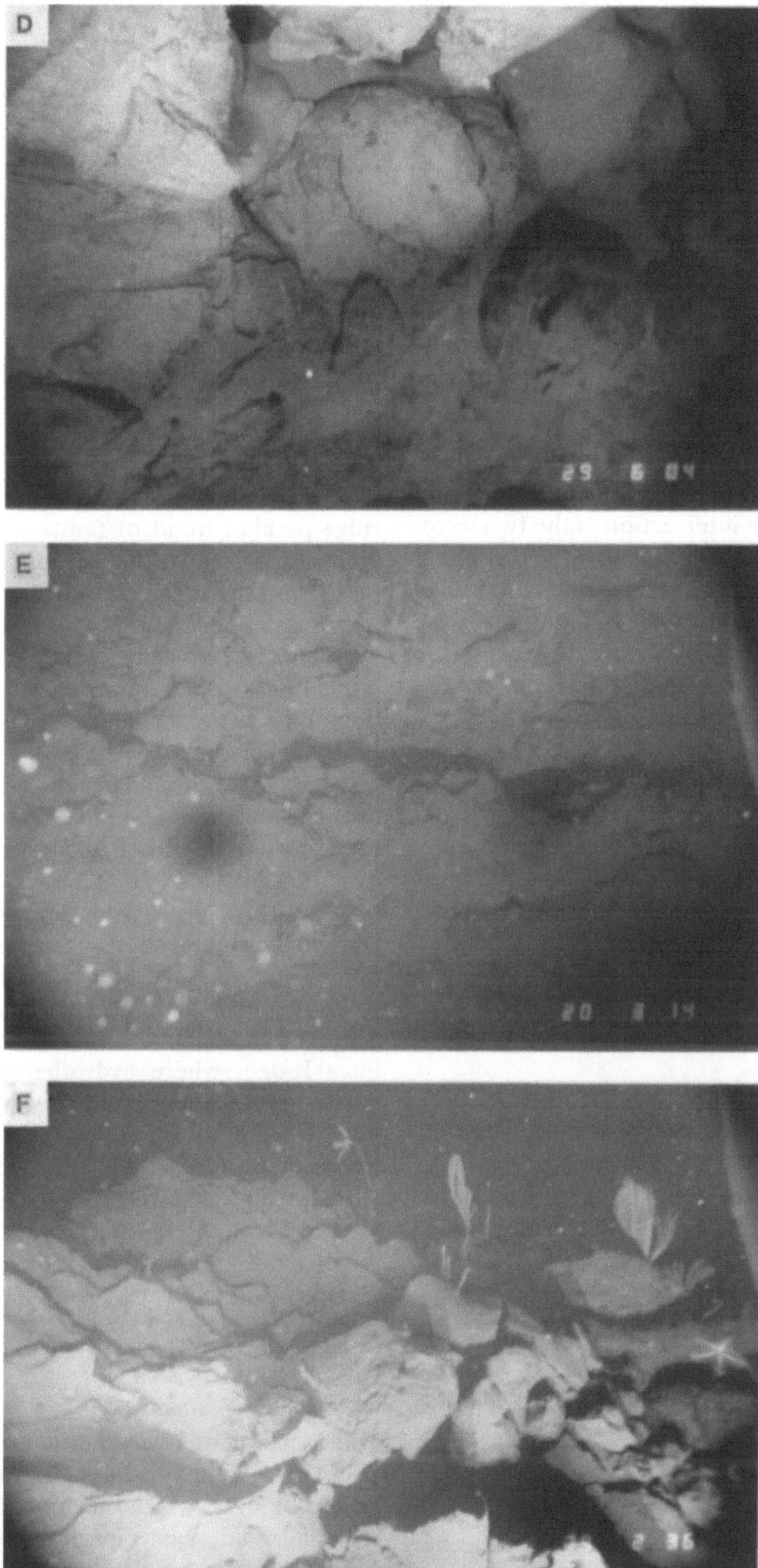

Fig. 3.13

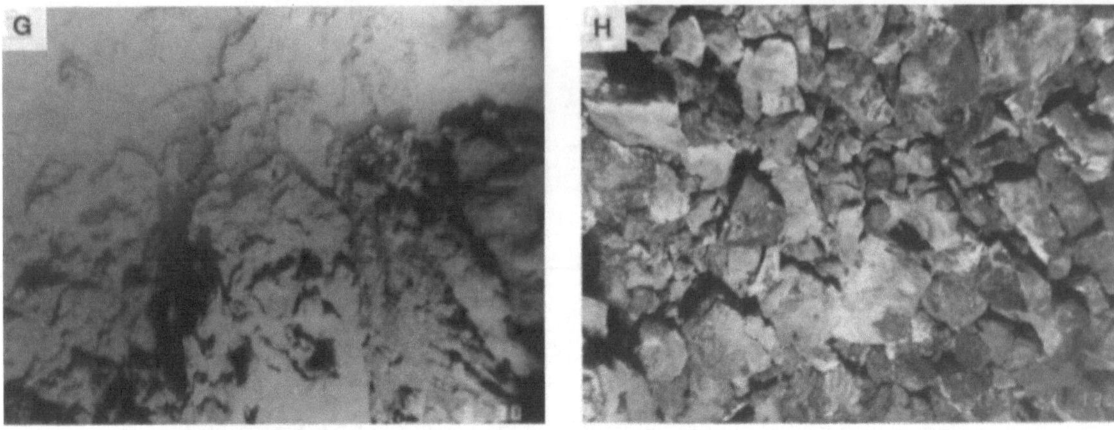

Fig. 3.13

may have been uplifted because of compression related to the intersection of the two tectonic trends. It is also possible that the linear ridge is a volcanically constructed medial ridge uplifted on the wall of the rift valley, or is the product of off-axis volcanism. However, the heavy sediment cover on and near the ridge is inconsistent with formation of the ridge by off-axis volcanism.

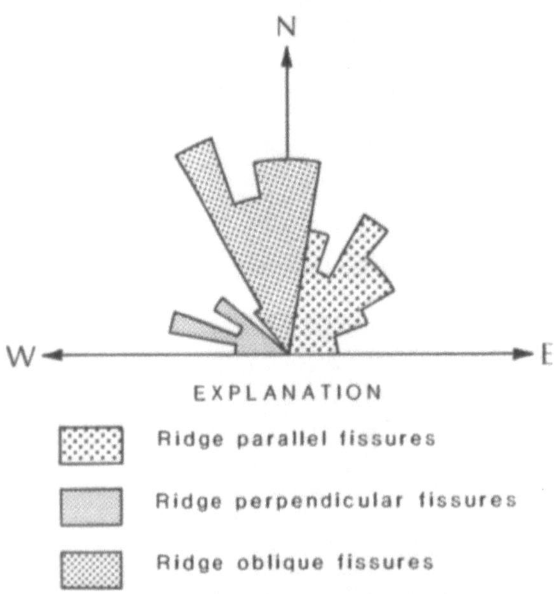

FIGURE 3.14. Rose diagram of the orientations of fissures in the axial valley. These data are compiled from camera stations 85-37, 85-41, 85-46, and the first part of 85-12. Most of the imaged fissures have the oblique 350° trend. The ridge-parallel trend of 022° is secondary. The ridge-perpendicular trend is represented by only a few fissures and small scarps of talus.

The GR14 site has two tectonic trends: the ridge-parallel trend of faults along which the elongate blocks of the rift mountains are uplifted, and the subsidiary oblique trend of fault scarps and open fissures oriented roughly 350°. The inferred site of active hydrothermal venting is off axis and coincides with the intersection of the two tectonic trends on the south end of the linear ridge. The role of faults along the margins of a rift valley as conduits for hydrothermal discharge has been emphasized previously (e.g., Rona, 1984; Malahoff, 1985). The area in the axial valley where hydrothermal deposits were photographed and dredged is near the intersection of a subsidiary cluster of 350° fissures and the ridge-parallel fissure trend. Similar structural relations have been observed at Axial Seamount on the Juan de Fuca Ridge, where hydrothermal sulfide deposits are located at the intersection of the fault bounding the caldera and an eruptive fissure system (Chase et al., 1985). Numerous on-land deposits are also thought to have formed at the intersections of two or more tectonic trends (e.g., Kutina, 1969; Mitchell and Garson, 1981).

Conclusions

The location of hydrothermal activity in the GR14 area of the northern Gorda Ridge is controlled by volcanic and tectonic features. At a regional scale, the volcanic budget along the ridge controls the location of hydrothermal vents so that the area of hydrothermal dis-

TABLE 3.1 Dredge locations and depths.

Dredge number	Start of dredge			End of dredge		
	Lat. (N)	Long. (W)	Depth (m)	Lat. (N)	Long. (W)	Depth (m)
L5-85-2D	42° 45.57′	126° 41.97′	2,745	42° 45.55′	126° 41.79′	2,726
L5-85-3D	42° 45.15′	126° 42.04′	2,682	42° 45.23′	126° 41.33′	2,571
L5-85-4D	42° 45.33′	126° 41.49′	2,675	42° 45.37′	126° 40.96′	2,542
L5-85-6D	42° 44.85′	126° 42.38′	2,690	42° 44.45′	126° 41.45′	2,549
L5-85-7D	42° 45.37′	126° 42.18′	2,594	42° 45.77′	126° 41.85′	2,791
L5-85-8D	42° 45.49′	126° 42.32′	2,734	42° 45.49′	126° 42.09′	2,734
L5-85-10D	42° 43.66′	126° 42.58′	2,719	42° 43.90′	126° 41.97′	2,542
L5-85-11D	42° 44.76′	126° 43.31′	2,720	42° 44.71′	126° 43.09′	2,717
L5-85-13D	42° 45.38′	126° 43.93′	3,075	42° 45.68′	126° 43.38′	2,868
L5-85-38D	42° 47.08′	126° 43.21′	3,045	42° 46.12′	126° 43.78′	3,038
L5-85-39D	42° 45.90′	126° 44.08′	3,031	42° 46.39′	126° 43.23′	2,956
L5-85-40D	42° 45.51′	126° 44.65′	3,026	42° 46.28′	126° 44.24′	3,045
L5-85-42D	42° 47.91′	126° 42.20′	3,141	42° 47.41′	126° 42.72′	3,097
L5-85-44D	42° 45.50′	126° 44.72′	2,994	42° 42.77′	126° 46.33′	2,823
L5-85-45D	42° 42.29′	126° 45.42′	3,145	42° 42.77′	126° 46.33′	3,034
L6-86-11D	42° 45.65′	126° 44.35′	3,067	42° 46.50′	126° 44.08′	3,067
L6-86-12D	42° 45.42′	126° 44.07′	3,075	42° 45.93′	126° 44.89′	2,980
L6-86-15D	42° 45.87′	126° 44.07′	3,075	42° 46.38′	126° 43.79′	3,075

charge occurs near the shallowest and volcanically most active section of the ridge. On a local scale, the presence of fissures and faults, and particularly their areas of intersection, controls the distribution of hydrothermal discharge.

Acknowledgments. We thank the captains and crews of the USGS R/V *S.P. Lee* and the NOAA ship *Discoverer* for their support. The Gorda Ridge Technical Task Force coordinated the cooperative work between government and academic groups. We acknowledge the efforts of Randy Koski and Andy Stevenson in photographing and dredging the east wall of the study area and of Debby Tatman in organizing and compiling photographic data. Chris Gutmacher and Stephanie Ross performed acoustic navigation under difficult conditions. We thank Alex Malahoff, Steve Hammond, and Bob Embley for multibeam bathymetry of the study area. Chris Fox and Bob Embley collaborated with P.A.R. in collection of the SeaMARC 1A data discussed in the text. Our thanks to Bill Normark, Stephanie Ross, Jan Morton, and Randy Koski for constructive comments. P.A.R. acknowledges the NOAA VENTS Program for support.

References

Atwater T, Mudie JD (1973) Detailed near-bottom geophysical study of the Gorda Ridge. J Geophys Res 78:8665–8686.

Baker ET, Massoth GJ, Collier RW, et al. (1987) Evidence for high-temperature hydrothermal venting on the Gorda Ridge, northeast Pacific Ocean. Deep-Sea Res 34:1461–1476.

Bolt BA, Lomnitz C, McElvilly TV (1968) Seismological evidence on the tectonics of central and northern California and the Mendocino Escarpment. Bull Seismol Soc Am 58:1725–1767.

Chandra U (1974) Seismicity, earthquake mechanisms, and tectonics along the western coast of North America from 42°N to 62°N. Bull Seismol Soc Am 64:1529–1549.

Chase RL, Delaney JR, Karsten JL, et al. (1985) Hydrothermal vents on an axis seamount of the Juan de Fuca Ridge. Nature 317:212–214.

Chezar H, Lee J (1985) A new look at deep-sea video. Deep-Sea Res 32:1429–1436.

Clague DA, Holmes ML (1987) Geology, petrology, and mineral potential of the Gorda Ridge. In Scholl DW, Grantz A, Vedder J (eds): Geology and Resource Potential of the Continental Margin of Western North America and Adjacent Ocean Basins—Beaufort Sea to Baja California: Circum-Pacific Council for Energy and Mineral Resources. Circum-Pacific Earth Sci Ser 6:563–580.

Collier RW, Holbrook SH, Robbins JM (1986) Studies of trace metals and active hydrothermal venting on the Gorda Ridge. Portland, Oregon, Department of Geology and Mineral Industries, Open-File Report 0-86-13: 36 p.

Davis AS, Clague DA (1987) Geochemistry, mineralogy, and petrogenesis of basalt from the Gorda Ridge. J Geophys Res 92: 10467–10483.

Davis AS, Clague DA (1989) Gabbroic xenoliths from the northern Gorda Ridge: Implications for magma chamber processes under slow-spreading centers. J Geophys Res (in press).

Detrick R, Buhl P, Vera E, et al. (1987) Multichannel seismic imaging of a crustal magma chamber along the East Pacific Rise. Nature 326: 35–41.

EEZ-Scan 1984 Scientific Staff (1986) Atlas of the Exclusive Economic Zone, western conterminous United States, U.S. Geological Survey Miscellaneous Investigations Ser I-1792: 152 p.

Feely RA, Lewison M, Massoth GL, et al. (1987) Composition and dissolution of black smoker particulates from active vents on the Juan de Fuca Ridge. J Geophys Res 92: 11347–11363.

Fox CG, Murphy KM, Embley RW (1988) Automated display and statistical analysis of interpreted deep-sea bottom photographs. Marine Geol 78: 199–216.

Francheteau J, Ballard RD (1983) The East Pacific Rise near 21°N, 13°N, and 20°S: Inferences for along-strike variability of axial processes of the mid-ocean ridge. Earth Planetary Sci Lett 64: 93–116.

Harper GD (1985) Tectonics of slow-spreading ridges and consequences of a variable depth to the brittle/ductile transition. Tectonics 4: 395–409.

Howard KJ, Fisk MR (1988) Hydrothermal alumina-rich clays and boehmite on the Gorda Ridge. Geochem Cosmochem Acta 52: 2269–2279.

Janecky DR, Seyfried WE Jr (1983) The solubility of magnesium-sulfate-hydrite in seawater at elevated temperatures and pressures. Am J Sci 283: 831–860.

Kadko D (1986) Radon-222 as a real-time tracer of hydrothermal activity on the Gorda Ridge. Portland, Oregon, Department of Geology and Mineral Industries, Open-File Report 0-86-14: 19 p.

Karlin R, Lyle M (1986) Sediment studies on the Gorda Ridge. Portland, Oregon, Department of Geology and Mineral Industries, Open-File Report 0-86-19: 76 p.

Koski RA, Shanks WC, Bohrson WA, Oscarson RL (1988) The composition of massive sulfide deposits from the sediment-covered floor of the Escanaba Trough, Gorda Ridge: Implications for depositional processes. Can Mineral 26: 655–673.

Kutina J (1969) Hydrothermal ore deposits in the western United States: A new concept of structural control of distribution. Science 165: 1113–1119.

Malahoff A (1985) Hydrothermal vents and polymetallic sulfides of the Galapagos and Gorda/Juan de Fuca ridge systems and of submarine volcanoes. Biol Soc Wash Bull 6: 19–41.

Mitchell AHG, Garson MS (1981) Mineral Deposits and Global Tectonic Settings. New York: Academic Press, 405 pp.

Morton JL, Holmes ML, Koski RA (1987) Volcanism and massive sulfide formation at a sedimented spreading center, Escanaba Trough, Gorda Ridge, northeast Pacific Ocean. Geophys Res Lett 14: 769–772.

Nelsen TA, Iricanin N, Trefry JH (1987) Suspended and bottom sediments from the Gorda Ridge. Portland, Oregon, May 11–13, 1987, Gorda Ridge Symposium, Program with Abstracts.

Riddihough RP (1980) Gorda plate motions from magnetic anomaly analysis. Earth Planetary Sci Lett 51: 163–170.

Rona PA (1984) Hydrothermal mineralization at seafloor spreading centers. Earth-Sci Rev 20: 1–104.

Rona PA, Clague DA (1986) Geologic setting of hydrothermal activity at the northern Gorda Ridge. EOS Trans Am Geophysic Union 67: 1028.

Rona PA, Clague DA (1989) Geologic controls of hydrothermal discharge on the northern Gorda Ridge. Geology (in press).

Seeber L, Barazangi M, Nowroozi A (1970) Microearthquake seismicity and tectonics of coastal northern California. Bull Seismol Soc Am 60: 1669–1699.

Stoddard PR (1987) A kinematic model for the evolution of the Gorda plate. J Geophys Res 92: 11524–11532.

Tobin DG, Sykes LR (1968) Seismicity and tectonics of the northeast Pacific Ocean. J Geophys Res 73: 3821–3845.

Wilson DS (1986) A kinematic model for the Gorda deformation zone as a diffuse southern boundary of the Juan de Fuca plate. J Geophys Res 91: 10259–10269.

Wilson DS (1989) Deformation of the so-called Gorda Plate. J Geophys Res (in press).

Wilson DS, Clague DA, Sleep NH, Morton JL (1988) Implications of magma convection for the size and temperature of magma chambers at fast spreading centers. J Geophys Res 93: 11974–11984.

4—Multistage Hydrothermal Systems in the Blanco Fracture Zone

ROGER HART, JOCHEN HOEFS, and DOUGLAS PYLE

Abstract

The deeper portions of paleohydrothermal systems are exposed in the Blanco Fracture Zone. Multistage hydrothermal alteration, brecciation, and sulfide deposition occurred over changing regimens of temperature and fluid chemistry. Oxygen isotope analyses and secondary mineral phases indicate that metagabbros formed at temperatures in excess of 500°C. Basalt and diabase breccias formed at temperatures between 350°C and 500°C. Quartz-chert-cemented breccias formed at temperatures below 200°C. Vein fillings of carbonate, prehnite, and zeolites formed at lower temperatures. The $^{87}Sr/^{86}Sr$ ratio of the fluids and the water/rock (W/R) ratio did not change appreciably during the hydrothermal events in the Blanco Fracture Zone. The temperature changes may have been the result of tectonic movement of the crust relative to heat sources. In the case of chert-cemented breccias the change of temperature rapidly followed brecciation and could not have been the result of either tectonic movement or transportation within the hydrothermal system. We propose that adiabadic cooling followed brecciation. The data suggest that the temperature and pressure of seawater increased as it circulated at depth in the Blanco Fracture Zone. The increase of temperature resulted in a decrease in the solubility of Mg and Cr. These elements were stripped from seawater in exchange for Si, Ca, and probably Al, which were leached from the crust and dissolved in the fluids at high temperatures. Cooling of the crust by hydrothermal solutions resulted in contraction and brecciation. The hydrothermal fluids expanded into the cracks in the crust and cooled adiabatically. The drop of temperature of the fluids caused a decrease in the solubility of Si, Ca, and Al. These elements precipitated in secondary minerals within the crack system. Ti was also mobilized locally within the crack system during this stage. The secondary minerals filled the cracks and curtailed hydrothermal circulation. This caused the crust to be reheated by equilibrating with the regional thermal gradient. The overall process was initiated again. These multiple episodes are now recorded in the hydrothermal phases. Sulfide deposition occurred in at least three stages: (1) disseminated pyrrhotite and minor pentlandite precipitated in the gabbros; (2) pyrite with minor chalcopyrite and sphalerite formed in veins in brecciated basalts; and (3) sphalerite, bornite, chalcopyrite, and galena formed in late stage drusy quartz.

Introduction

Fracture zones offer two excellent opportunities to investigate submarine hydrothermal systems. First of all, they act as conduits for concentrating fluid flow. Second, through tectonic uplift, they expose deeper sections of the hydrothermal system not accessible at spreading ridge vent sites.

Rona (1976) suggested that the discharge from ridge crest geothermal systems is concentrated in the fractures at rift margins. In Phanerozoic and Archaean ophiolites, massive sulfide deposits occur within fracture zones that transect the sheeted dikes and lower pillow basalts (Franklin et al., 1981; Harper et al., 1987). Evidence of active hydrothermal activity within oceanic fracture zones is extensive. Bonatti et al. (1976) found pyrite concretions in the Romanche Fracture Zone. Jenkins and

Clarke (1976) detected a helium anomaly near the Charlie Gibbs transform fault. Inactive hydrothermal vents have been identified in Fracture Zone A at 37° N on the Mid-Atlantic Ridge (Hoffert et al., 1978; Lalou et al., 1977). Small positive temperature anomalies were measured in the Siqueiros Transform Fault (Crane, 1976). Metalliferous sediments are present in the Blanco depression of the Blanco Fracture Zone (Selk, 1977). Large methane anomalies have been detected in both the Blanco depression and the Cascadia depression of the Blanco Fracture Zone (Massoth, personal communication, 1985). Thus, fracture zones are known focal points of hydrothermal activity and massive sulfide deposition.

In this chapter we describe brecciated greenstones, concentrically zoned hydrothermally altered basalts, quartz-chert-cemented breccias, and hydrothermally altered gabbros as evidence of multistage hydrothermal activity within the Blanco Fracture Zone.

Results and Discussion

The greenstones and hydrothermal breccias discussed here were dredged from two different localities. The bulk of the samples, more than 20 kg, were recovered in 1985 by the USGS R/V *S.P. Lee* at water depths of 4,100 to 2,900 m from the northern wall of the Blanco Fracture Zone where it crosses the Gorda Depression (site A, Fig. 4.1). Precise locations and sample descriptions are presented in Hart et al. (1986). Another sample was dredged from the southern wall of the Blanco Fracture Zone by W.G. Melson (see Melson, 1968, and site B, Fig. 4.1) and procured on loan from the Smithsonian Institution of Washington. Nineteen samples were cut into polished thin sections for scanning electron microscopy (SEM) and electron microbeam (EM) analysis. One hundred fifty-six microprobe analyses of individual sulfide and silicate mineral grains were collected. Billets of the same samples were powdered and analyzed by instrumental neutron activation analysis (INA), atomic adsorption spectroscopy using both flame (FASS) and heated graphite atomization (HGA), and X-ray diffraction. Strontium isotopic analyses of the samples were carried out at the Max-Planck Institute, Mainz, Germany, and oxygen isotope analyses were done at the Geochemical Institute at the University of Gottingen. Details of the chemistry of the samples and analytical methods are discussed in detail by Hart et al. (1986). Selected analyses of whole rock samples are presented in Table 4.1, and oxygen and strontium isotopic compositions are presented in Table 4.2.

Hydrothermally Brecciated Basalts

Hydrothermal alteration of pillow basalts under greenschist conditions has been reported in both dredged and drilled samples of oceanic crust (Melson et al., 1968; Melson and Van Andel, 1966; Humphris and Thompson, 1978a; Stakes and O'Neil, 1982; Honnorez et al., 1985; Alt et al., 1985). Brecciation and reopening of veins by pressurized fluids are characteristic of this hydrothermal process (Honnorez et al., 1985; Alt et al., 1985; Delaney et al., 1987). In this section, we describe evidence of brecciation of pillow lavas by hydrothermal fluids in the Blanco Fracture Zone. Discussion of quartz-chert-cemented breccias is deferred to a later section.

Primary textures of the basalts indicate they were in some cases holohyaline, hypohyaline, and tachylitic glassy pillow basalts (RHGR-3, RHGR-21); however, most were phaneritic pillow interiors or intrusive diabasic sills and dikes with porphyritic plagioclase (labradorite) in tabular laths and variolitic growths (RHGR-5H and V, RHGR-24, RHGR-26). Subophitic plates of augite are in some cases Ti-rich. Opaque minerals include magnetite and ilmenite along with disseminated pyrite and chalcopyrite, which comprises as much as 2% of the ground mass.

The $\partial^{18}O$ composition (Table 4.2) of the brecciated basalts ranges from 5.95 to 9.65 and indicates that even those samples such as RHGR-26 without obvious alteration mineralogy have exchanged oxygen with hydrothermal fluids. The $^{87}Sr/^{86}Sr$ ratio of the freshest sample

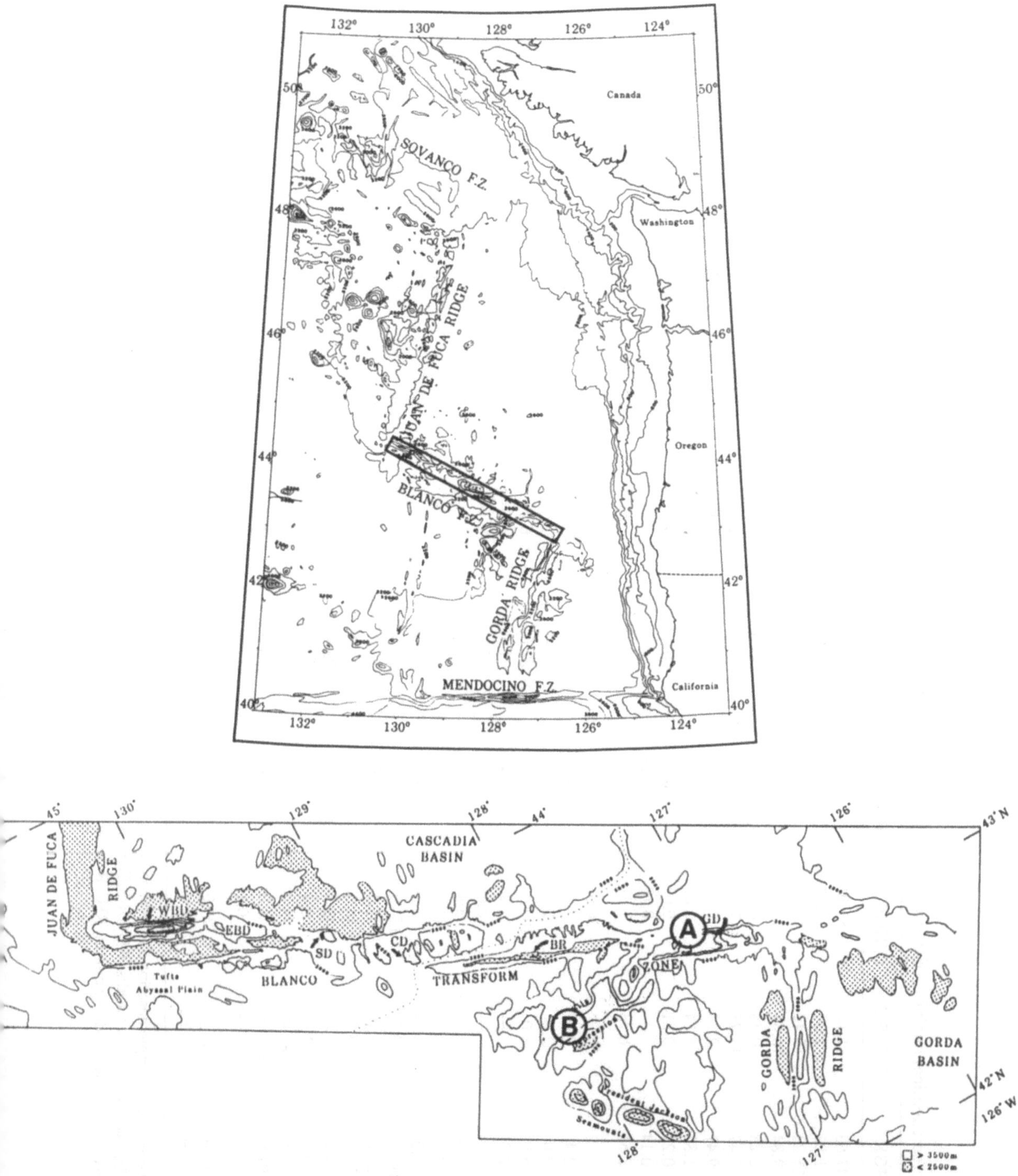

FIGURE 4.1. Location maps for hydrothermal samples from the Blanco Fracture Zone (from Embley et al., 1987). Site A is the location of samples dredged in 1985; site B is the location of samples reported in Melson (1968).

Table 4.1. Selected chemical analyses of hydrothermally altered Blanco Fracture Zone samples.

	RHGR-2	RHGR-3	RHGR-26	RHGR-5H	RHGR-5V	RHGR-6	RHGR-11	RHGR-16	RHGR-21	RHGR-24
SiO_2 (%)	47.60	41.10	50.60	46.50	39.00	83.00	74.60	51.60	44.90	53.50
TiO_2	1.82	2.04	1.87	1.56	1.01	0.41	0.87	1.61	1.8	1.92
Al_2O_3	15.20	15.42	14.10	14.70	17.50	3.04	7.28	12.87	16.33	13.34
Fe_2O_3	9.84	16.87	11.24	10.47	13.97	6.11	5.35	9.57	14.01	10.84
MnO	0.28	0.38	0.42	0.33	0.36	0.20	0.08	0.20	0.37	0.52
MgO	7.47	11.76	7.79	8.73	5.69	2.60	3.19	8.23	10.71	8.65
CaO	9.47	2.24	9.44	9.28	12.60	0.45	1.23	7.23	3.88	9.69
Na_2O	1.85	1.54	1.86	1.83	0.96	0.07	1.24	2.28	1.78	2.56
K_2O	0.21	0.21	0.07	0.14	0.17	0.01	0.05	0.21	0.40	0.29
P_2O_5	0.14	—	0.13	0.13	—	—	—	—	—	—
H_2O	3.55	6.73	2.32	3.60	5.45	1.65	3.25	3.92	5.76	3.14
Ba (ppm)		30.0	30.0	—	—	3.0	—	—	—	—
Li	2.4	9.6	3.1	4.6	3.9	24.5	41.5	2.4	7.3	7.6
Rb	2.0	2.0	2.0	2.0	—	1.0	1.0	2.0	1.0	2.0
Sr	155.0	46.0	109.0	121.0	82.0	6.0	33.0	122.0	25.0	141.0
Hf	2.3	—	2.9	1.9	1.4	—	1.5	2.1	2.6	2.8
U	0.2	—	0.6	—	0.1	—	—	0.3	0.2	0.3
La	2.9	—	3.8	2.4	1.9	—	1.6	2.7	3.8	5.5
Nd	8.9	12.1	12.1	7.2	6.8	—	6.5	9.8	11.1	10.1
Eu	1.0	—	1.3	1.0	1.4	—	0.8	0.9	1.3	1.8
Cr	243.0	410.0	154.0	377.0	251.0	93.0	119.0	383.0	148.0	163.0
Ni	61.0	53.0	63.0	92.0	53.0	—	63.0	48.0	56.0	83.0
Co	35.0	43.0	45.0	39.0	44.0	25.0	17.0	40.0	39.0	37.0
Sc	41.6	—	42.3	42.8	28.9	—	18.4	42.9	39.3	43.2
Pb	0.1	1.0	0.4	0.2	0.5	0.5	0.6	2.0	0.2	4.4
Cu	91.8	183.0	76.4	61.4	550.0	3,610.0	31.4	2.7	4.0	24.3
Zn	91.0	197.0	124.0	174.0	678.0	621.0	68.0	27.0	212.0	221.0
Sb	0.1	—	0.1	—	—	—	0.2	0.1	0.1	0.2
Ag (ppb)	10.0	30.0	30.0	10.0	40.0	810.0	20.0	—	—	—
Au	—	—	1.4	—	—	—	—	—	1.0	—

[a] Analytical uncertainties and sample descriptions given in Hart et al. (1986).
Rock types are as follows: Brecciated basalts, RHGR-3, RHGR-26, RHGR-5 (H = Host, V = Vein), RHGR-24; concentrically zoned basalt, RHGR-21; chert-cemented breccia, RHGR-11; drusy quartz, RHGR-6; gabbros, RHGR-2, RHGR-16.

TABLE 4.2. Oxygen and strontium isotopic composition of Blanco Fracture Zone metaigneous rocks.

	RHGR-2	RHGR-3	RHGR-26	RHGR-5H	RHGR-5V
$\delta^{18}O$	5.83	5.96	7.04	7.05	7.48
$^{87}Sr/^{86}Sr$	0.703165 (±6)	0.703917 (±107)	0.702588 (±24)	0.703839 (±13)	0.703672 (±17)

TABLE 4.2. *Continued*

	RHGR-6	RHGR-11	RHGR-16	RHGR-21	RHGR-24
	9.45	12.83	5.93	9.69	7.67
	0.704825 (±27)	0.703686 (±14)	0.704015 (±15)	0.703911 (±19)	0.704004 (±14)

(RHGR-26) is 0.702588, similar to that of fresh MORB. The $^{87}Sr/^{86}Sr$ ratios for brecciated basalts and vein material range from 0.703672 to 0.704056. These values reflect a small variation either in fluid composition or in the W/R ratio; large-scale variations of either are ruled out.

Secondary minerals are abundant in most samples. In general, plagioclase is albitized; augite is typically rimmed with actinolite. The groundmass is altered to chlorite-saponite assemblages; ilmenite is altered to leucoxene. An exceptional example of alteration spots of chlorite surrounding actinolite is shown in Figure 4.2A. Vein and vug fillings include saponite, chlorite, epidote, iron hydroxides, and, later, infillings of carbonate (both ankerite and calcite), prehnite, and zeolites.

Brecciation occurred on a variety of scales ranging from complex nets of hairline fractures (RHGR-3, RHGR-21, RHGR-24) to large-scale venation of samples up to 23 cm in diameter (RHGR-5H and V). In some cases the brecciation may be the result of magmatic or tectonic processes; in most samples, however, evidence of hydrothermal brecciation, as shown in Figure 4.2, is abundant. First, schistosity and flow banding are absent. Second, basalt fragments show evidence that they were shattered, transported, and rounded by fluids under pressure. Third, fragments are hydrothermally altered on the rims and supported in a matrix of alteration minerals. Lastly, pyrite, chalcopyrite, and sphalerite in the alteration matrix show evidence of fragmentation during precipitation. Overall, the textural evidence shows that fragments reacted chemically with fluids as they broke apart. This observation suggests that the brecciation was caused by pressurized hydrothermal fluids and that precipitation of alteration minerals was induced by a change of pressure and/or temperature that accompanied the brecciation process.

The basalt breccia RHGR-5 (H = host, V = vein) offers an opportunity to compare the chemistry of a hydrothermally brecciated vein with the host rock. Chemical analyses (Table 4.1), show that Al_2O_3, FeO, and CaO are enriched in the vein relative to the host rock. These enrichments reflect the chlorite-epidote-dominated mineralogy of the vein. Since the vein material fills cracks rather than replaces host basalt, the Al, Fe, and Ca must have precipitated from hydrothermal fluids. Sphene, with TiO_2 concentrations as high as 33% (see Table 4.3), is present as a trace phase in the matrix, suggesting TiO_2 has also been mobilized in hydrothermal fluids.

Concentrically Zoned Hydrothermal Alteration

Concentrically zoned hydrothermal alteration is common in fragmented basalts. Bands of alteration minerals reflect several conditions including (1) change of temperature, (2) change of W/R ratio, (3) change of fluid chemistry, and (4) diffusion across chemical concentration gradients. Cann (1969) described fragments of

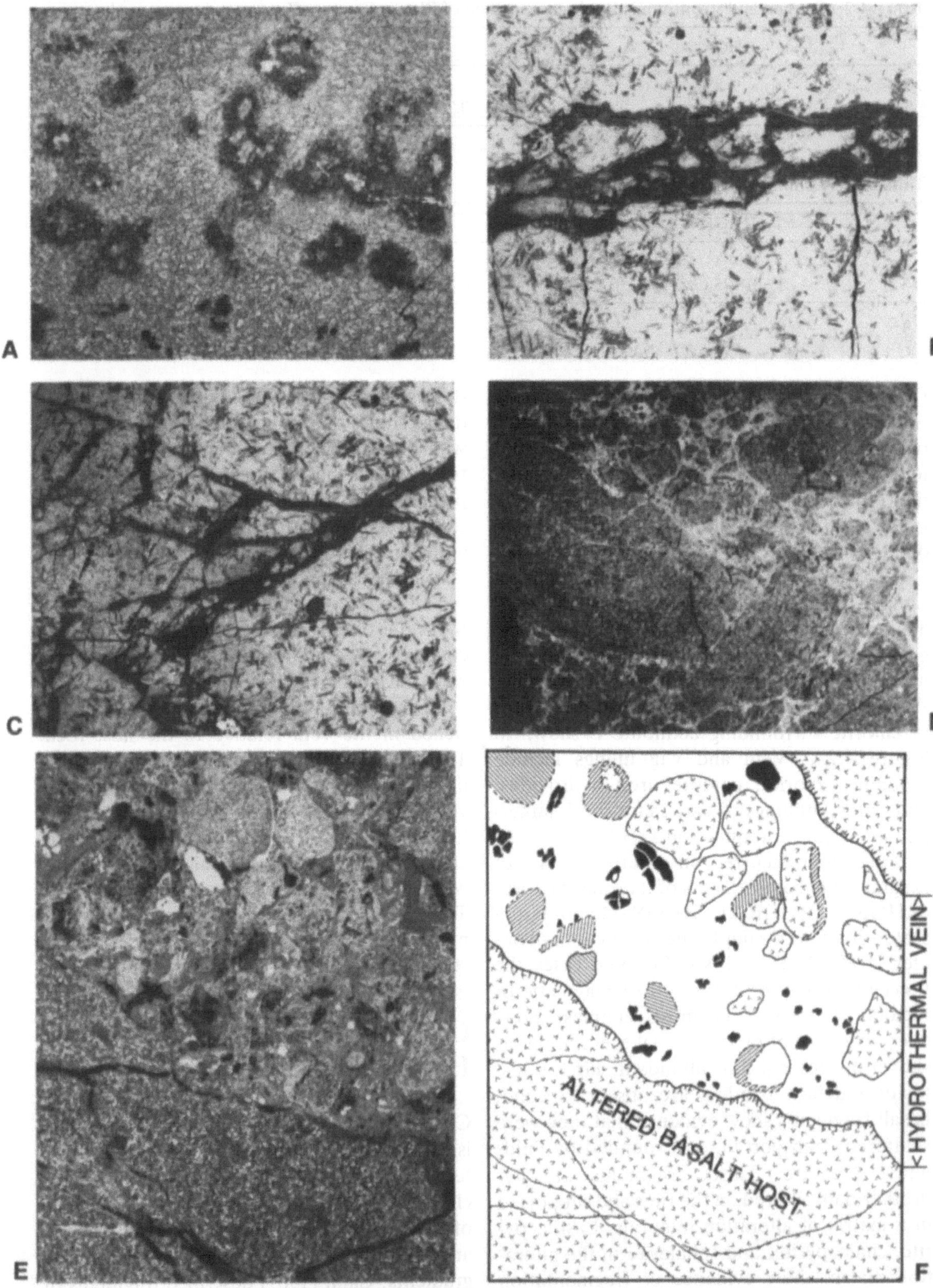
A
B
C
D
E
F
ALTERED BASALT HOST
HYDROTHERMAL VEIN

spilitized basalts from the Carlsberg Ridge with cores of actinolite-albite and rims of chlorite rich assemblages. Cann (1969) and Hart (1973) attributed the zonation to changing temperature regimens and suggested that the actinolite-albite formed at high temperatures near the ridge axis and the chlorite-rich rims formed at lower "retrograde" temperatures as the crust moved away from the source of heat at spreading ridges. Multistage alteration, as a result of change of temperature regimen, has been documented in numerous dredge samples and at DSDP site 504b (Stakes and Scheidegger, 1981; Alt et al., 1985). On the basis of experimental seawater-basalt interaction, Seyfried et al. (1978) suggested that chemical zonation can result from changes in W/R ratios; albite-actinolite assemblages formed at W/R less than 10, and chlorite-rich assemblages formed at W/R ratios greater than 10. A similar zonation was predicted by the theoretical modeling of Reed (1983), who used thermodynamic parameters to simulate seawater-basalt interaction. Most recently, Embley et al. (1987) have shown, on the basis of $^{87}Sr/^{86}Sr$ ratios, that the cores of the least altered sections of a stockwork pipe in the Galapagos Rift interacted with seawater with a W/R ratio of 4 whereas concentric zones around the core formed with W/R ratios as high as 30.

Change in the chemistry of hydrothermal fluids is another possible cause for concentric zoning of hydrothermal alteration. Silica-cemented greenstone breccias represent examples of deposits formed in fluids that changed from Mg-rich fluids to Si-rich fluids, perhaps as a result of ion exchange with the wall rocks as the fluids circulate through the crust. Small-scale variations of the Fe/Mg ratio in chlorites in greenstones reflect variations in the Fe/Mg ratio of the fluids, perhaps as the result of pulsing admixtures of Mg-rich cold seawater (Franklin, 1987).

Concentrically zoned, hydrothermally altered samples from the Blanco Fracture Zone occur as individual hand-size specimens (RHGR-21) and as clasts in the chert-cemented breccia discussed below (RHGR-11). The hand specimens are mottled; dark areas of grayish-green chlorite surround centimeter-size lighter areas. In thin section, four to six alternating dark and light bands (~0.1 mm) are apparent in circular, triangular, and polygonal shapes (Fig. 4.3). The circular shapes may enclose vesicle infillings (Fig. 4.3a). In places the banding parallels relict pla-

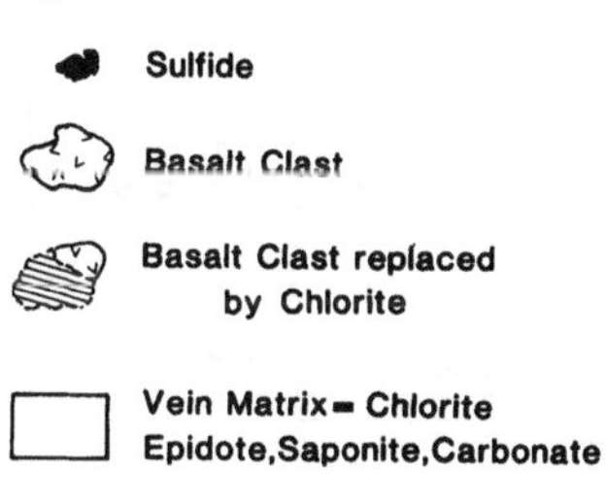

FIGURE 4.2. Hydrothermally altered and brecciated basalts are depicted in all panels. (A) Reverse-image photomicrograph of hydrothermally brecciated basalt, RHGR-24, with spots of chlorite surrounding actinolite (white) cores. The spots are 0.5 mm in diameter. (B) Reverse-image photomicrograph of brecciated basalt, RHGR-24, depicting a vein with angular brecciated basalt fragments in a matrix of chlorite cut by prehnite veinlets. An opaque groundmass (white) surrounds phenocrysts of laboradorite (black). The width of the vein varies from 1 to 1.5 mm. (C) A network of thin (less than 0.5 mm) veins dissect porphritic basalt, RHGR-24. The veins are filled with early stage chlorite, pyrite, and chalcopyrite and cut by later calcite. (D) This reverse-image photomicrograph of RHGR-3 shows a network of saponite and chlorite veins enclosing semirounded basalt clasts. (E) This is a reverse-image photomicrograph of brecciated vein RHGR-V. A matrix of epidote, chlorite, carbonate, and sphene surrounds and partially replaces well-rounded fragments of phaneritic host basalt entrained in the vein. Opaque grains of pyrite, chalcopyrite, and sphalerite are depicted white. (F) Schematic map of Figure 4.2e. The chalcopyrite, sphalerite, and pyrite (black areas) show evidence of in situ fragmentation, suggesting they precipitated in a dynamic environment. Corners of basalt fragments show rounding due to attrition by dissolution in hydrothermal fluids; the fragments have been partially or wholly replaced by chlorite (stippled areas).

TABLE 4.3. Microprobe analyses of concentric zoned hydrothermal alteration in traverse A–B from Figures 4.4C and D and 4.5.

	C	C	A	S	A	C	A	S	A	C	C	C	A	S	S	A	C	C	A
SiO_2	29.11	29.33	30.62	29.31	36.41	30.55	49.16	30.21	42.15	29.21	29.30	26.99	40.75	26.64	35.02	31.69	35.40	29.58	27.29
TiO_2	0.92	0.04	11.38	26.86	2.01	0.86	0.59	36.55	0.09	0.83	0.14	0.04	0.11	35.13	5.95	33.39	5.66	0.04	0.04
Al_2O_3	15.24	16.50	4.39	4.58	11.29	14.20	4.58	5.61	14.47	16.51	13.59	17.69	12.86	8.68	9.61	12.17	12.02	16.19	18.05
FeO	18.68	19.31	8.79	3.48	16.89	18.50	13.70	4.31	23.09	18.51	14.28	21.16	21.14	10.24	8.91	0.72	18.90	19.11	21.35
MnO	0.43	0.42	0.36	0.05	0.52	0.41	0.25	0.08	0.04	0.44	0.40	0.42	0.05	0.06	0.05	0.34	0.04	0.41	0.44
MgO	19.01	20.86	9.15	1.09	15.75	19.71	15.45	1.12	13.85	19.30	19.93	18.71	13.91	1.45	9.74	13.61	3.93	20.91	18.57
CaO	0.09	0.11	6.16	22.39	5.32	0.21	11.59	20.20	1.64	0.11	6.34	0.16	3.35	17.31	0.37	0.49	0.33	0.13	0.14
Na_2O	0.27	0.25	0.57	0.14	0.73	0.27	1.01	0.07	0.43	0.23	0.24	0.06	0.36	0.41	0.37	0.49	0.33	0.29	0.14
K_2O	0.26	0.29	0.46	0.15	0.29	0.25	0.18	0.15	0.18	0.26	0.13	0.10	0.14	0.13	0.27	0.18	0.17	0.23	0.10
Extra oxygen	16.01	12.89	28.12	11.44	10.78	15.01	3.49	1.70	4.06	15.04	15.50	14.66	7.06	0.00	20.39	7.32	7.86	13.11	13.77

Key: C, chlorite; A, amphibole; S, sphens S.

TABLE 4.4. Microprobe analyses of fragments in siliceous breccias.

	Chlorite in chert breccia matrix	Chlorite in chert breccia	Chlorite in quartz-cemented breccia	Dark chlorite in quartz-cemented breccia	Light chlorite in quartz-cemented breccia	Light chlorite in quartz-cemented breccia	Light chlorite in quartz-cemented breccia	Dark chlorite in quartz-cemented breccia	Dark chlorite in quartz-cemented breccia	Actinolite interlayered with chlorite in quartz-cemented breccia	Albite vein in chlorite fragment in quartz-cemented breccia	Albite Phenocryst in basalt fragment	Albite Phenocryst in basalt fragment
SiO_2	28.83	30.92	27.98	29.43	29.17	27.35	35.14	27.92	28.28	46.81	65.03	68.49	67.10
TiO_2	0.04	0.04	0.03	0.03	0.05	0.05	0.03	2.28	0.03	0.05	0.04	0.04	0.04
Al_2O_3	15.95	24.69	13.60	13.23	15.99	17.47	16.03	11.31	11.32	14.59	19.72	19.45	18.45
FeO	19.86	18.01	15.63	16.31	19.99	23.06	20.17	12.90	13.62	14.26	0.46	0.06	0.22
MnO	0.34	0.32	0.35	0.34	0.34	0.37	0.33	0.40	0.34	0.34	0.34	0.05	0.05
MgO	21.03	17.05	18.15	18.33	19.24	18.00	16.51	15.40	19.04	12.11	0.28	0.06	0.06
CaO	0.08	0.42	0.56	0.30	0.34	0.14	0.22	1.85	0.52	4.41	0.82	0.21	0.09
Na_2O	0.20	0.10	0.05	0.09	0.05	0.06	1.76	0.10	0.07	0.06	11.60	11.38	11.69
K_2O	0.13	0.13	0.26	0.32	0.20	0.17	0.34	0.26	0.28	0.27	0.14	0.09	0.06
Extra oxygen	12.47	8.32	23.30	21.62	14.62	13.34	9.46	27.49	26.49	7.10	1.47	0.19	2.24

gioclase phenocrysts (Fig. 4.3b). In other areas the banding traces longer lineations (Fig. 4.3c, d), suggesting control by microfractures such as reported in highly altered basalt from DSDP site 417A (Juteau et al., 1980; Alt and Honnorez, 1979) and from an alteration pipe beneath massive sulfides in the Galapagos Rise (Embley et al., 1987). The texture of the microfracture network in the Blanco Fracture Zone samples is different from that reported for deep sea hyoclastites (Embley et al., 1987) and may be the result of hydrothermal brecciation.

The banding reflects alternating concentrations of chlorite, amphibole (probably actinolite), and sphene. In thin section chlorite dominates; thin semiopaque bands of fine-grained sphene define the concentric shapes. Amphibole needles, identified by microprobe to be of actinolite composition, crosscut the chlorite-sphene contacts. Some of the sphene is intergrown with leucoxene that has replaced primary ilmenite; however, most of the sphene bands do not conform to ilmenite crystal shapes. Bands of sphene in circular and peninsular shapes are shown in Figure 4.4 and are precisely matched by high concentrations of titanium as mapped by long count time SCEM (Fig. 4.4a, c). An EM traverse of the bands (Table 4.3; Fig. 4.5) shows TiO_2 concentrations as high as 36.55% correlated with CaO concentrations as high as 22.39%. The TiO_2 is anticorrelated with FeO, indicating the lack of a genetic relationship between sphene and ilmenite. We propose that titanium and calcium are incompatible with chlorite and therefore trapped in bands around centers of chlorite nucleation and then combine with available silica to form sphene.

Quartz-Chert-Cemented Breccias

Quartz-chert-cemented breccias are common in the stockwork section of ophiolites, especially at the contact between the sheeted dikes and pillow basalts at subseafloor depths of 400 to 800 m (Spooner and Fyfe, 1973; Roberts and Reardon, 1978; Richards et al., in press; Lydon and Galley, 1986; Collinson, 1986). Quartz and chert have been recovered from several localities in hydrothermally altered sea floor (Rona et al., 1980; Hekinian et al., 1983; Alt et al., 1986; Delaney et al., 1987). The breccias from the Blanco Fracture Zone are similar, with noteworthy differences, to those dredged from the Kane Fracture Zone as described by Delaney et al. (1987). These authors note that rounded and sorted high-Fe chlorite and sulfide clasts are surrounded by haloes of euhedral quartz and suspended in a matrix of microcrystalline quartz. Fluid inclusion homogenization studies indicate that the quartz-cemented breccias formed at temperatures between 195 and 280°C (Delaney et. al., 1987). Based on sorting and rounding, these authors propose that the fragments were transported upward in fluid zones at velocities in the range 50 to 100 cm/sec. Based on texture relationships, Delaney et al. (1987) proposed at least three hydrothermal stages in the Kane Fracture Zone: (1) prebrecciation chloritization, silicification, and sulfide deposition; (2) episodic brecciation and subsequent quartz cementation; and (3) late-stage quartz veining.

The same three hydrothermal stages are evident in the Blanco Fracture Zone breccias and are described below.

Prebrecciation Alteration and Mineralization

Figure 4.6a, b shows a quartz-cemented breccia from site B with subround clasts of basalt and chlorite-rich fragments. Microprobe analyses of the alteration mineralogy are given in Table 4.4. Chlorite is the dominant alteration mineral in the basalt clasts, indicating that the solutions were Mg-rich in this stage. The feldspar is ubiquitously albitized, indicating the solutions were Na-rich. Variations in the color of chlorite reflect variations in the Mg/Fe ratio, suggesting the chemistry of the fluids changed during chloritization.

The site A chert-cemented breccias contain clasts of the concentrically zoned, hydrothermally altered basalt described above (Fig. 4.6d), indicating that at least one period of

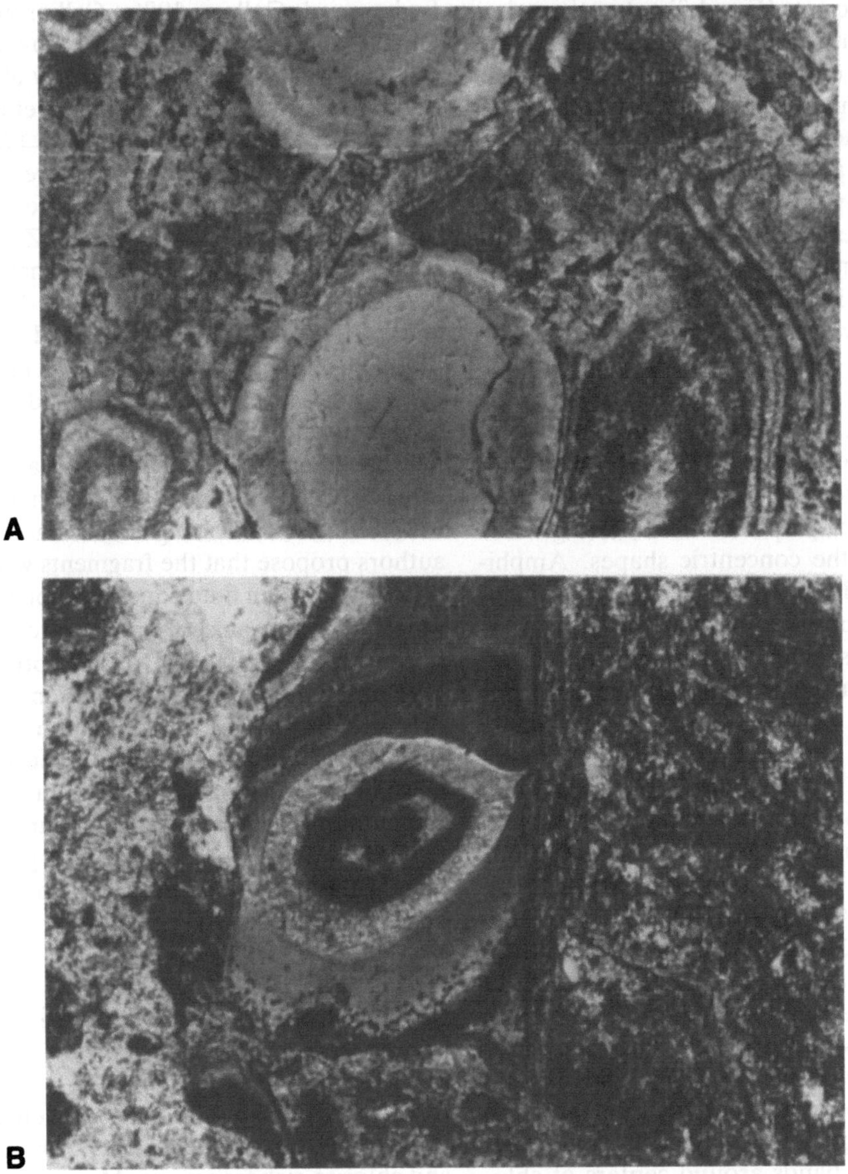

FIGURE 4.3. All panels are photomicrographs of concentric zoning in sample RHGR-21. (A) This photomicrograph shows concentrically zoned vesicles filled with bands of chlorite (light gray center) and opaque sphene aggregates (black bands), rimmed with bands of amphibole, most likely actinolite (light gray bands adjacent to sphene bands). Vesicles are 0.5 mm in diameter. (B) Concentrically zoned hydrothermal banding conforms to the edge of a phenocryst of albitized plagioclase on the right. (C, D) Concentrically zoned alteration conforms to a network of lineations, perhaps microfractures developed during hydrothermal activity.

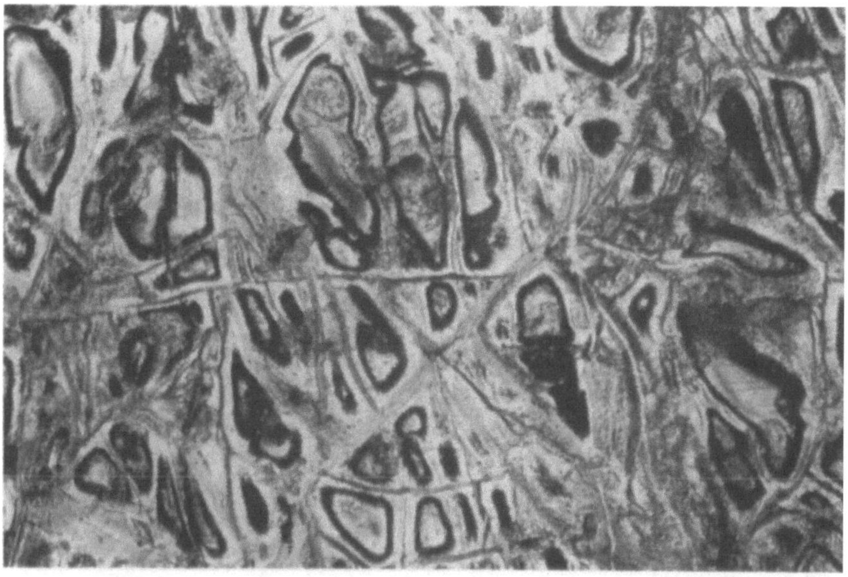

C

Fig. 4.3 D

brecciation and alteration preceded the fragmentation of the clasts. All of the clasts were altered under greenschist conditions; chlorite, albite, and sulfide are abundant. The sulfide minerals (mainly pyrite) are concentrated in the clasts rather than in the siliceous matrix, suggesting that sulfide precipitation occurred in this first stage; however, a fragment of drusy quartz (RHGR-6), possibly vein material, shows copper and zinc sulfide mineralization that may have occurred at a much later stage.

Brecciation and Cementation

The texture of the quartz-chert-cemented breccias suggests that multiple events of fragmentation took place with concurrent dissolution and cementation. The quartz-breccia from the southern Blanco Fracture Zone (site B) exhibits euhedral quartz crystals in layers and haloes. A well-rounded clast composed of angular fragments of chlorite (Fig. 4.6a, b) is cemented by quartz. The angular fragments

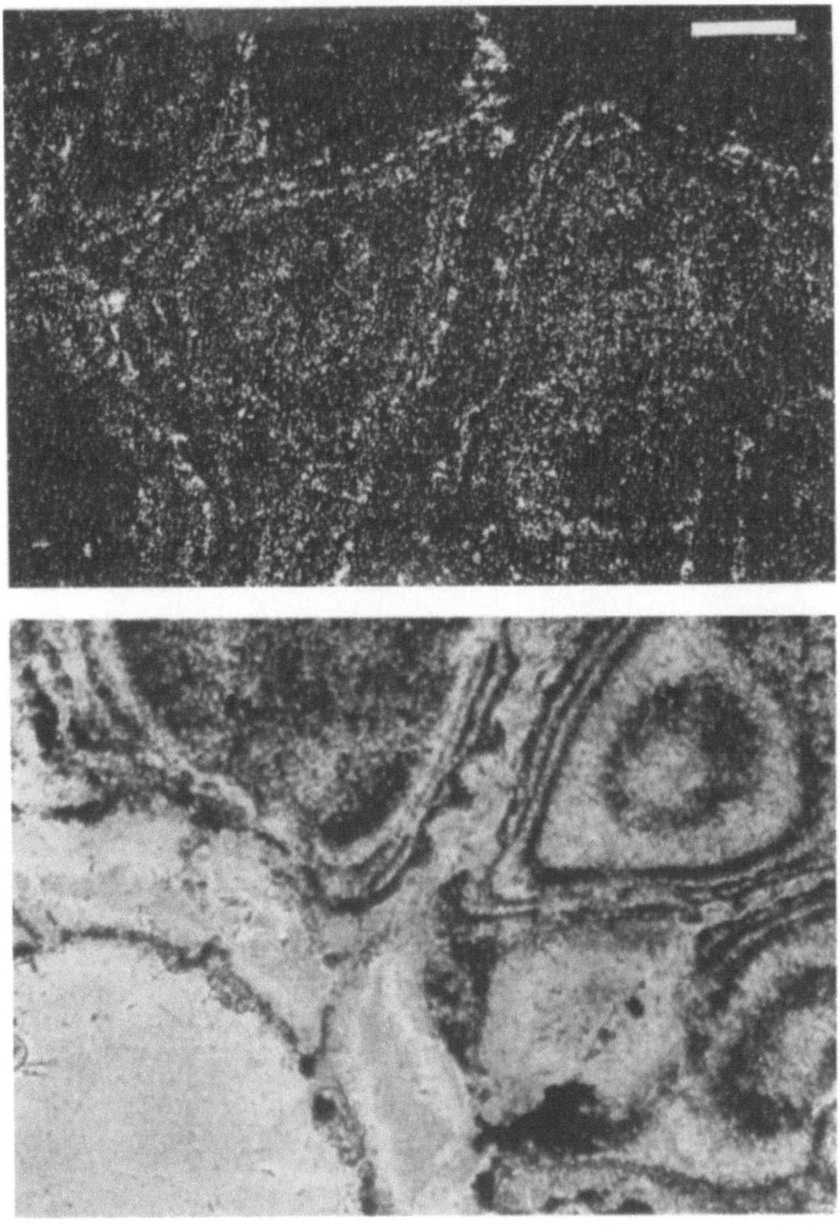

A

B

FIGURE 4.4. Concentric banding in sample RHGR-21 with Ti maps (A and B) collected by long count time SEM. Areas of high Ti concentration are indicated by the white areas on the maps (4C and 4D) and correspond to opaque aggregates (black areas) in the photomicrographs on the right. EM analyses of the aggregates (Table 4.3 and Figure 4.5) indicate that they are of sphene composition. This figure shows that the areas of high Ti concentration that surround chlorite do not conform to shapes expected for pseudomorphs of ilmenite.

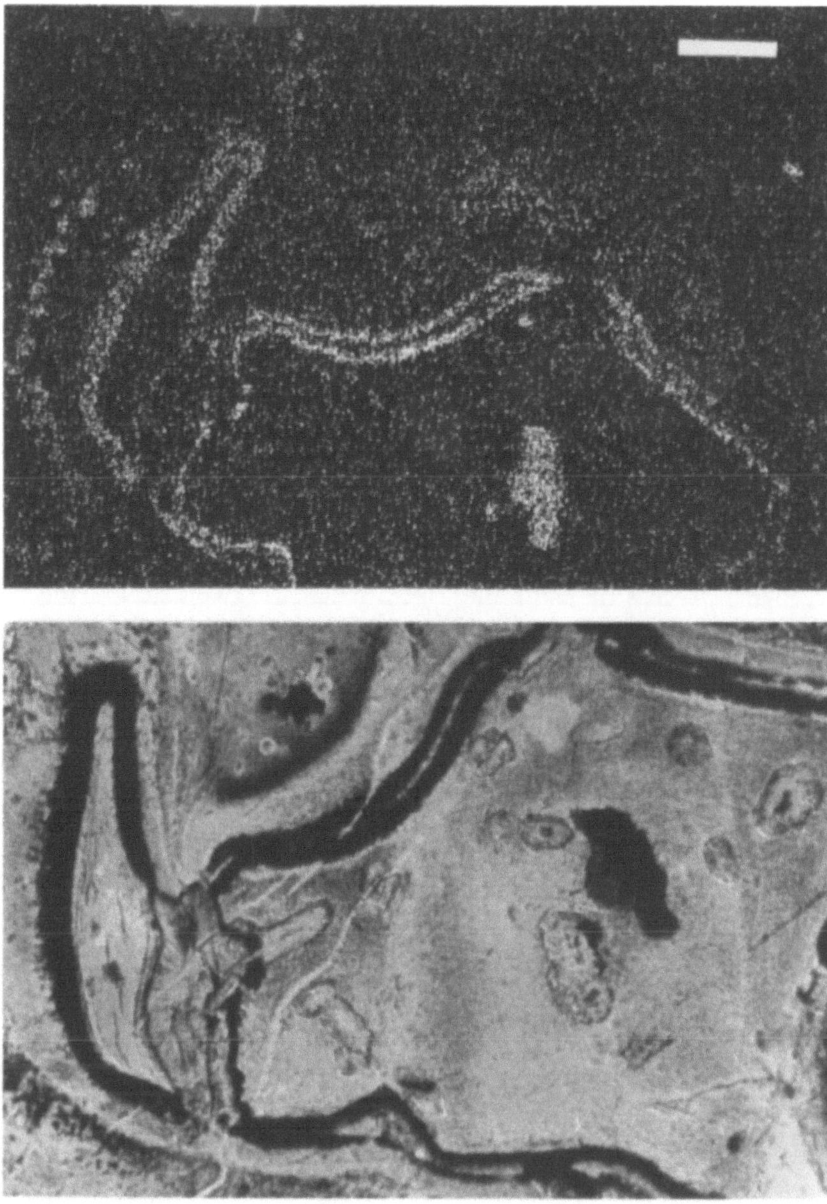

C

D

Fig. 4.4

must have aggregated into the larger clast soon after they were fractured, or they would show evidence of rounding. The larger clast is well rounded, suggesting transport in fluids at high velocities similar to those from the Kane Fracture Zone (Delaney et al., 1987) The transportation and rounding of the clast must have occurred after the angular chlorite fragments were cemented together by silica but before the quartz matrix precipitated. These observations suggest that episodic brecciation and cementation occurred in the Blanco Fracture Zone.

The chert-cemented breccias from site A present a different story. Only a few of the largest clasts are rounded, and sorting is poor (Fig. 4.6c–f). In the bottom middle of Figure 4.6c, chert haloes surround two basalt clasts; veins of chert cut through the clasts. Abundant angular and unsorted fragments are suspended in the chert matrix as if in a viscous gel. These observations suggest that (1) pebbles were par-

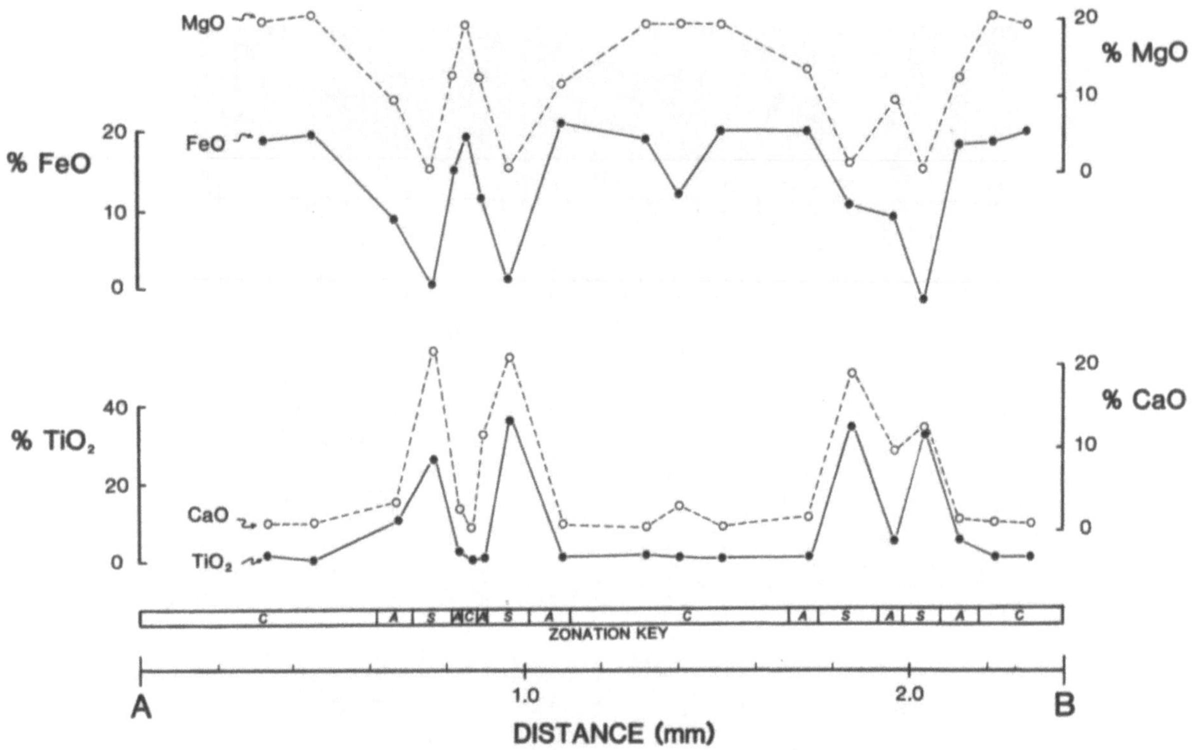

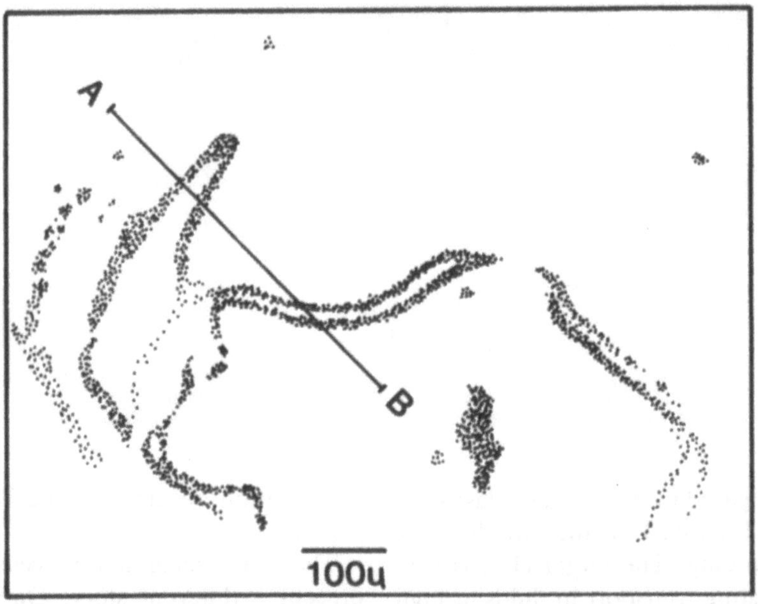

FIGURE 4.5. Element distribution cross section of microprobe analysis from traverse A–B. The data are from Table 4.2, and the zonation key is as follows: C = chlorite-rich zone; A = amphibole-rich zone; S = sphene-rich zone. The insert shows the location of the traverse across the section shown in Figure 4.4C.

tially dissolved during fragmentation, (2) the entrained fragments were not transported in fluids, and (3) the fragments were cemented together shortly after fragmentation. In general, the fragment density in the matrix is variable; clear areas of pure chert cut through cloudy fragment-rich areas, suggesting sluggish flow in a viscous fluid. High viscosity may have resulted from the formation of solutions supersaturated in silica polymers (colloids) at high pH, or it may be the result of a dense fluidized bed load.

Late-Stage Fracturing and Vein Filling

In contrast to the previous stages, wherein the fragments were entrained in a fluid or plastic flow, the late-stage fracture event is characterized by brittle fracture and subsequent infilling of fragment-free quartz. As shown in Figures 4.6e and f, the late-stage fractures cut across both the matrix and the basalt clasts, indicating the chert matrix had hardened before this stage. Fracturing took place by decompressional or tensional release. Several small pieces of pure quartz (RHGR-6) with disseminated bornite, chalcopyrite, and sphalerite probably formed during this late-stage event.

Overall, the siliceous breccias demonstrate a complex interrelationship between fluid flow, fragmentation, and precipitation of silica. The $\partial^{18}O$ of silicified samples RHGR-6 and RHGR-11 are 9.45 and 12.83. Assuming the samples are pure SiO_2 and using the Epstein-Knauth equilibrium curve, the quartz-chert-cemented breccias probably formed at temperatures in the range of 180 to 200°C. This range of temperatures is consistent with the fluid inclusion results on quartz breccias from the Kane Fracture Zone (Delaney et al., 1987). The $^{87}Sr/^{86}Sr$ ratios of 0.703825 and 0.703729 overlap with those of the brecciated basalts and indicate that neither the strontium isotopic composition of the fluids nor the W/R ratio changed significantly between brecciation of the basalt fragments and precipitation of the siliceous matrix. The quartz-chert-cemented breccias probably formed when silica-laden hydrothermal fluids cooled within an upflow zone.

Pyroxene Metagabbros

Metagabbros occur in ophiolites below the sheeted dike complex and are commonly recovered from the ocean floor in the vicinity of fracture zones (Cann and Funnell, 1967; Chann, 1971; Bonatti et al., 1975; Myashiro et al., 1971; Engel and Fisher, 1975; Ito and Anderson, 1983; Stakes and Vanko, 1986; Kelley and Delaney, 1987; Mével, 1987). Isotopic and fluid inclusion studies indicate that most metagabbros were hydrothermally metamorphosed by seawater in multiple stages at temperatures between 350 and 700°C and at W/R ratios between 0.1 and 2 (Ito and Clayton, 1983; Stakes and Vanko, 1986; Kelley and Delaney, 1987). Gabbros recovered from fracture zones display evidence of microfracturing, ductile deformation, and schistose recrystallization at temperatures up to 700°C and depths of 3 km below the sea bed (Stakes and Vanko, 1986; Kelley and Delaney, 1987). Multistage hydrothermal alteration of gabbros includes (1) amphibolitization of pyroxene and albitization of the plagioclase, (2) overprinting of high-temperature phases by chlorite and epidote, and (3) late fracturing and vein infilling of chlorite, epidote, carbonates, and zeolites. These stages may reflect progressive cooling and tectonic uplift of the oceanic crust.

Two fragments of gabbro (RHGR-2 and RHGR-16) were recovered by dredging at site A of the Blanco Fracture Zone (Hart et al., 1986). Both samples have well-preserved igneous textures free from schistosity (Fig. 4.7a, b). Laths of plagioclase (An 63) as long as 4.5 mm are present in subophitic plates of augite up to 2.5 mm in length. Minor ilmenite, magnetite, carbonate, pyrrhotite, and sarce pentlandite are disseminated throughout the samples. The disseminated sulfides are readily visible in hand specimens and vary in abundance from 1.5% to 2%. The association of the sulfides with alteration minerals suggests they are of hydrothermal origin. Microprobe analyses of secondary phases in gabbro are given in Table 4.5. As shown in Figure 4.7, three stages of alterations of the gabbros are apparent in thin section: (1) Actinolite replaces augite, plagioclase is albitized, and leucoxene replaces

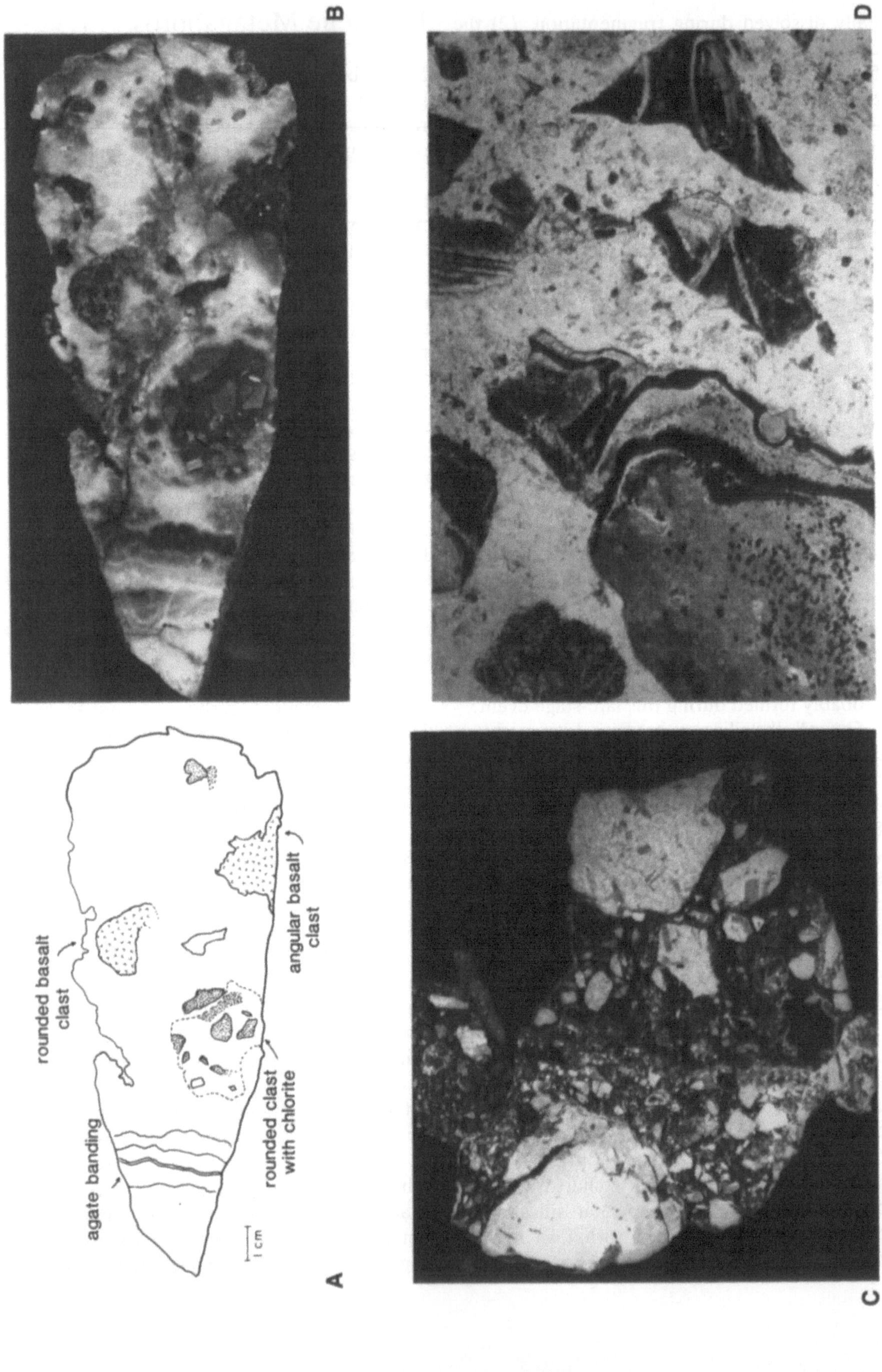
A
B
C
D
agate banding
rounded basalt clast
rounded clast with chlorite
angular basalt clast
1 cm

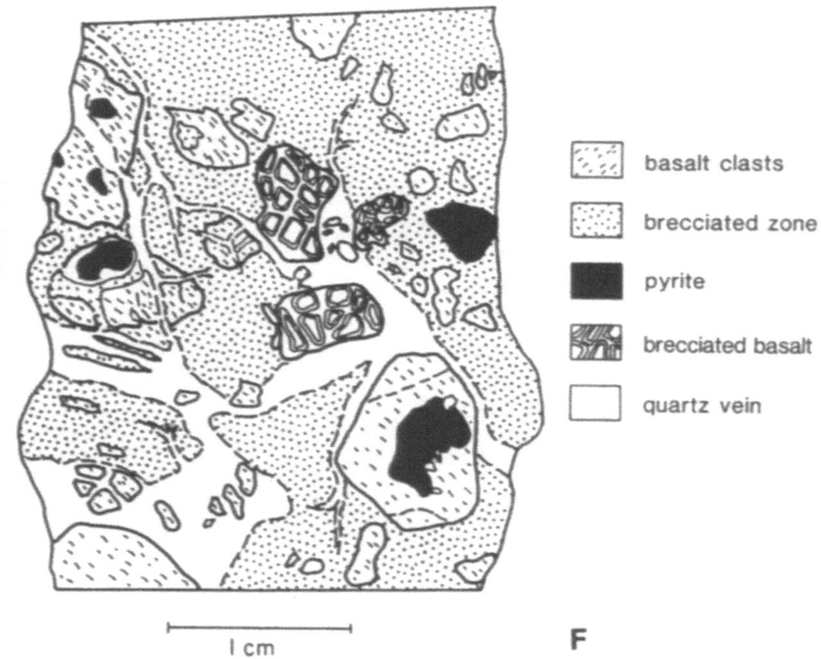

FIGURE 4.6. All panels show quartz-chert-cemented breccias from the Blanco Fracture Zone. (A) The schematic map shows the quartz-cemented breccia dredged from site B of the Blanco Fracture Zone (Melson, 1968) and depicted in the next panel. (B) Photograph of the quartz-cemented breccia from site B. Note the rounded clast of angular chlorite-rich fragments. These fragments were cemented together into the clast that must have then been transported and rounded prior to precipitation of the quartz matrix. (C) Reverse-image photomicrograph of chert cemented breccia RHGR-11. The larger clasts (1 mm) are hydrothermal altered basalt. The white interior areas are sulfides. Note that unlike the quartz-cemented breccia, the clasts in the chert-cemented breccia are angular and unsorted with respect to size. Note that some basalt clasts such as the one at middle bottom show haloes and veins of partially dissolved fragments. These fragments are suspended in the chert matrix, indicating that fragmentation, dissolution of the fragments, and precipitation of the chert matrix were closely related in time. (D) This is a photomicrograph of angular clasts of concentrically zoned hydrothermally altered basalt similar to those depicted in Figures 4.3 and 4.4. This photograph suggests that at least one episode of fracturing and hydrothermal activity preceded the formation of the chert-cemented breccias. (E) Reverse-image photomicrograph of chert-cemented breccias showing that brittle fracture of clast-rich chert matrix extends into a basalt clast with pyrite crystal (white). This photograph suggests that the final episode of fracturing occurred after the chert matrix had solidified. (F) Schematic representation of Figure 4.6e, showing fragment basalt clast, brecciated matrix with angular particles of basalt, opaque pyrite crystals, concentrically zoned basalt, and late-stage quartz vein filling brittle fractures.

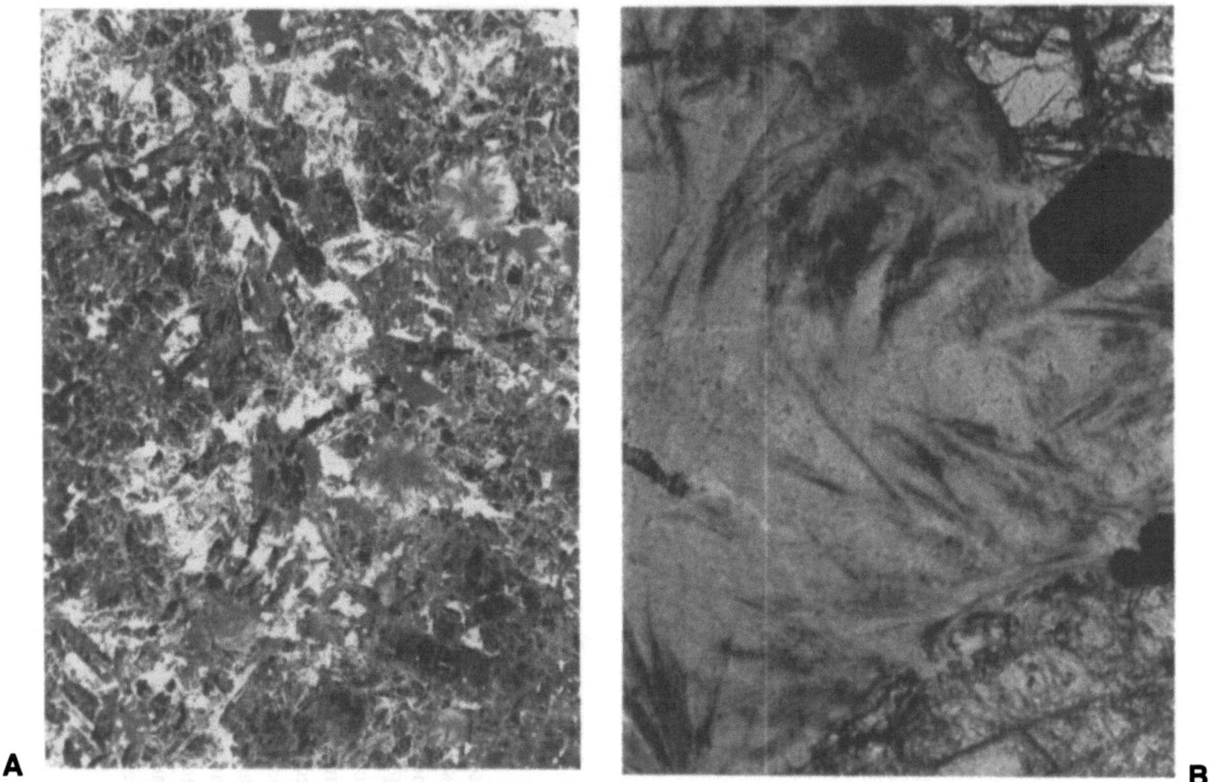

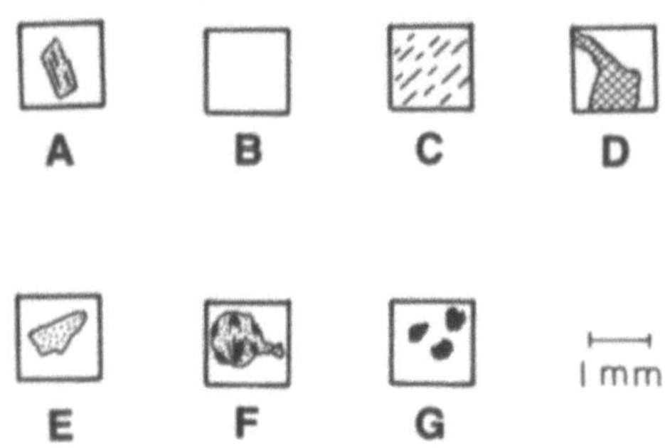

FIGURE 4.7. All panels depict hydrothermally altered gabbro RHGR-2. (A) Reverse-image photomicrograph of gabbro RHGR-2. (B) Photomicrograph of chlorite spot with actinolite needles and opaque sulfides on periphery. (C) Schematic representation of Figure 4.5a showing lathes of plagioclase (*A*) now largely albitized (An63); subophitic augite-ilmenite-magnetite matrix (*B*); hydrothermally altered pyroxene matrix with alteration mineralogy of leucoxene, actinolite, and sphene (*C*); vein albite (*D*) infilling a vug with chlorite (*E*) lining; and chlorite spots with actinolite needles rimmed by pyrrhotite and pentlandite (*F*).

TABLE 4.5. Microprobe analyses of secondary phases in gabbros.

	Albite vein	Albite vein	Albite replacing labradorite	Chlorite spot	Chlorite spot	Actinolite replacing augite	Actinolite vein cutting augite	Actinolite needle in chlorite spot	Actinolite needle in chlorite spot	Actinolite needle in chlorite spot	Late-stage analcite vein
SiO_2	68.49	63.72	56.48	31.48	29.15	48.62	44.47	48.02	50.21	42.15	54.89
TiO_2	0.04	0.04	0.04	0.04	0.04	0.30	0.15	0.15	0.16	0.09	0.04
Al_2O_3	19.90	20.65	25.19	15.24	11.88	5.26	9.26	8.00	10.07	14.47	23.02
FeO	0.06	1.12	0.06	18.28	24.39	13.19	11.67	24.48	20.34	23.09	0.06
MnO	0.04	0.04	0.05	0.04	0.04	0.22	0.25	0.41	0.44	0.44	0.05
MgO	0.09	1.24	0.05	18.14	16.76	15.71	12.24	9.54	9.38	13.85	0.06
CaO	0.46	0.29	6.85	3.02	6.85	10.65	9.56	7.36	6.23	1.64	0.06
Na_2O	11.30	12.77	8.25	0.18	8.25	1.15	1.54	0.56	1.93	0.43	11.66
K_2O	0.08	0.12	0.31	0.26	0.20	0.09	0.13	0.09	0.18	0.18	0.06
Extra oxygen	0.00	0.00	2.71	14.33	2.71	4.81	10.73	4.38	1.06	3.60	10.08

ilmenite. (2) Spheroidal spots of chlorite with radiating bundles of actinolite needles partially replace augite (see Fig. 4.7a–c). To the best of our knowledge this unique configuration of "spotted" alteration has not been previously reported in oceanic gabbros. (3) The chlorite-antinolite spots are cut and sometimes filled by albite, prehnite, and a colorless zeolite (probably thomsonite).

Although these metagabbros are strongly altered, their chemical and oxygen isotopic compositions are not significantly different from fresh gabbros or basalts. This observation supports the suggestion of Stakes and Vanko (1986) that hydrothermal alteration of gabbros mainly induces redistribution of chemical elements within the whole rock without significant chemical exchange between rock and fluid. This may be in part a result of low W/R ratios; the $^{87}Sr/^{86}Sr$ ratios of the Blanco Fracture Zone gabbros, however, are similar to those of the breccias, indicating similar W/R ratios. Amphibolites with up to twice the H_2O contents of the Blanco Fracture Zone samples show enrichments in MgO and K_2O, perhaps related to the formation of biotite (Cann, 1971), and depletion in SiO_2, which may be transported by hydrothermal solutions to overlying stockwork.

Element Mobility and Sulfide Formation in the Blanco Fracture Zone Hydrothermal System

Analyses of dredge samples from the Blanco Fracture Zone indicate a hydrothermal system in which magmatic rocks underwent multistage alteration and fracturing within a restricted time frame. The complex relationships of changing temperature, W/R ratio, and fluid chemistry are characteristic of sub-sea-floor hydrothermal zones described on the ocean floor and in ophiolites (Alt et al., 1986; Delaney et al., 1987; Stakes and Vanko, 1987; Juteau et al., 1980; Embley et al., 1987).

In this section we attempt to constrain the physical properties of the hydrothermal fluids in the Blanco Fracture Zone in terms of (1) temperature, (2) W/R ratio, and (3) fluid chemistry.

Temperature

Estimates of temperature are constrained by the mineralogy of metamorphic facies and by oxygen isotope compositions. Actinolite-rich assemblages as documented in the Blanco Fracture Zone gabbros are traditionally assigned to the upper greenschist (lower actinolite) facies with temperatures in excess of 500°C (Elthon, 1981). The oxygen isotopic composition of the Blanco Fracture Zone metagabbros is compared to that of other

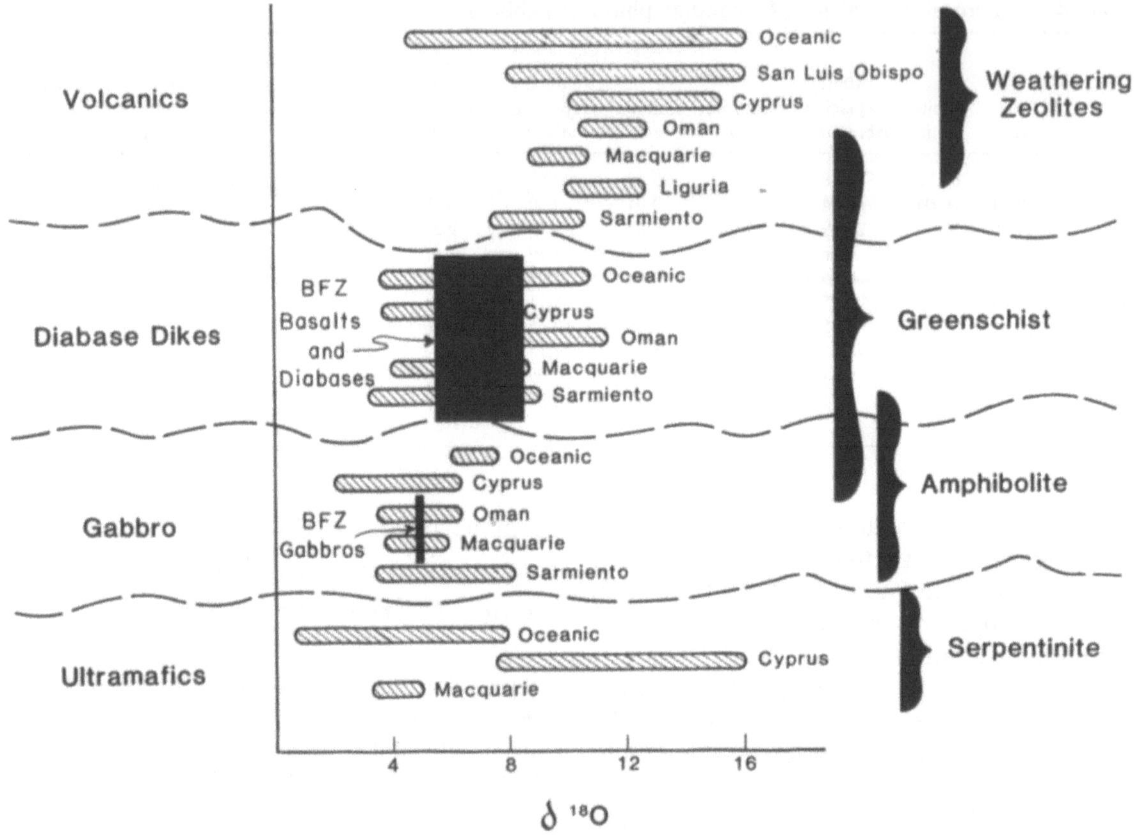

FIGURE 4.8. The oxygen isotopic analysis of Blanco Fracture Zone samples compared to pseudostratigraphy of oceanic crust and ophiolites.

oceanic and ophiolitic metagabbros in Figure 4.8. The oxygen isotope results support the mineralogical evidence that the Blanco Fracture Zone gabbros were hydrothermally altered at temperatures in excess of 500°C. The chlorite-albite-actinolite assemblages of the brecciated basalts correspond to temperatures in the range 350 to 550°C. The $\partial^{18}O$ of Blanco Fracture Zone samples altered under greenschist facies range from +5.83 to +7.76 and are consistent with the range of temperature suggested by the mineralogy (see Fig. 4.8).

The silica-cemented breccias with $\partial^{18}O$ as high as +12.83 suggest they formed at temperatures between 100 and 200°C. Since the siliceous breccias contain fragments of basalt altered at higher temperature, a drop in temperature of several hundred degrees occurred between the alteration of the basalt clasts and their cementation in the breccia. In the case of the quartz-cemented breccia shown in Figure 4.6b, the change of temperature may reflect transportation of the clasts in an up-welling zone before cementation. In the case of the chert-cemented breccia the fragments are angular and could not have been transported. Since the fragments are partially corroded and suspended in chert; we propose that an adiabatic drop of temperature accompanied brecciation and induced precipitation of the chert in the matrix. Zeolites, analcite, laumontite, saponite, carbonate and prehnite in late-stage veins may have formed at temperatures similar to those of the siliceous breccias. Overall, metamorphosed igneous rocks from the Blanco Fracture Zone provide evidence of hydrothermal activity in dynamic temperature regimens.

Water/Rock Ratio

$^{87}Sr/^{86}Sr$ ratios are common indicators of W/R ratio. The $^{87}Sr/^{86}Sr$ ratios of hydrothermal altered igneous rocks from the Blanco Fracture Zone fall between the values of 0.703165 and 0.704056. These values correspond to W/R ratios between 10 and 100. Higher W/R ratios as expected from open-system flushing of seawater through a fracture system are ruled out. The concentrically zoned basalt (RHGR-21), with an H_2O content of 5.76%, is the most severely hydrated of the Blanco Fracture Zone samples. The banded succession of mineralogy in this sample suggests distinct pulses of hydrothermal fluids through the fracture system.

Fluid Composition

The chemical composition of the hydrothermal fluids is reflected by the minerals that precipitate from them. The chemistry of fluids evolves in response to variations of the solubility of dissolved species. For example, the solubility of Mg (OH) in aqueous solution decreases with temperature, whereas at pressures typical of the sub-seabed, the solubility of SiO_2 increases (Reed, 1983). Thus, at high temperatures, hydrothermally evolving seawater is expected to lose magnesium and pick up silica. Experimental basalt-seawater interactions have substantiated this hypothesis (Seyfried et al., 1978). Some features of the chemistry of hydrothermally altered igneous rocks from the Blanco Fracture Zone can be explained in terms of the relationship between the solubilities of magnesium and silica. As shown in Figure 4.9, those samples altered at temperatures in excess of 350°C are enriched in MgO, as expected from precipitation of magnesium from seawater. On the other hand, samples rich in SiO_2 are those formed at lower temperatures, apparently from solutions depleted in magnesium and enriched in silica. These relationships are also demonstrated by the oxygen isotope composition as shown in Figure 4.10. This plot also suggests Cr may be stripped from seawater along with magnesium at high temperatures.

Overall the data suggest that as seawater circulates at depth, it heats up in response to the thermal gradient while picking up silica, calcium, and aluminum in exchange for magnesium. The fluids undergo thermal expansion and rapidly cool the crust, which contracts and brecciates. The volume of the crust decreases owing to cooling; the volume of the fluids increases as it fills the cooling cracks in the crust. The fluid expansion may result in rapid adiabatic cooling. Precipitation of quartz and chert follows as a result of the change in pressure and temperature.

In addition, at least one other phase of solution chemistry is indicated by epidote-chlorite-carbonate-rich veins in the greenstone breccias (RHGR-5V). Epidote-rich zones have been reported in the sheeted dikes and lower pillow lavas of the Josephine and Trodoos Ophiolites, and their formation is attributed to precipitation to Ca-rich mature fluids in hydrothermal outflow zones (Harper et al., 1987). The veins in the Blanco Fracture Zone breccias are enriched by 30% to 40% in CaO over the host rock, indicating they must have formed in contact with Ca-rich fluids. They are also 20% to 30% enriched in Al_2O_3. The solubility of Al in hydrothermal fluids has not been studied adequately and is generally assumed to be negligible, despite the evidence from Al-rich hydrothermal mounds and sediments off ridge axis in the South Pacific that suggest considerable mobility of Al in the fluid phase. The effects of temperature, pressure, and pH on Al solubility are unknown, but being a major constituent of chlorite, its association with Ca-rich epidote and carbonate suggests it may precipitate from solution in the well-known Ca solubility minimum between 250 and 350°C. If this is the case, then a model for change of fluid composition in the Blanco Fracture Zone could be as follows: Mg-rich seawater penetrates the crust at depth and precipitates Mg-rich chlorites and actinolites at temperatures in excess of 350°C while leaching Ca, Si, and Al from the host rocks. As the result of pressure and temperature drop on the up-welling limb of a convection cycle, the solutions precipitate Ca and Al rich epidote-chlorite-carbonate phases. Upon further relief of pressure by fracturing and by adiabatic cooling, supersaturated solutions precipitate silica as quartz and chert which cement the hydrothermal breccias.

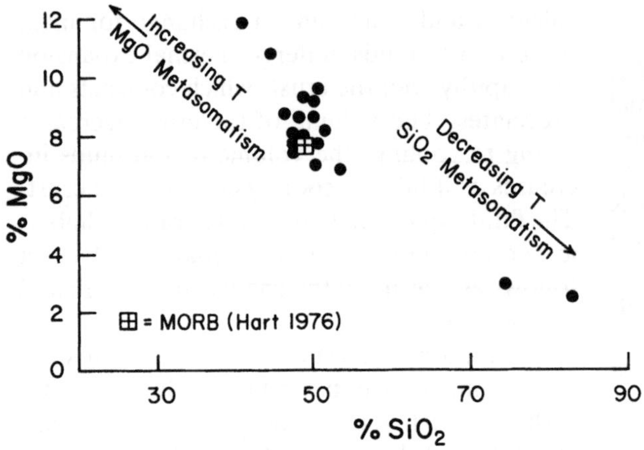

FIGURE 4.9. A plot of weight percent MgO vs. weight percent SiO_2 in hydrothermally altered samples from the Blanco Fracture Zone.

The sulfides that form at each stage of solution evolution are distinct. Disseminated Fe-depleted pyrrhotite and pentlandite are associated with the highest-temperature Mg-rich phases in the gabbro. Pyrite, chalcopyrite, and sphalerite form at the intermediate stage associated with Ca and Al rich epidote-chlorite assemblages. The silica cemented breccias contain no sulfides in the matrix, but a few fragments of drusy quartz, perhaps originally vug fillings in the stockwork breccia, contain copper and zinc sulfides.

Conclusions

Gabbros, diabases, and basalts dredged from the Blanco Fracture Zone have been hydrothermally altered under greenschist conditions. Multistage fragmentation, brecciation, and mineral precipitation accompanied the hydrothermal activity.

The basalt and diabase breccias are characterized by rounded and altered basalt clasts surrounded by a matrix of alteration minerals. Precipitation of sulfides (pyrite, chalcopyrite, and sphalerite) accompanied first-stage vein fillings. Veins are enriched in aluminum (owing to chlorite formation), indicating the hydrothermal fluids dissolve and reprecipitate aluminum. Later vein fillings occurred at lower temperature under prehnite-pumpyllite conditions and are characterized by precipitation of carbonate. Some basalt samples show concentric banding of chlorite-actinolite-sphene around microfracture nets. Sphene is also present in vein fillings, indicating Ti was mobilized by hydrothermal fluids.

Fragmented, silicified, and altered basalt clasts are included in quartz-chert-cemented breccias. Cross-cutting relationships indicate silicification occurred in several episodes. In the quartz-cemented breccia, angular chlorite-rich fragments were rapidly cemented together into a clast that was then rounded by transportation in the up-welling limb of a hydrothermal system. In contrast, partially dissolved angular fragments are suspended in the chert-cemented matrix, suggesting the hydrothermal fluids existed in a reactive, stagnant, and viscous phase. Late-stage veins of crystalline microquartz are free of suspended dissolved basalt fragments, suggesting they precipitated from a fluid phase that penetrated brittle fractures. Hydrothermal fluids evidently scavenged the silica from basalt during chloritization. Brecciation then induced a physical-chemical change that resulted in precipitation of silica in the matrix. Oxygen isotope analyses indicate that the siliceous phases formed at lower temperatures than chlorite-rich clasts. The strontium isotope composition of the siliceous phases is similar to that of the brecciated basalts, suggesting they formed at similar W/R ratios from fluids of similar strontium isotopic composition.

The gabbros were altered at the highest temperature. They are characterized by round millimeter sized chlorite spots that cut the igneous texture, and contain needles of actino-

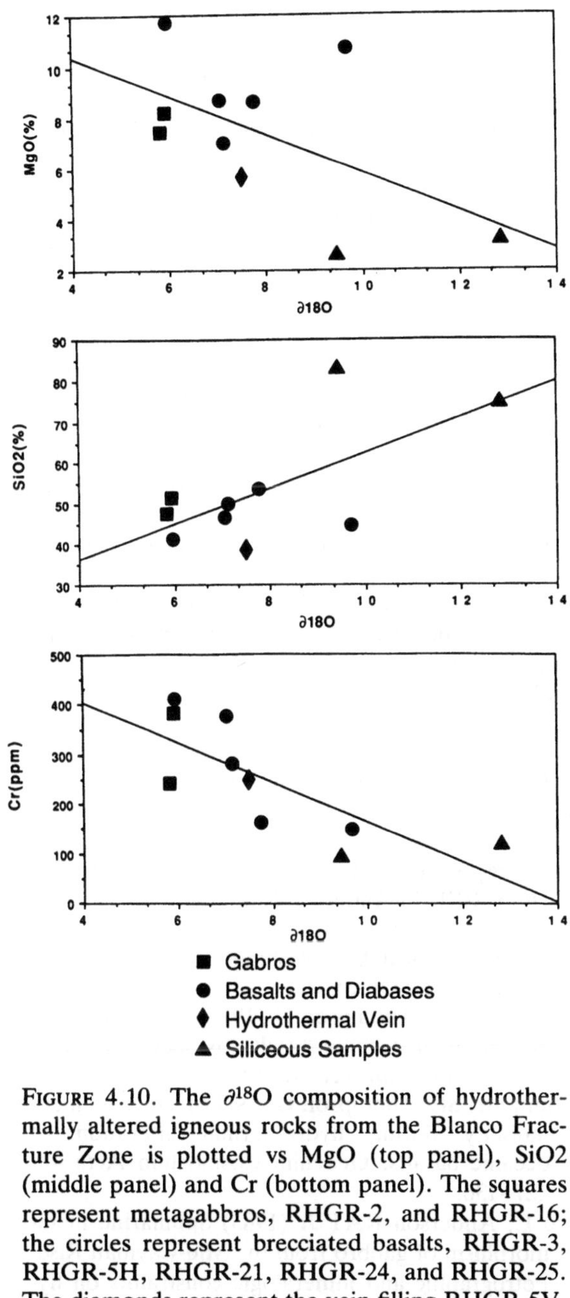

FIGURE 4.10. The $\partial^{18}O$ composition of hydrothermally altered igneous rocks from the Blanco Fracture Zone is plotted vs MgO (top panel), SiO2 (middle panel) and Cr (bottom panel). The squares represent metagabbros, RHGR-2, and RHGR-16; the circles represent brecciated basalts, RHGR-3, RHGR-5H, RHGR-21, RHGR-24, and RHGR-25. The diamonds represent the vein filling RHGR-5V. The triangles represent siliceous samples RHGR-6 and RHGR-11.

lite, and are rimmed with sulfides. Veins of albite with cores of prehnite formed during a later stage at lower temperatures.

Acknowledgments. This work was funded by the Gorda Ridge Technical Task Force through a grant by the Oregon Department of Geology and Mineral Industries. R.H. is grateful for a grant from the Max Planck Society that enabled determination of the strontium isotopic compositions presented in this chapter. We thank Drs. Greg McMurray and Don Hull for expediting the public presentation of this material at Task Force meetings and in various publications. We thank the USGS Marine Geology Branch for dredging operations carried out under the supervision of Dr. David Clague and Mr. Andrew Stevenson.

SEM analyses were provided by Drs. George Dooley and Charles Dellenbach of the U.S. Bureau of Mines and Al Soeldner, Department of Plant Pathology at Oregon State University. We are grateful to Greg Campi, Bobbi Conard, James Robbins, and Sandy Moore of Oregon State University and to Michael Shaffer of the University of Oregon for technical assistance in the analyses. We thank Dr. William G. Melson of the Smithsonian Institution for providing a sample of the quartz-cemented breccia.

The authors benefited from discussion with Drs. Greg McMurray, Martin Fisk, John Delaney, Ross Heath, and Robert Embley. We thank Robert Embley and the NOAA-PMEL VENTS program for providing the location and SEABEAM maps on which the dredging was based. We thank Sue Pullen for preparing the manuscript and Dave Reinert and Brendan Hart for their help in preparing the figures.

References

Alt JC, Emmerman R (1983) Geochemistry of hydrothermally altered basalts: DSDP hole 504B, leg 83. In Anderson RN Honnorez J, Becker K, et al. (eds): Initial Reports of the Deep Sea Drilling Project 83. Washington: U.S. Government Printing Office, pp 249–262.

Alt JC, LaVerne C, Muehlenbachs K (1985) Alteration of the upper oceanic crust: Mineralogy and processes in DSDP hole 504B, leg 83. In Anderson RN, Honnorez J, Becker K et al. (eds): Initial Reports of the Deep Sea Drilling Project, 83. Washington: U.S. Government Printing Office, pp 217–247.

Alt JD, Honnorez J, Laverne C, Emmermann R (1986) Hydrothermal alteration of 1-km section through the upper oceanic crust, Deep Sea Drilling Project hole 504B: Mineralogy, chemistry, and evolution of seawater-basalt interactions. J Geophys Res 91:10309–10336.

Bonatti E, Honnorez J, Kirst P, Radiciati F (1975) Metagabbros from the Mid-Atlantic Ridge at 06°N: Contact-hydrothermal-dynamic metamorphism beneath the axial valley. J Geol 83:61–78.

Bonatti E, Guerstein-Honnorez BM, Honnorez J (1976) Copper-iron sulfide mineralizations from the equatorial Mid-Atlantic Ridge. Econ Geol 71:1515–1525.

Cann JR (1969) Spilites from the Carlsberg Ridge, Indian Ocean. J Petrol 10:1–19.

Cann JR (1971) Petrology of basement rocks from Palmer Ridge, N.E. Atlantic. Phil Trans R Soc Lond Ser A 268:605–617.

Cann JR, Funnell BM (1967) Palmer Ridge: A section of the upper part of the ocean crust? Nature 213:6661–6664.

Collinson TB (1986) Hydrothermal mineralization and basalt alteration in stockwork zones of the Bayda and Lasail massive sulfide deposits, Oman ophiolite. MA Thesis, University of California, Santa Barbara, 164 p.

Crane K (1976) The intersection of the Siqueiros Transform Fault and the East Pacific Rise. Marine Geol 21:25–46.

Delaney JR, Mogk DW, Mottl MJ (1987) Quartz-cemented, sulfide-bearing greenstone from Mid-Atlantic Ridge; sample of a high-temperature, high-salinity hydrothermal upflow zone. J Geophys Res 92:9175–9192.

Elthon D (1981) Metamorphism in oceanic spreading centers. In Emiliani C (ed): The Sea, Vol 7. New York: John Wiley and Sons, pp 285–303.

Embley RA, Kulm LD, Massoth G, Abbott D, Holmes M (1987) Morphology, structure and resource potential of the Blanco Transform Fault Zone. In Scholl DW et al. (eds): Geological Resource Potential of Western North America and Adjacent Ocean Basins—Beaufort Sea to Baja California. AAPG Memoir Ser 6:549–561.

Engel CG, Fisher RL (1975) Granitic to ultramafic rock complexes of the Indian Ocean Ridge System, Western Indian Ocean. Geol Soc Am Bull 86:1553–1578.

Franklin JM (1987) Volcanic-associated massive sulfide deposits—an update. In Andrew CJ, Crowe RWA, Finley S, Pennell WM, Pyne JF (eds): Geology and Genesis of Mineral Deposits in Ireland. Dublin: Irish Association of Economic Geologists.

Franklin JM, Lydon JW, Sangster DF (1981) Volcanic-associated massive sulfide deposits. Econ Geol 75th anniversary volume: 485–627.

Harper GD, Bowman JRJ, Kuhns R (1988) Field chemical and isotopic aspects of submarine hydrothermal metamorphism of the Josephine Ophiolites, northwest California. J Geophys Res 93:4625–4655.

Hart RA (1973) A model for the chemical exchange in the basalt-seawater system of oceanic layer 2. Can J Earth Sci 10:799–816.

Hart RA, Pyle D, Robbins J (1986) Hydrothermal sulfides, breccias and greenstones from the Gorda Depression. State of Oregon Department of Geology and Mineral Industries Open-File Report 0-86-16; 31 p.

Hekinian R, Fevrier MN, Avedik F, et al. (1983) East Pacific Rise near 13°N: Geology of new hydrothermal fields. Science 219:1321–1324.

Hoffert M, Perseil A, Hekinian R, et al. (1978) Hydrothermal deposits sampled by diving saucer in Transform "A" near 37°N on the Mid-Atlantic, Famous Area. Oceanol Acta 1:72–86.

Honnorez J, Alt JC, Honnorez-Guertsein BM, et al. (1985) Stockwork-like sulfide mineralization in young oceanic crust: Deep Sea Drilling Project Hole 504B. In Anderson RN, Honnorez J, Becker K, et al. (eds): Initial Reports of the Deep Sea Drilling Project 83. Washington: U.S. Government Printing Office.

Humphris SE, Thompson G (1978a) Hydrothermal alteration of oceanic basalts by seawater. Geochim Cosmochim Acta 42:107–125.

Humphris SE, Thompson G (1978b) Trace element mobility during hydrothermal alteration of oceanic basalts. Geochim Cosmochim Acta 42: 127–136.

Ito E, Anderson RAT Jr (1983) Submarine metamorphism of gabbro from the Mid-Cayman Rise: Petrographic and mineralogic constraints on hydrothermal processes at slow spreading ridges. Contrib Mineral Petrol 82:371–388.

Ito E, Clayton RN (1983) Submarine metamorphism of gabbros from the Mid-Cayman Rise: An oxygen isotopic study. Geochim Cosmochim Acta 47:535–546.

Jenkins WJ, Clarke WB (1976) The distribution of ^{3}He in the western Atlantic Ocean. Deep-Sea Res 23:481–494.

Juteau T, Noack Y, Whitechurch H, Courtois C (1980) Mineralogy and geochemistry of alteration products in holes 417A and 417D basement samples (Deep Sea Drilling Project, Leg 51). In Donnelly T, Franchetau J, et al (eds): Initial Reports of the Deep Sea Drilling Project, 53. Washington: U.S. Government Printing Office, pp 1273–1279.

Kelley DS, Delaney JR (1987) Two-phase separation and fracturing in mid-ocean ridge gabbros at temperatures greater than 700°C. EPSL 83:53–66.

Lalou C, Brichet E, Ku TL, Jehanno C (1977) Radiochemical, scanning electron microscope (SEM) and X-ray dispersive energy (EDAX) studies of a FAMOUS hydrothermal deposit. Marine Geol 24:245–258.

Lydon JW, Galley A (1986) The chemical and mineralogical zonation of Mathiati alteration pipe, Cyprus, and genetic significance. In Gallapher MJ, Ixer RA, Neary CR, Prichard HM (eds): Metallogeny of Basic and Ultrabasic Rocks. London: Institute of Mining and Metallogeny, pp 49–68.

Mével C (1987) Evolution of oceanic gabbro from DSDP leg 82: Influence of the fluid phase on metamorphic crystallizations. Earth Planetary Sci Lett 83:67–79.

Melson WG, Van Andel TH (1966) Metamorphism in the Mid-Atlantic Ridge, 22°N latitude. Mar Geol 4:165–186.

Melson WG, Thompson G, Van Andel TJH (1968) Volcanism and metamorphism in the Mid-Atlantic Ridge, 22°North latitude. J Geophys Res 73:5925–5941.

Miyashiro A, Shido F, Ewing M (1971) Metamorphism in the Mid-Atlantic Ridge near 24 and 30°N East Pacific Rise site. Earth Planet Sci Lett 67:297–308.

Reed M (1983) Seawater-basalt reaction and the origin of greenstones and related ore deposits. Econ Geol 78:80–90.

Richards HG, Cann JR, Jensensius J (1988) Mineralogical and metasomatic zonation of the alteration zone of Cyprus sulfide deposits. Econ Geol 83:71–89.

Roberts RG, Reardon EJ (1978) Alteration and ore-forming processes at Mattagami Lake Mine, Quebec, Can. J Earth Sci 15:1–21.

Rona PA (1976) Pattern of hydrothermal mineral deposition: Mid-Atlantic Ridge Crest at Latitude 26°N. Marine Geol 21:59–66.

Rona PA, Bostrom K, Epstein S (1980) Hydrothermal quartz vug from the Mid-Atlantic Ridge. Geology 8:569–572.

Selk BW (1977) The manganese-enriched sediments of the Blanco Trough: Evidence for hydrothermal activity in a fracture zone. MSc Thesis, Oregon State University, Corvallis.

Seyfried WE, Mottl MJ, Bischoff JL (1978) Seawater/basalt/ratio effects on the chemistry and mineralogy of spilites from the ocean floor. Nature 275:211–213.

Spooner ETC, Fyfe WS (1973) Sub-seafloor metamorphism, heat and mass transfer. Contrib Mineral Petrol 42:287–304.

Stakes DS, O'Neil JR (1982) Mineralogy and stable isotope geochemistry of hydrothermally altered oceanic rocks. Earth Planet Sci Lett 57:285–304.

Stakes DS, Scheidegger KF (1981) Temporal variations in secondary minerals from Nazca Plate basalts. In Nazca Plate: Crustal Formation and Andean Convergence. Geological Society of America Ann Arbor 154, pp 109–130.

Stakes DS, Vanko DA (1986) Multistage hydrothermal alteration of gabberoic rocks from the failed Mathematician Ridge. Earth Planet Sci Lett 79:75–92.

5—Distribution and Composition of Massive Sulfide Deposits at Escanaba Trough, Southern Gorda Ridge

JANET L. MORTON, RANDOLPH A. KOSKI, WILLIAM R. NORMARK, and STEPHANIE L. ROSS

Abstract

Massive sulfide deposits were photographed and sampled at two sites (southern or SESCA and northern or NESCA) on the sediment-covered Escanaba Trough segment of Gorda Ridge, a slow-spreading midocean ridge located offshore of northern California within the U.S. Exclusive Economic Zone. Massive sulfide ledges, mounds, and chimneys in various stages of oxidation and erosion are spatially associated with the margins of small (1 km wide, 100 m high) sediment hills at SESCA and NESCA, and with faulted sediment on the north flank of a volcanic hill at NESCA. The sediment hills are uplifted above volcanic edifices that rise through sediment filling the axial valley. Two types of massive sulfide have been dredged from Escanaba Trough. Porous aggregates of pyrrhotite-rich sulfide contain small amounts of intergrown isocubanite, chalcopyrite, sphalerite, and loellingite. In some samples, pyrrhotite is partly to completely replaced by marcasite. Pyrrhotite-rich massive sulfide generally contains low amounts of base and precious metals, but barite encrustations on some of the samples are enriched in Zn, Pb, Ag, and Au. This sulfide type is largely derived from mound and talus deposits associated with the sediment hills at SESCA and NESCA. Zoned polymetallic sulfide samples from a single dredge at the NESCA site comprise the second massive sulfide type. These polymetallic samples are fragments from the wall of a sulfide structure with a large fluid channelway. Proceeding from the inner to outer wall, the mineral zonation includes isocubanite, sphalerite, arsenopyrite, galena, tetrahedrite, a Pb-Sb sulfosalt, and argentite. Polymetallic sulfide samples are enriched in Pb, As, Sb, Ag, Hg, and Sn relative to massive sulfide recovered from other spreading axes. Other dredge samples include sulfide-cemented sediment-clast breccia and sediment containing abundant interstitial galena and hydrocarbon. The composition of massive sulfide from both areas indicates extensive interaction between hydrothermal fluids and the sediment in Escanaba Trough. Pyrrhotite-rich sulfide probably forms above sites of diffuse, sluggish discharge, whereas polymetallic sulfide structures formed in response to focused, high-velocity discharge at a "black smoker" type of vent.

Introduction

The majority of ridge-crest hydrothermal systems that have been studied to date occur at sediment-free spreading axes. Most of these known bare-rock hydrothermal deposits have limited areal extent. Sediment-covered spreading centers are relatively rare, but have recently become the focus of considerable interest because of the possibility for hosting large massive sulfide deposits analogous to some ancient sediment-hosted ore deposits. Sediment-hosted hydrothermal deposits located at oceanic spreading centers have been observed at Guaymas Basin, Gulf of California (Lonsdale et al., 1980; Koski et al., 1985), and at Middle Valley on the northern Juan de Fuca Ridge (Davis et al., 1987).

Escanaba Trough at the southern end of Gorda Ridge is a spreading center covered by as much as 500 m of turbiditic and interbedded hemipelagic sediment. Geological and geophysical investigations conducted in 1985 and 1986 aboard the U.S. Geological Survey re-

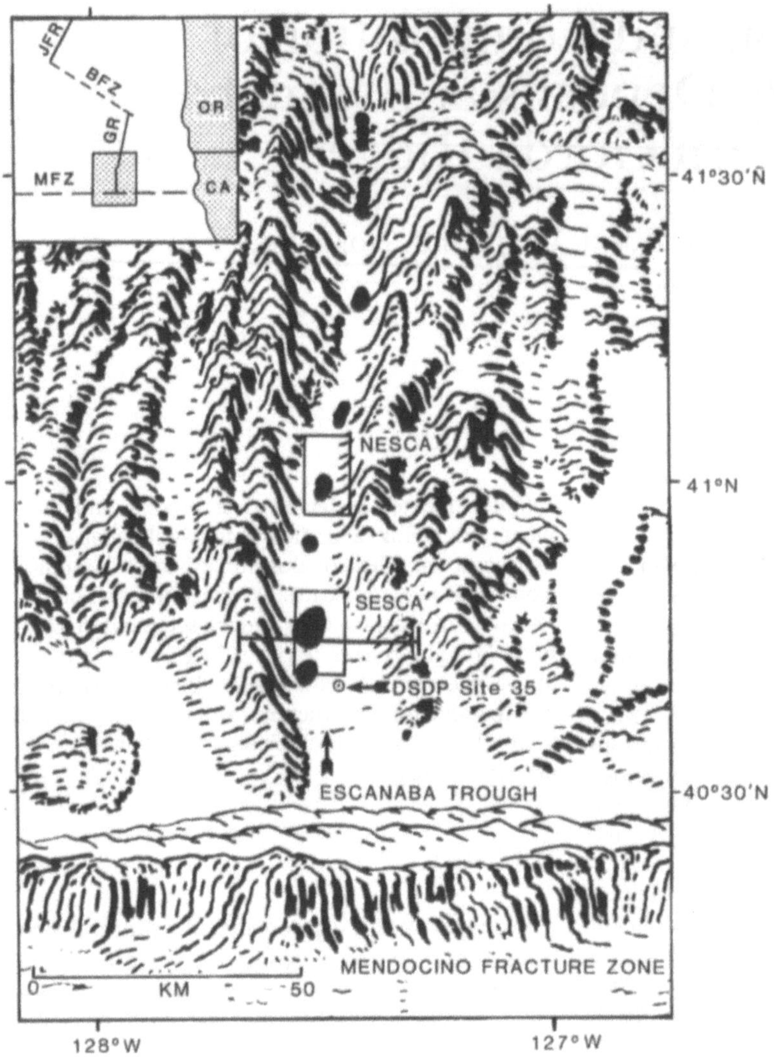

FIGURE 5.1. Physiographic map of Escanaba Trough showing volcanic edifices (black spots), location of seismic-reflection profile 7 (Fig. 5.2), SESCA site (Fig. 5.3), NESCA site (Fig. 5.4), and location of DSDP site 35. Adapted from compilations by T. Chase, B. Seekins, and K. Lund (unpublished). Inset shows location of Gorda Ridge (GR) between Blanco Fracture Zone (BFZ) and Mendocino Fracture Zone (MFZ). Juan de Fuca Ridge (JFR) is ridge segment to the north.

search vessel *S.P. Lee* mapped and sampled several inactive sulfide deposits at two sites on Escanaba Trough. Subsequent dives with the submersible *Sea Cliff* obtained additional sulfide samples from some of the deposits (see Holmes and Zierenberg, this volume). This chapter presents the preliminary results on the distribution and composition of the sulfide deposits based on photographic, dredging, and coring operations conducted during *S.P. Lee* cruises L1-86-NC and L2-86-NC and is in part a summary of material presented in two U.S. Geological Survey Open-File Reports (Morton et al., 1987a; Benninger and Koski, 1987).

Geologic Setting

Gorda Ridge is located in the northeast Pacific Ocean 200 to 300 km west of northern California and southern Oregon (see Fisk and Howard, this volume, for a discussion of the regional geologic setting of Gorda Ridge) and is within the U.S. Exclusive Economic Zone (EEZ). Escanaba Trough comprises the southernmost segment of Gorda Ridge, extending 130 km northward from the intersection of the ridge with the Mendocino Fracture Zone (at 40° 25′ N latitude, 127° 30′ W longitude) to a

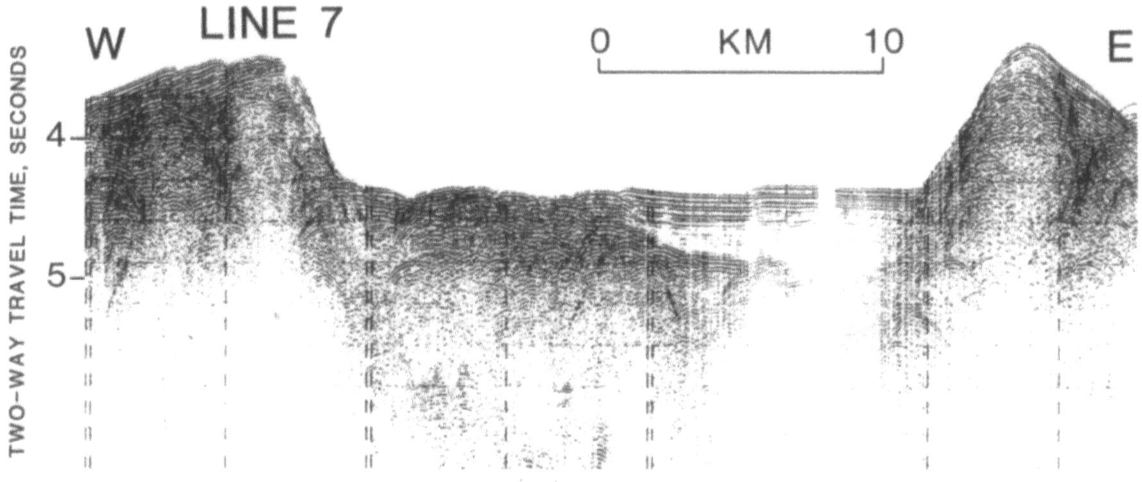

FIGURE 5.2. Single-channel seismic-reflection profile 7 over SESCA site. The volcanic edifice is near the west side of the axial valley, and faults offset the turbidite sediment in the eastern half of the axial valley. Vertical exaggeration is 6.4. Location of profile is shown in Figure 5.1.

right-lateral offset of the ridge axis near 41° 35′ N latitude, 127° 25′ W longitude (Fig. 5.1). The total spreading rate along Escanaba Trough segment is about 2.3 cm/year (Atwater and Mudie, 1973; Riddihough, 1980), which is characteristic of a slow-rate spreading center. South of 41° 15′ N latitude, the axis of Escanaba Trough is covered by turbidite and hemipelagic sediment (Moore, 1970; Vallier et al., 1973; Karlin and Lyle, 1986). A drill hole at DSDP site 35 (Fig. 5.1) penetrated 390 m of primarily Pleistocene turbidites without reaching basement (McManus et al., 1970).

Reconnaissance seismic-reflection surveys suggested that igneous rocks intrude the sedimentary fill at several localities near the axis of Escanaba Trough (Moore and Sharman, 1970; Clague and Holmes, 1987). A detailed seismic-reflection survey conducted in 1985 revealed that the igneous bodies are discrete volcanic edifices with both intrusive and extrusive components and are spaced at approximately 15-km intervals along the axis (Fig. 5.1; Holmes and Morton, 1986; Morton et al., 1987b). The sedimentary fill of the trough above some edifices has been uplifted as much as 100 m (Fig. 5.2). Locally, fresh basalt flows are exposed on the sea floor. The areas of the volcanic edifices appear on GLORIA side-scan records as nearly circular, highly reflective zones within the less reflective axial-valley sediment (EEZ-SCAN 84 Scientific Staff, 1986). Reconnaissance station work conducted on the 1985 cruise suggested that the volcanic centers are sites of active or recent hydrothermal activity. Heat-flow values in excess of 1200 mW/m^2 were measured adjacent to two of the volcanic centers (Abbott et al., 1986). Pyrrhotite-rich massive-sulfide samples were dredged from the flanks of the volcanic center near 40° 45′ N latitude, 127° 30′ W longitude (Holmes and Morton, 1986; Morton et al., 1987b). These sulfide deposits were the first ones identified on a midocean ridge spreading axis within the U.S. EEZ. One of the sulfide samples contained nonbiodegraded asphaltic petroleum of hydrothermal origin (Kvenvolden et al., 1986).

Based on the geophysical studies and sampling results in 1985, two of the larger volcanic edifices were chosen for detailed work in 1986. The southern Escanaba Trough or SESCA site is located at 40° 45′ N latitude, 127° 30′ W longitude; the northern Escanaba Trough or NESCA site is located at 41° 00′ N latitude, 127° 30′ W longitude (Figs. 5.1, 5.3, and 5.4).

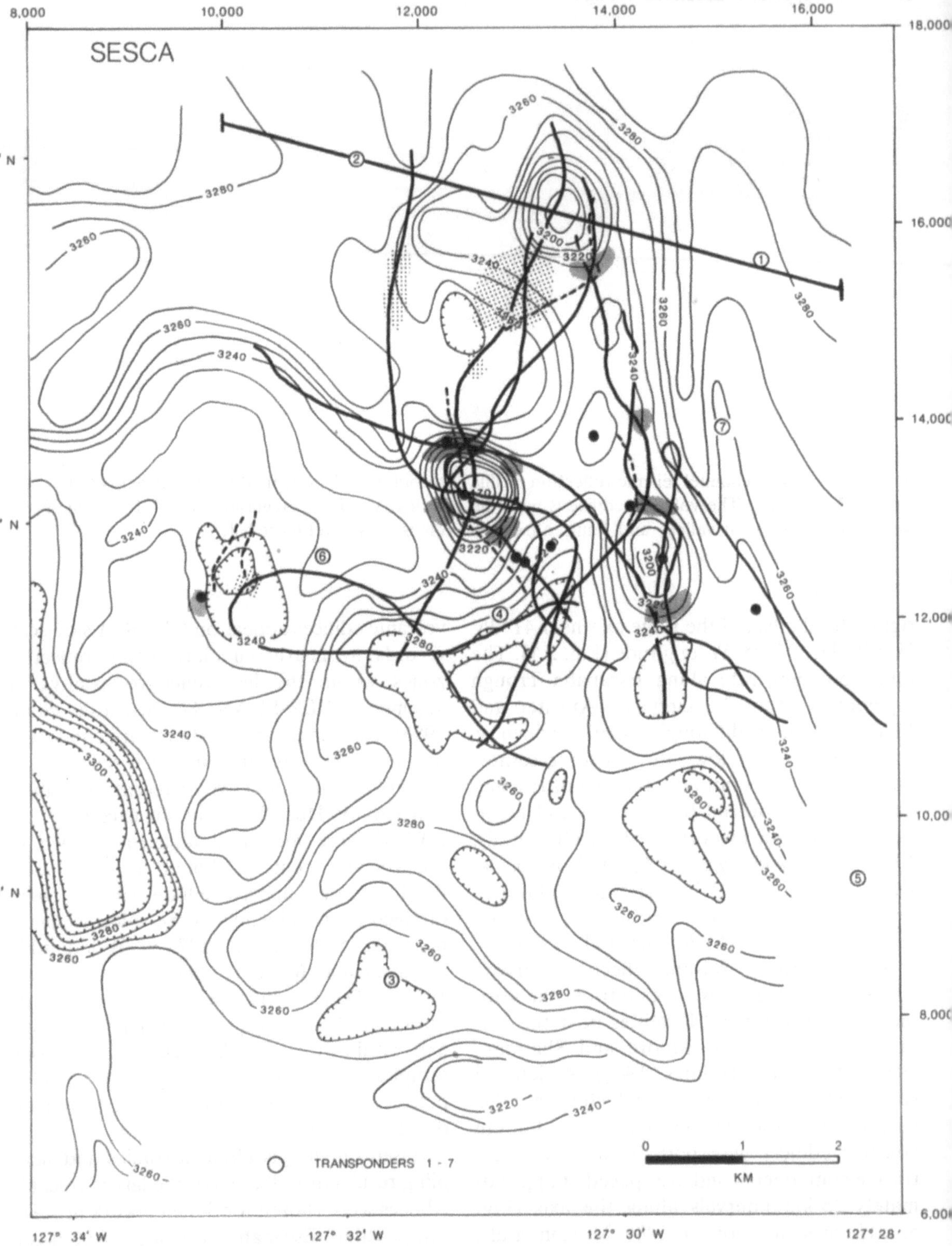

FIGURE 5.3. Bathymetric map of the SESCA site (10-m contour interval) showing location of photographed and/or sampled sulfide deposits (gray) and approximate extent of outcropping volcanic rock (stippled pattern). Also shown are location of 3.5-kHz profile shown in Figure 5.5, location of camera (solid lines), dredge (dashed lines), and gravity core (solid dots) stations, and location of sea floor transponders (circled numbers). Location of map is shown in Figure 5.1.

Methods

An acoustic-transponder positioning system provided a common base for all sample stations and geophysical lines as well as the positions for the *Sea Cliff* dives. An integrated satellite (NNSS)-Doppler sonar-Loran C navigation system and a satellite global positioning system (GPS) provided supplemental navigation during deployment and calibration of the transponder net and near the fringes of positioning available from the transponder net. A relay transponder was attached to the wire above the camera or sampling device to provide direct position information for each station. When travel times to three or more sea floor transponders were obtained, position errors were generally less than 3 to 5 m, sufficiently accurate for reoccupation of station sites.

Underway geophysical equipment consisted of a 12-kHz echosounder and a 3.5-kHz subbottom profiler, both hull mounted. In addition, 160 km of single-channel seismic reflection data were collected utilizing an 80-in^3 water gun as an energy source. A bathymetric survey of the SESCA site, with track lines spaced at 350-m intervals, was conducted during a 1986 *USNS Narragansett* cruise preceding the *S.P. Lee* cruises; a similar survey of the NESCA site, with tracklines spaced at approximately 300-m intervals, was conducted during the *S.P. Lee* cruises. The bathymetric data were digitized and contoured to produce the base maps for all sampling work (Figs. 5.3 and 5.4).

The primary activities during both 1986 *S.P. Lee* cruises were sea floor photography, dredging, and gravity coring (Figs. 5.3 and 5.4). In addition, we conducted two hydrocast stations and two fish trap deployments (see Morton et al., 1987a, for additional information on sample collection procedures).

Twelve camera tows were made in the NESCA and SESCA areas. The photographic system consists of both a 35-mm still camera and a self-contained color video system mounted in a rugged steel frame (Chezar and Lee, 1985). The sled was generally towed 3 to 5 m above the sea floor and yielded sea floor images 3 to 4 m wide. The still camera obtained one exposure every 14 sec on Ektachrome color film. Output from a sled-mounted temperature sensor was acoustically telemetered to the ship during camera stations, providing an indication of relative changes in bottom water temperature (i.e., true temperature values are not provided).

In addition to the 4.5 kg of massive sulfide dredged from the SESCA site in 1985 (Morton et al., 1987b), 44.1 kg of sulfide-bearing material was dredged from SESCA and 17.8 kg from NESCA in 1986 (Benninger and Koski, 1987; Koski et al., 1988). The petrographic and microchemical characterization of the sulfides was accomplished by a combination of incident-light microscopy, scanning electron microscopy (SEM), and electron microprobe (EM) analysis on doubly polished thin sections. Chemical analyses of bulk sulfide- and sulfate-rich samples were obtained by quantitative dc-arc optical-emission spectroscopy (Fe, Cu, Zn, Pb, Au, Bi, Cd, Sb, Se, Hg), semiquantitative emission spectroscopy (Ag, As, Ba, Mn, Sn), and LECO total S analysis (S). Arsenic greater than 0.2 wt% was analyzed by inductively coupled plasma emission spectroscopy. Organic carbon was determined by subtracting carbonate, analyzed by a coulometric technique, from total carbon analyzed with the LECO total-C analyzer.

Distribution of Sulfide Deposits

The bathymetry and location of major sulfide deposits of the SESCA and NESCA sites are shown in Figures 5.3 and 5.4, respectively. The distribution and structural/geomorphic setting of sulfide deposits differs between the two study areas, which will be discussed separately.

SESCA

The SESCA site is characterized in bathymetric and seismic-reflection data by a broad, low uplift over much of the volcanic edifice relative to the surrounding sediment of the axial valley. Geophysical profiles and photographic results indicate that most of the volcanic complex is sediment covered; only three small areas of ex-

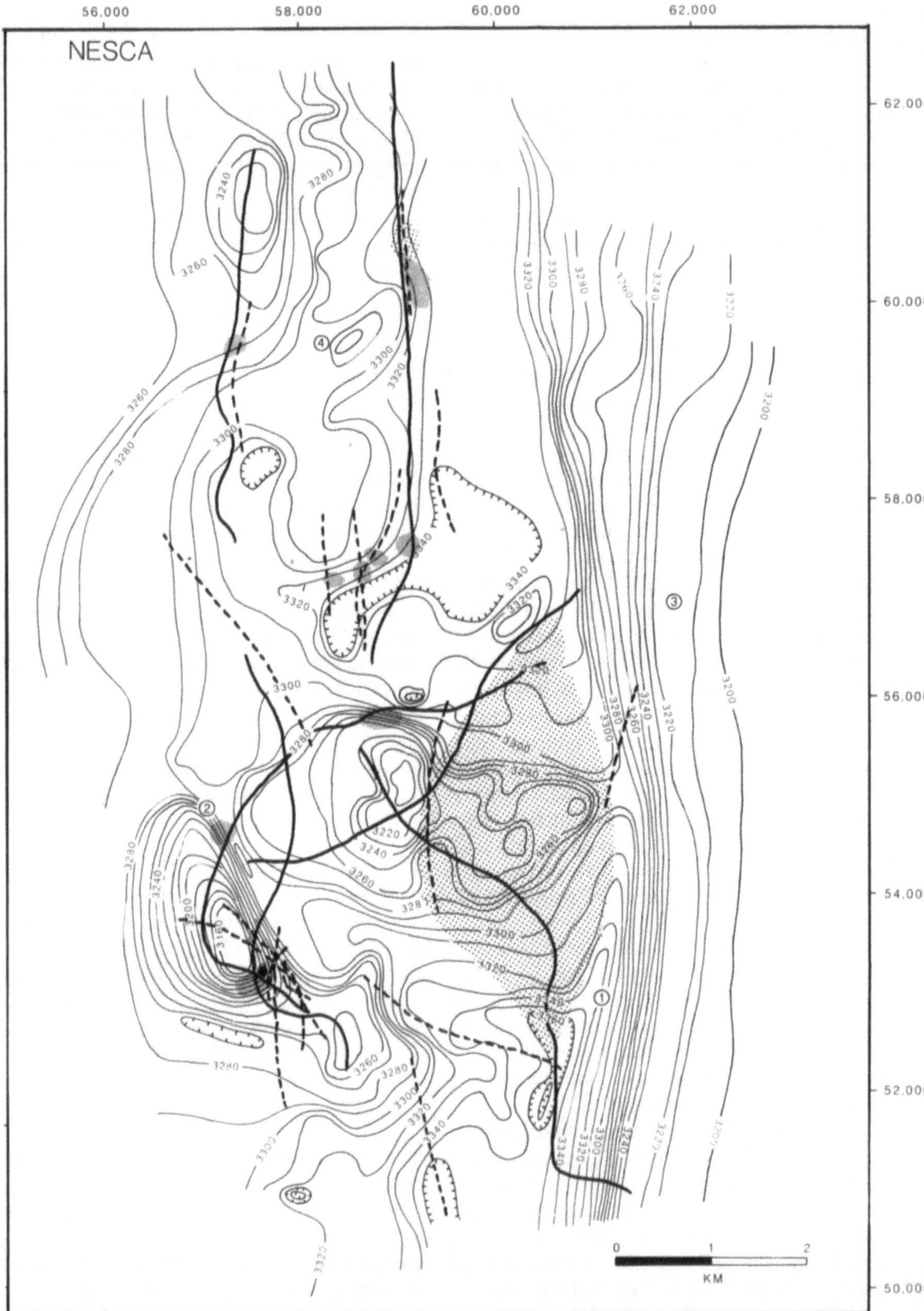
NESCA
56,000
58,000
60,000
62,000
62,000
60,000
58,000
56,000
54,000
52,000
50,000
41° 02 N
41° 00 N
40° 58 N
127° 32 W
127° 30 W
127° 28 W
0
1
2
KM

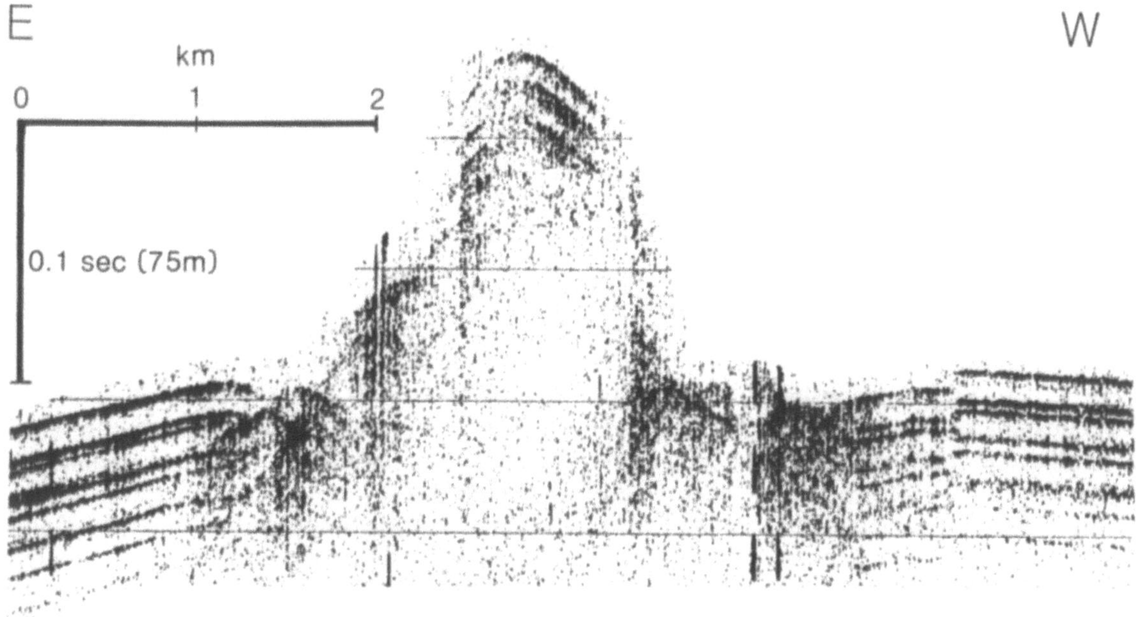

FIGURE 5.5 East-west 3.5-kHz profile across northern hill in SESCA area. Separation of major reflectors within the sediment cover is quite similar on both flanks and crest of the hill. Vertical exaggeration is 20. Location of profile is shown in Figure 5.3.

posed basaltic rock, primarily fissured pillow basalt, were photographed at SESCA (Fig. 5.3). Gravity cores obtained from sediment overlying the volcanic edifice consist predominantly of gray-green hemipelagic mud with thin layers of turbiditic silt and sand (Morton et al., 1987a).

The most prominent physiographic features of the SESCA area are three circular hills that rise as much as 100 m above the sediment-covered volcanic complex. The hills are bounded by steep slopes with angle-of-repose sediment and talus. Near the summit of the hills, slopes steeper than the angle of repose of talus expose outcropping sediment. Seismic-reflection and 3.5-kHz profiles indicate that the hills are composed largely of sediment, although resolution of the profiles is not sufficient to determine the nature of the core of the hills. A 3.5-kHz reflection profile (Fig. 5.5) across the northernmost hill in the SESCA area shows that the sediment under the summit has an internal layering nearly identical to the surrounding lower-relief areas. If much of this sediment cover is turbiditic (i.e., gravity-driven flow deposits) in origin, it suggests relatively recent uplift of the hill. The form of the southwestern hill, which has a nearly horizontal summit area, is that of a butte with short near-vertical slopes around the summit flanked by steep talus slopes. No exposures of volcanic rock were observed during submersible surveys or sampled during dredging on the sides of the hills. No surficial faults or fissures around the base of the hills were observed during camera or submersible studies; however, recent faults were observed on single-channel and 3.5-kHz reflection profiles at or near the base of the slopes bounding the sediment hills. Thus, the available observations are consistent with at

FIGURE 5.4. Bathymetric map of the NESCA site (10-m contour interval) showing location of photographed and/or sampled sulfide deposits (gray), approximate extent of outcropping volcanic rock (stippled pattern), and location of camera and sampling stations and sea floor transponders (symbols same as in Fig. 5.4). Location of map is shown in Figure 5.1.

least two mechanisms of formation of the circular hills. The hills may be cored by bysmalithic intrusions (Lonsdale and Lawver, 1980) that uplifted their sediment caps during intrusion. The presence of pillow basalt exposed at the sea floor near the base of the northern hill (Fig. 5.3) is consistent with this hypothesis. Alternatively, the hills may have formed by mass wasting of a larger area of uplifted sediments, perhaps on a horst block formed during recent extension of the axial valley or during doming above the volcanic edifice.

Several large sulfide deposits up to 100 m in lateral extent occur around the base of the three hills at SESCA (Fig. 5.3). The larger deposits consist of constructional ledges and mounds often surrounded by sulfide talus aprons (Fig. 5.6). The deposits range in age from young, unweathered deposits with chimneylike projections to older, partly oxidized deposits locally degraded by mass wasting. Blocks of recently disturbed white sediment and encrustations of low-relief white material were photographed in several places, notably in the basin near transponder 4 at SESCA and on the southern and western slopes of the southwestern hill (Fig. 5.3). Hydrothermal talc and white mudstone altered to a chlorite-talc aggregate in dredges from the same region suggest that the photographed white material is of hydrothermal origin at some locations.

The schematic east-west geologic section in Figure 5.7a summarizes the major observations on the distribution of hydrothermal deposits in the SESCA area. The major sulfide deposits are exposed at the base of the hills and on the slopes of the bench on which the hills stand. The white deposits occur within the uplifted sediment sections of the hills and at or near the sea floor immediately south and west of the hills.

NESCA

The axial valley floor at NESCA is flanked on either side by broad, sedimented terraces that have been uplifted 100 to 150 m above the valley floor (only the eastern terrace is shown on Fig. 5.4). Two prominent hills sit atop the NESCA volcanic edifice. The hill to the southwest is sediment capped and bounded by a steep scarp on the eastern side. No volcanic rocks have been photographed on or dredged from the slopes of this hill, similar to the smaller hills at SESCA. At the larger hill near the center of the NESCA site, fresh pillow basalt is locally exposed on the sea floor (Fig. 5.4). The north flank of the largest peak of the central volcanic hill is formed by a series of fault-bounded north-dipping terraces. Volcanic rocks are not exposed by the faults, and seismic records indicate a considerable thickness (approximately 150 m) of sediment flanking this side of the hill. Circular features (up to a few meters in diameter) resembling sediment-covered collapse pits observed on photographs from the elongate basin north of the central hill suggest that the basin might be a thinly sedimented lava lake.

Large sulfide deposits occur at NESCA near the base of the sediment hill and in the faulted sediment section north of the central volcanic hill. The deposits near the base of the sediment hill are similar in form and setting to the SESCA deposits. Observations from the submersible *Sea Cliff* indicate that massive sulfide on the north flank of the central volcanic hill is exposed over a 30-m vertical thickness and is capped by chimneylike spires several meters high (Holmes and Zierenberg, this volume). Dredged sulfide samples from this location are fresh and unoxidized. Small isolated chimneys and sulfide crusts occur along the western slope bounding the elongate basin north of the central volcanic hill. The presence of clam shells and dead vestimentifera (tube worms) at the sulfide deposit on the north flank of the volcanic hill (Holmes and Zierenberg, this volume) also suggests recent venting.

Figure 5.7b is a north-south schematic geologic section that summarizes the main observations on the distribution of massive sulfide deposits north of the central volcanic hill in the NESCA area. Although active venting was not observed at either the NESCA or SESCA sites, small increases in bottom water temperature were recorded over the sediment hill at NESCA, suggesting that some diffuse fluid discharge is occurring. Active hydrothermal venting has subsequently been observed and

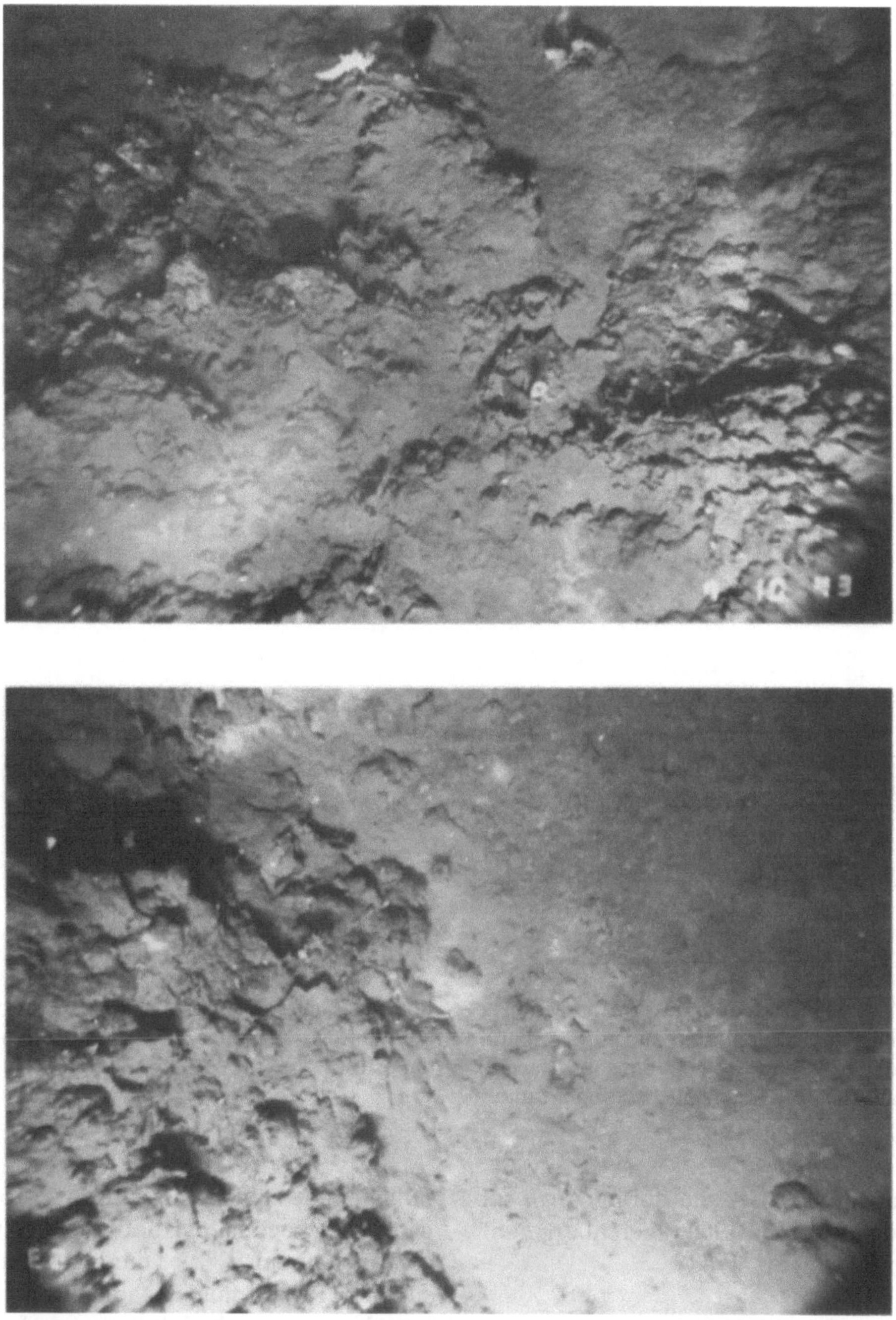

FIGURE 5.6. Photographs of weathered sulfide mounds on the sea floor at SESCA site. Width of each photo is about 3 m.

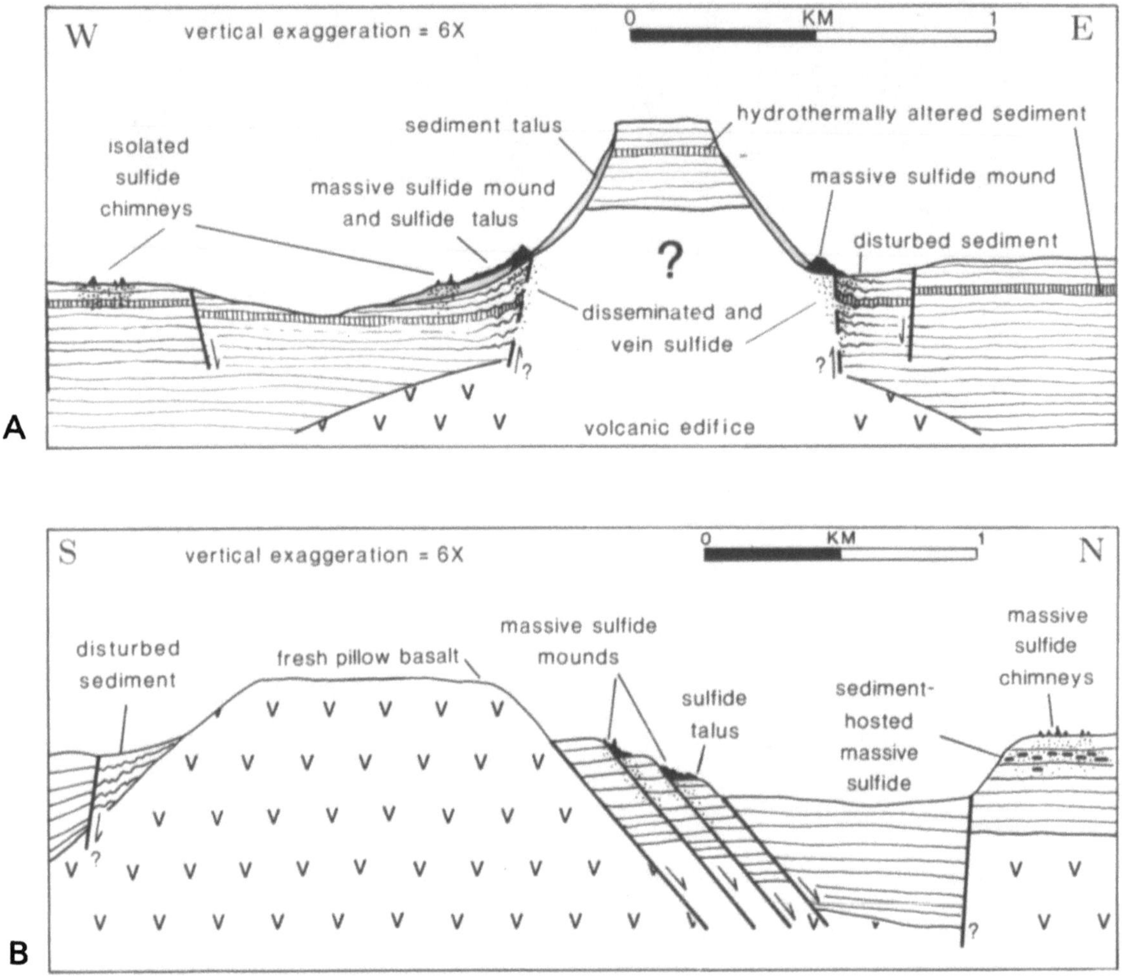

FIGURE 5.7. (A) Schematic east-west section of SESCA area summarizing the type and distribution of sulfide deposits, and the possible subsurface extent of stratigraphically controlled sediment alteration. (B) Schematic north-south section of NESCA area summarizing the type and distribution of sulfide deposits.

sampled at the NESCA site during *Alvin* submersible dives in June 1988 (Zierenberg et al., 1988; Campbell et al., 1988).

Composition of Sulfide Deposits

The location and general description of sulfide-bearing samples recovered from Escanaba Trough in 1986 are presented in Table 5.1. Most of the sulfide samples can be grouped into two rock types: homogeneous, pyrrhotite-rich massive sulfide (Fig. 5.8a); and zoned polymetallic sulfide (Fig. 5.8b). Porous aggregates of hexagonal pyrrhotite and/or marcasite comprise about 80% of the recovered massive sulfide. Fossil fluid channelways in many samples are marked by zones of randomly oriented, coarse-grained pyrrhotite plates surrounded by pyrrhotite with radial orientation. In some samples, marcasite has partially or completely replaced pyrrhotite. Pyrite is concentrated along the outer surfaces of some samples but is a volumetrically minor phase. Small amounts of isocubanite ($CuFe_2S_3$), chalcopyrite ($CuFeS_2$), and sphalerite (ZnS) occur

TABLE 5.1. Location and results of dredge, gravity core, and camera stations that recovered samples of massive sulfide, hydrothermally altered sediment, sulfide-bearing sediment, and sulfide-bearing basalt[1].

Station[2] Number	Start N Latitude	Start W Longitude	End N Latitude	End W Longitude	Study Site	Sample weight[3] or length[3]	General Descriptions
L1 86 NC							
3D	40°45.47′	127°30.46′	40°46.08′	127°31.11′	SESCA	0.257	Sulfide-bearing sediment, 0.156 kg; altered sediment 0.092 kg; pyrrhotite-rich massive sulfide, 0.009 kg.
9C	40°45.20′	127°30.51′	40°48.05′	127°30.49′	SESCA	0.001	Sulfide-bearing sediment, 0.001 kg.
15D	40°46.74′	127°31.24′	40°46.06′	127°31.06′	SESCA	2.170	Massive talc and talc-bearing sediment, 1.156 kg; sulfide-bearing sediment and sediment breccia, 0.972 kg; altered sediment, 0.027 kg; massive sulfide, 0.015 kg.
16D	40°46.46′	127°29.97′	40°45.90′	127°29.98′	SESCA	5.050	Pyrrhotite-rich massive sulfide and massive sulfide with attached barite crusts, 3.515 kg; barite crusts, 1.100 kg; unsorted sulfide-sulfate rubble, 0.435 kg.
22D	40°47.00′	127°30.87′	40°47.77′	127°30.18′	SESCA	0.350	Sulfide-bearing sediment breccia, 0.315 kg; pyrrhotite-rich massive sulfide with barite, 0.035 kg.
24D	40°45.81′	127°30.89′	40°46.10′	127°30.74′	SESCA	36.298	Pyrrhotite-rich massive sulfide and massive sulfide with barite, 21.407 kg; pyrrhotite-rich massive sulfide with talc overgrowths, 3.365 kg; sulfide-bearing sediment and sediment breccia, 1.456 kg; altered sediment, 0.225 kg; barite crusts, 0.195 kg; unsorted rubble, 9.650 kg.
27G	40°53.35′	127°31.65′			MESCA[4]	33.0	Pyrrhotite-rich sulfide interlayered with gray sand from 72 cm to base of core.
29D	40°58.34′	127°29.88′	40°58.97′	127°30.96′	NESCA	*	Pyrrhotite-rich massive sulfide, <0.150 kg; hydrocarbon-bearing and sulfide-bearing sediment.
30D	40°59.94′	127°30.12′	41°01.10′	127°31.18′	NESCA	*	Pyrrhotite-rich massive sulfide, <0.020 kg.
34D	40°58.56′	127°30.14′	40°58.98′	127°30.74′	NESCA	0.726	Pyrrhotite-rich massive sulfide, 0.726 kg.
35D	40°58.59′	127°30.48′	40°58.84′	127°30.30′	NESCA	*	Pyrrhotite-rich massive sulfide; sulfide-bearing sediment.
L2 86 NC							
9G	40°45.59′	127°33.05′			SESCA	27.0	27 cm of fragmental massive sulfide overlies chloritized mud and is capped by red-brown metalliferous mud.
14D	41°00.68′	127°29.76′	41°01.08′	127°29.62′	NESCA	17.580	Polymetallic massive sulfide, 11.835 kg; hydrocarbon-bearing and sulfide-bearing sediment, 5.630 kg; unsorted rubble, 0.015 kg.
16D	41°00.64′	127°29.86′	41°01.16′	127°29.91′	NESCA	3.875	Sulfide-bearing sediment with barite, 0.043 kg; talc-bearing sediment, 0.056 kg; sediment, 3.775 kg.
18D	40°57.69′	127°29.33′	40°58.33′	127°29.44′	NESCA	0.152	Pyrrhotite-rich massive sulfide, 0.152 kg.
24D	41°02.44′	127°29.44′	41°03.09′	127°29.44′	NESCA	2.773	Sulfide-bearing plagioclase phyric basalt, 2.773 kg.
25D	40°58.49′	127°30.24′	40°59.29′	127°31.05′	NESCA	2.225	Sulfide-bearing sediment breccia, 1.245 kg; pyrrhotite-rich massive sulfide, 0.980 kg.

1. Many dredges recovered unlithfied sediment and unmineralized basalt which is not included in this table, therefore the listed dredge weights are not representative of the total dredge recovery. 2. Station codes: C = camera, D = dredge, G = gravity core. 3. Weight in kg for dredge and camera samples; length in cm of mineralized zone in gravity cores. 4. Middle Escanaba Trough, edifice approximately half way between SESCA and NESCA sites. * Sample weights not available.

A

B

FIGURE 5.8. (a) Pyrrhotite-rich massive sulfide sample from dredge L1-86-NC-24D, SESCA site. (b) Polymetallic massive sulfide sample from dredge L2-86-NC-14D, NESCA site.

TABLE 5.2. Bulk composition of massive sulfide, barite crust, and sulfide-rich sediment

	Pyrrhotite-rich sulfide					Barite crust	Polymetallic sulfide		Sulide-rich sediment
	L186NC-24D-1	L286NC-18D	L287NC-25D	L685NC-32D-2	L186NC-16D-1	L186NC-16D-2	L286NC-14D-1	L286NC-14D-2	L286NC-14D-6
Wt %									
As	0.03	0.55	N	0.50	0.20	0.72	1.70	2.70	2.75
Ba	0.04	<0.0005	<0.004	0.10	2.70	23.0	2.50	0.016	0.35
Cu	1.21	1.01	0.42	0.97	4.62	0.49	1.82	0.94	0.26
Fe	42.0	35.2	57.0	40	33.3	6.4	33.1	16.9	22.9
Pb	0.20	<0.02	<0.02	0.50	0.65	4.38	7.60	1.27	13.7
Zn	0.34	0.07	1.10	1.60	2.30	17.3	20.3	43.2	7.10
S	45.2	42.1	36.6	N	37.5	21.1	32.6	32.5	24.0
C*	0.32	0.13	0.11	5.6	0.14	0.15	0.07	0.06	2.24
ppm									
Ag	14	<3	<3	40	67	387	681	184	339
Au	<0.2	<0.2	<0.2	N	1.4	2.0	<0.2	<0.2	<0.2
Bi	25	27	12	N	44	1.9	40	100	9.7
Cd	4.4	4.8	4.5	50	270	830	630	1850	150
Mn	100	7	7	3000	50	500	15	50	7000
Sb	24	132	172	N	530	2700	960	1000	8700
Se	180	380	380	200	95	<10	1000	720	<10
Sn	300	500	200	500	70	150	700	700	>1500

* organic
Analysts: R. Mays, J. Consul, C. Heropoulos, N. Elsheimer, J. Kent, S. Neil, P. LaMothe
Additional chemical data are reported in Benninger and Koski (1987)
N = Not analyzed

in the interstices between pyrrhotite grains. Loellingite ($FeAs_2$), with minute (less than 2 μm) inclusions of native bismuth, is concentrated around fluid channelways. Crusts of drusy barite up to 1.5 cm thick and vuggy encrustations as much as 10 cm thick of massive talc and chlorite occur on many pyrrhotite-rich samples from the SESCA site. The form of the pyrrhotite-rich massive sulfide samples suggests that they were probably recovered from sulfide mounds and surrounding talus aprons.

Bulk chemical analyses for five pyrrhotite-rich samples and one barite crust are given in Table 5.2. The pyrrhotite-rich sulfide is generally low in base and precious metal content compared to massive sulfide from sediment-free spreading axes, but similar to pyrrhotite-rich samples from Guaymas Basin, Gulf of California (Koski et al., 1985; Peter, 1986). The high C content of sample L685NC-32D-2 reflects the presence of asphaltic petroleum, which is interlayered with sulfide and sediment in the sample (Kvenvolden et al., 1986). Barite crusts (sample L186NC-16D-2 and part of sample L186NC-16D-1) are enriched in Zn (to 17%), Pb (to 4%), As (to 0.7%), Sb (to 2700 ppm), Ag (to 390 ppm), and Au (to 2 ppm) relative to pyrrhotite-rich massive sulfide. Hannington et al. (1986) demonstrated a similar enrichment of Au with increased Pb, Ag, As, and Sb in sulfide samples from Axial Seamount on Juan de Fuca Ridge and from southern Explorer Ridge.

About 12 kg of zoned, dark-gray massive sulfide were recovered from dredge L286NC-14D at the NESCA site. The largest sample, a bowl-shaped fragment from a sulfide vent structure, is divided into three zones on the basis of mineralogical and textural variations. The 0.5- to 1-cm-thick inner wall, which formed in direct contact with the hydrothermal fluid, is composed of coarse-grained Cu-Fe sulfide (isocubanite) + sphalerite. The 3-cm-thick middle wall contains alternating pyrrhotite-rich layers and layers of isocubanite + sphalerite + arsenopyrite (with fine-grained inclusions of native bismuth). Galena occurs in the outer part of the middle wall. The sulfide assemblage in the 0.5-cm-thick outer wall, which was in contact with seawater, consists of sphalerite, galena, tetrahedrite, a Pb-Sb sulfosalt phase, and argentite. The polymetallic sulfide fragment represents part of the wall of a sulfide vent structure with a large central conduit.

Bulk chemical analyses of two polymetallic-sulfide samples are shown in Table 5.1. The samples have high Zn, Pb, As, Sb, and Ag contents. The Pb (7.6%), As (2.7%), and Sb (1,000 ppm) concentrations are the highest reported values for bulk sulfide samples from either sediment-free or sediment-covered spreading centers. It is noteworthy that despite the high amounts of Zn, Pb, As, Sb, and Ag, the Au content is below the detection limit (0.2 ppm). Cu, Pb, As, Ag, Cd, Ba, Sb, and Sn are also zoned in the polymetallic samples owing to the mineralogical zonation (Koski et al., 1988). The inner wall of sample L286NC-14D-1 is enriched in Cu (4.6%), As (5.4%), Zn (35%), Cd (1,800 ppm), and Se (1,400 ppm), whereas the outer wall is enriched in Pb (21%), Ag (3,000 ppm), Sb (16,000 ppm), Ba (1,500 ppm), Hg (130 ppm), and Sn (>1,500 ppm) (Benninger and Koski, 1987).

Approximately 10% of the dredge samples consist of indurated sediment with interstitial, fracture-controlled, and disseminated sulfide. Angular clasts of altered siltstone and mudstone are cemented by pyrrhotite or marcasite-rich massive sulfide in samples from both the SESCA and NESCA sites. The polymetallic sulfide assemblage galena + sphalerite + pyrrhotite + loellingite forms the cement for an indurated coarse-sand sample from the NESCA site. This sediment sample has the highest Pb (13.7%), As (2.75%), and Sb (8,700 ppm) of the bulk samples analyzed, and also contains high Zn, Ag, Mn, and Sn (Table 5.2). The presence of hydrocarbon in the sediment is indicated by the high C content (2.24%).

One gravity core at SESCA (L286NC-9G) penetrated red-brown metalliferous mud overlying sand-size sulfide sediment and large nodules composed predominately of pyrrhotite and Cu-Fe sulfide. Another gravity core (L186NC-27G) located halfway between the SESCA and NESCA sites recovered pyrrhotite-rich massive sulfide interlayered with coarse sand (Table 5.1).

Dredge L286NC-24D from the NESCA site recovered a cobble-size block of plagioclase-phyric basalt that contains pyrrhotite, smectite, and chlorite along fractures.

Conclusions

On the basis of photographic and dredge coverage, sulfide deposits appear to be widespread at the SESCA and NESCA sites at Escanaba Trough. All 13 camera stations conducted on cruises L1-86-NC and L2-86-NC photographed hydrothermal deposits, and sulfide was recovered in 14 of 23 dredge stations. In total, eight major deposits, up to 100 m in lateral extent and 30 m in vertical exposure, and many small deposits were photographed.

All of the observed sulfide deposits are sediment hosted. The larger deposits occur at the base of the steep slopes of the sediment hills at SESCA and NESCA, along the fault scarps north of the central hill at NESCA, and along the north-south-trending slopes (probably fault controlled) along the west side of the lava lake terrain at NESCA. These faults probably act as primary conduits for the upflowing hydrothermal fluids. The recovery of sulfide-cemented sediment breccia and interlayered massive sulfide and turbiditic sand in dredge samples and a gravity core also suggests that hydrothermal fluids may move laterally through porous sediment. Dredge samples, photographs, and submersible observations all indicate that there may be extensive stratigraphic intervals of mineralized and altered sediment.

The high concentration of Pb and As, as well as the ratios of sulfur (Koski et al., 1988) and lead isotopes (LeHuray et al., 1988) of massive sulfide, and the presence of asphaltic petroleum in recovered sulfide samples, indicate significant interaction between hydrothermal fluid and sediment at depth. The two types of massive sulfide probably formed under different venting regimes. The numerous fossil fluid channelways and the chemical homogeneity of the pyrrhotite-rich samples suggest slow, diffuse discharge and low-temperature gradients (Koski et al., 1988). The low hydrothermal fluid velocity probably enhances subsurface deposition of base-metal sulfide in sediment beneath the pyrrhotite-rich mounds. The distinct zoning and composition of the polymetallic sulfide samples suggest that they formed in response to steep temperature gradients at a site of focused, high-velocity fluid discharge, analogous to the "black smoker" vents at sediment-free ridge axes.

Results of investigations in Escanaba Trough through 1987 indicate that massive sulfide deposits on the sediment-covered sea floor are widespread and larger in volume than sulfide deposits on most sediment-free ridge axes of the northeast Pacific Ocean. Most of the materials recovered from sulfide mounds and talus associated with the sediment hills are porous accumulations of Fe sulfide with relatively low base-metal content; however, Au appears to be concentrated in barite-rich encrustations deposited with pyrrhotite-rich massive sulfide. Zoned Cu-Zn-Pb-Ag sulfide deposits have been dredged from only one location at NESCA, but this type of mineralization suggests that high-temperature, high-velocity "black smoker" venting has occurred in Escanaba Trough. The bimodal composition of these deposits, the presence of fracture-controlled and interstitial sulfide in sediment, and the recovery of massive sulfide layers in sediment cores suggest that a considerable amount of sulfide may have been deposited in the sediment column below the sea floor. In addition, the consistent zonation patterns in many ancient volcanic- and sediment-hosted massive sulfide deposits (Franklin et al., 1981; Lydon, 1984) lead us to predict that the underlying sediment should be a sink for Cu sulfide mineralization. Similar predictions have been made for hydrothermal systems in Guaymas Basin (Von Damm et al., 1985; Koski et al., 1985) and the Middle Valley segment of the Juan de Fuca Ridge (Davis et al., 1987).

The Escanaba Trough segment of Gorda Ridge is entirely within the U.S. EEZ, but only a small fraction of its length (the SESCA and NESCA sites) has been subject to detailed investigation. Fundamental questions that remain include the actual size, grade, and distribution of deposits in the axial valley; the relationship of deposits to structure and stratigraphy; the depth of penetration and flow patterns of hydrothermal fluids in the sediment and underlying volcanic basement; the interplay among volcanic, sedimentary, and hydrothermal processes in this slow-spreading environment; and the composition of hydrothermal fluids. Our continuing research program has two objectives: (1) to improve our understanding of ore-forming processes in sedimentary environments within oceanic rifts, and (2) to develop more quantitative measurements of the size and metal content of massive sulfide accumulations in Escanaba Trough. The data base necessary to achieve these objectives should be derived from further geophysical, photographic, and sampling work by surface ships and submersibles at the known areas of mineralization and elsewhere along the 100-km length of Escanaba Trough, complemented by carefully targeted drilling to document the critical third dimension of the seabed.

Acknowledgments. We thank Captain Al McClenaghan and the officers, crew, and scientific party members of the *S.P. Lee* cruises L1-86-NC and L2-86-NC for their hard work that made these studies possible. Valuable discussions with *Sea Cliff* dive participants, especially Mark Holmes and Robert Zierenberg, and critical reviews by Robert Zierenberg and Tracy Vallier improved this manuscript.

References

Abbott DH, Morton JL, Holmes ML (1986) Heat-flow measurements on a hydrothermally-active, slow spreading ridge: The Escanaba Trough. Geophys Res Lett 13:678–680.

Atwater T, Mudie JD (1973) Detailed near-bottom geophysical study of the Gorda Ridge. J Geophys Res 78:8665–8686.

Benninger LM, Koski RA (1987) Preliminary descriptions of sulfide samples dredged in 1986 from Escanaba Trough, southern Gorda Ridge. U.S. Geological Survey Open-File Report 87-375-B, 26 pp.

Campbell AC, German C, Palmer MR, Edmond JM (1988) Preliminary report of the chemistry of hydrothermal fluids from the Escanaba Trough (abstract). EOS, Trans Am Geophys Union 69:1271.

Chezar H, Lee J (1985) A new look at deep-sea video. Deep-Sea Res 32:1429–1436.

Clague DA, Holmes ML (1987) Geology, petrology, and mineral potential of the Gorda Ridge. In: Scholl DW, Grantz A, Vedder JG (eds): Geology and Resource Potential of the Continental Margin of Western North America and Adjacent Ocean Basins—Beaufort Sea to Baja California. Houston: Circum-Pacific Council for Energy and Mineral Resouces Earth Science Series, Vol 6, pp 563–580.

Davis EE, Goodfellow WD, Bornhold BD, et al. (1987) Massive sulfides in a sedimented rift valley, northern Juan de Fuca Ridge. Earth Planetary Sci Lett 82:49–61.

EEZ-SCAN 84 Scientific Staff (1986) Atlas of the Exclusive Economic Zone, Western Conterminous United States. U.S. Geological Survey Miscellaneous Investigations Series I-1792 152 pp, scale 1:500,000.

Franklin JM, Lydon JW, Sangster DM (1981) Volcanic-associated massive sulfide deposits. In Economic Geology 75th Anniversary Volume, pp 485–627.

Hannington MD, Peter JM, Scott SD (1986) Gold in sea-floor polymetallic sulfide deposits. Econ Geol 81:1867–1883.

Holmes ML, Morton JL (1986) Massive sulfide deposits from Escanaba Trough, Gorda Ridge (abstract). Geol Soc Am Abstr Program 18:118.

Karlin R, Lyle M (1986) Sediment studies on the Gorda Ridge. State of Oregon, Department of Geology and Mineral Industries, Open-File Report O-86-19, 76 pp.

Koski RA, Lonsdale PF, Shanks WC, Berndt ME, Howe SS (1985) Mineralogy and geochemistry of a sediment-hosted hydrothermal sulfide deposit from the southern trough of Guaymas Basin, Gulf of California. J Geophys Res 9:6695–6707.

Koski RA, Shanks WC III, Bohrson WA, Oscarson RL (1988) The composition of massive sulfide deposits from the sediment-covered floor of Escanaba Trough, Gorda Ridge: Implications for depositional processes. Can Mineral 26:655–673.

Kvenvolden KA, Rapp JB, Hostettler F, Morton JL, King JD, Claypool GE (1986) Petroleum associated with polymetalic sulfide on the Gorda Ridge, offshore Northern California. Science 234:1231–1234.

LeHuray AP, Church SE, Koski RA, Bouse RM (1988) Pb isotopes in sulfides from mid-ocean ridge hydrothermal sites. Geology 16:362–365.

Lonsdale PF, Lawver LA (1980) Immature plate boundary zones studied with a submersible in the Gulf of California. Geol Soc Am Bull 91:555–569.

Lonsdale PF, Bischoff JL, Burns VM, Kastner M, Sweeney RE (1980) A high-temperature hydrothermal deposit on the seabed at a Gulf of California spreading center. Earth Planetary Sci Lett 49:8–20.

Lydon JW (1984) Ore deposit models—8. Volcanogenic massive sulphide deposits. I. A descriptive model. Geosci Can 11:195–202.

McManus DA et al. (1970) Site 35. In McManus DA et al. (eds): Initial Reports of the Deep Sea Drilling Project, Vol 5. Washington: U.S. Government Printing Office, pp 165–172.

Moore GW (1970) Sea-floor spreading at the junction between Gorda Rise and Mendocino Ridge. Geol Soc Am Bull 81:2817–2824.

Moore GW, Sharman GF (1970) Summary of scan site 4. In McManus DA et al. (eds): Initial Reports of the Deep Sea Drilling Project, Vol 5. Washington: U.S. Government Printing Office, pp 761–773.

Morton JL, Normark WR, Ross SL, et al. (1987a) Preliminary report, cruises L1-86-NC and L2-86-NC; Escanaba Trough, Gorda Ridge. U.S. Geological Survey Open-File Report 87-375-A, 13 pp.

Morton JL, Holmes ML, Koski RA (1987b) Volcanism and massive sulfide formation at a sedimented spreading center, Escanaba Trough, Gorda Ridge, northeast Pacific Ocean. Geophys Res Lett 14:769–772.

Peter JM (1986) Genesis of hydrothermal vent deposits in the Southern Trough of Guaymas Basin, Gulf of California: A mineralogical and geochemical study. MS Thesis, University of Toronto, Toronto, Ontario, Canada, 164 pp.

Riddihough RP (1980) Gorda Plate motions from magnetic anomaly analysis. Earth Planetary Sci Lett 51:163–170.

Vallier TL, Harold PJ, Girdley WA (1973) Provenances and dispersal patterns of turbidite sand in Escanaba Trough, northeastern Pacific Ocean. Marine Geol 15:67–87.

Von Damm KL, Edmond JM, Measures CI, Grant B (1985) Chemistry of submarine hydrothermal solutions at Guaymas Basin, Gulf of California. Geochim Cosmochim Acta 49:2221–2237.

Zierenberg RA, Koski RA, Ross SL, Shanks WC III, Slack JF (1988) Preliminary results of ALVIN dives on active sediment-hosted massive sulfide deposits in the Escanaba Trough, southern Gorda Ridge (abstract). EOS, Trans Am Geophys Union 69:1488.

6—Submersible Observations in Escanaba Trough, Southern Gorda Ridge

MARK L. HOLMES and ROBERT A. ZIERENBERG

Abstract

A series of eight dives using the U.S. Navy submersible *Sea Cliff* was conducted in July and August 1986 to map hydrothermal sulfide deposits and determine the distribution of benthic fauna at two study areas in Escanaba Trough at the southern end of Gorda Ridge. Five dives examined three seafloor hills that had been surveyed and sampled during previous surface ship programs, including one at latitude 41° N, longitude 127° 30′ W (the NESCA site), and two at latitude 40° 45′ N, longitude 127° 30′ W (the SESCA site). The hills, approximately 700 m in diameter and 100 m high, are complex polygenetic features produced by a combination of intrusive and extrusive volcanic processes. The flanks of the hills are characterized by steep, scarp-like pitches separated by gently sloping terraces. The small hills are mostly sediment covered, although the steep flanks locally expose stratified mudstone. Extensive deposits of iron-rich massive sulfide were encountered on the mid and lower flanks of all three hills, in water depths ranging from 3,180 to 3,230 m. Some of these deposits were several tens of meters wide and more than 100 m long, suggesting sustained hydrothermal discharge.

Two dives investigated areas several hundred meters from the domes where surface ship measurements indicated a change in both bathymetry and bottom morphology. The seafloor at these two sites is characterized by closely spaced ridges up to 10 m high. Some of the ridge slopes expose mudstone slabs that have broken, contributing talus into the intervening swales. Hydrothermal sulfide deposits are located on the crests of some of these ridges. These deposits differ from those on the hills, occurring as discrete mounds, broken plates, and small chimneys spaced approximately 1 to 2 m apart. The nature of these deposits indicates diffuse hydrothermal circulation and sulfide deposition within the uppermost sediment layers.

The easternmost flank of the central hill at NESCA was studied during a single dive. This area, near one of the north-trending fault zones that bounds Escanaba Trough, is the site of extensive extrusive volcanism. Lightly sedimented basalt pillows up to 3 m in diameter cover the seafloor; glassy rinds were occasionally seen.

No signs of active hydrothermal discharge were seen during the 41 hours of bottom observations conducted during the dive program. The generally weathered nature of the sulfide deposits on the hills indicates that active venting slowed some time ago, at least in the areas visited by the submersible. The extensive but much less massive sulfide deposits observed away from the hills might be somewhat younger. The subsurface growth of sulfide appears to have lifted and deformed the most recently deposited sediment.

Introduction

The signing of the Exclusive Economic Zone Proclamation in 1983 added more than 3.9 billion acres to the land area over which the United States exercises sovereignty in the exploitation of mineral resources. The proclamation provided impetus to a number of federally funded programs to conduct studies on the geological framework and mineral resource potential of this vast region, with particular emphasis being given to certain high-priority areas. One of these areas was Gorda Ridge (Fisk and Howard, this volume), a segment of the midocean ridge system along which new seafloor crust is being generated through volcanic processes. Located 170 to 265 km west of northern California and southern Oregon, Gorda Ridge is the only segment of this spreading center system that lies within the U.S. EEZ.

This paper discusses the results of a series of eight dives conducted in Escanaba Trough on southern Gorda Ridge (Fig. 6.1) using the U.S. Navy's deep submersible vehicle *Sea Cliff* (Appendix A) in the summer of 1986. The dive program was part of a program, begun in 1983, whose objective was to conduct multidisciplinary studies of deep-sea mineral deposits on Gorda Ridge.

Previous Studies in Escanaba Trough

Atwater and Mudie (1973) conducted detailed near-bottom geophysical measurements in northern Escanaba Trough which documented a distinct change in full spreading rate from 7.4 cm/year to 2.4 cm/year about 2 mA. Their studies also showed the pronounced block-faulted nature of the rift valley floor. In 1984, the U.S. Geological Survey began a systematic survey of the U.S. EEZ using the GLORIA (Geological LOng-Range Inclined Asdic) mapping system. GLORIA imagery from the Escanaba Trough area (EEZ-SCAN 84 Scientific Staff, 1986) revealed irregular zones of anomalous sea floor backscatter, and seismic reflection records collected along the same tracks showed faults and strong reflections from subsurfaces structures in the 500-m-thick sediment fill. Similar features had been seen almost 20 years previously—in 1965, during the site survey work for hole 35 (Fig. 6.1) of the Deep Sea Drilling Project (McManus et al,. 1970).

U.S. Geological Survey and Oregon State University scientists, the latter working under the auspices of the Gorda Ridge Techanical Task Force (McMurray, this volume) utilized the USGS research vessel *S.P. Lee* in 1985 to conduct a variety of reconnaissance geophysical and geological studies over Escanaba Trough (Abbott, 1986; Bibee, 1986; Karlin and Lyle, 1986). The seismic reflection data and bottom photographs collected in the early phases of this work revealed a series of buried volcanic edifices spaced about 15 to 20 km apart along the axis of Escanaba Trough in water depths of 3,200 to 3,300 m (Fig. 6.1). High heat flow values measured at two of these domelike features suggested the possibility of hydrothermal venting on or near the edifices (Abbott, 1986; Abbott et al., 1986). Following the geophysical work, approximately 6 kg of iron-rich metallic sulfides were dredged from the flank of a small hill associated with a subsurface volcanic edifice (later to become known as the SESCA site) near latitude 40° 45′ N, longitude 127° 30′ W. This discovery represented the first known occurrence of hydrothermal sulfide deposits within the U.S. EEZ (Holmes and Morton, 1986; Morton et al., 1987a; Kvenvolden et al., 1986). In addition to the geological and geophysical data, baseline information on the benthic communities of Escanaba Trough was obtained using camera observations made in 1985 from the *S.P. Lee* (Carey et al., 1986).

To evaluate the size and composition of the sulfide deposits, the Gorda Ridge Technical Task Force coordinated a series of expeditions involving researchers from several federal and state agencies to obtain additional samples, geophysical data, bathymetric information, and photographic images during the period from May through July 1986 (Morton et al., 1987b). The work was carried out aboard vessels of the National Oceanic and Atmospheric Administration (*Surveyor*), the U.S. Navy (*Narragansett*), and the U.S. Geological Survey (*S.P. Lee*). These multidisciplinary efforts (Morton et al., 1987a, this volume) concentrated on the two largest intrusive zones at latitude 40° 45′ N, longitude 127° 30′ W (SESCA), and latitude 41° 00′ N, longitude 127° 30′ W (NESCA). The data collected during these cruises were used to develop sites and transects for the program of dives using the submersible *Sea Cliff* (DSV-4).

Geologic Setting of the NESCA and SESCA Sites in Escanaba Trough

The research cruises to Escanaba Trough in 1985 and 1986 established the basic geologic and tectonic setting of the NESCA (Northern

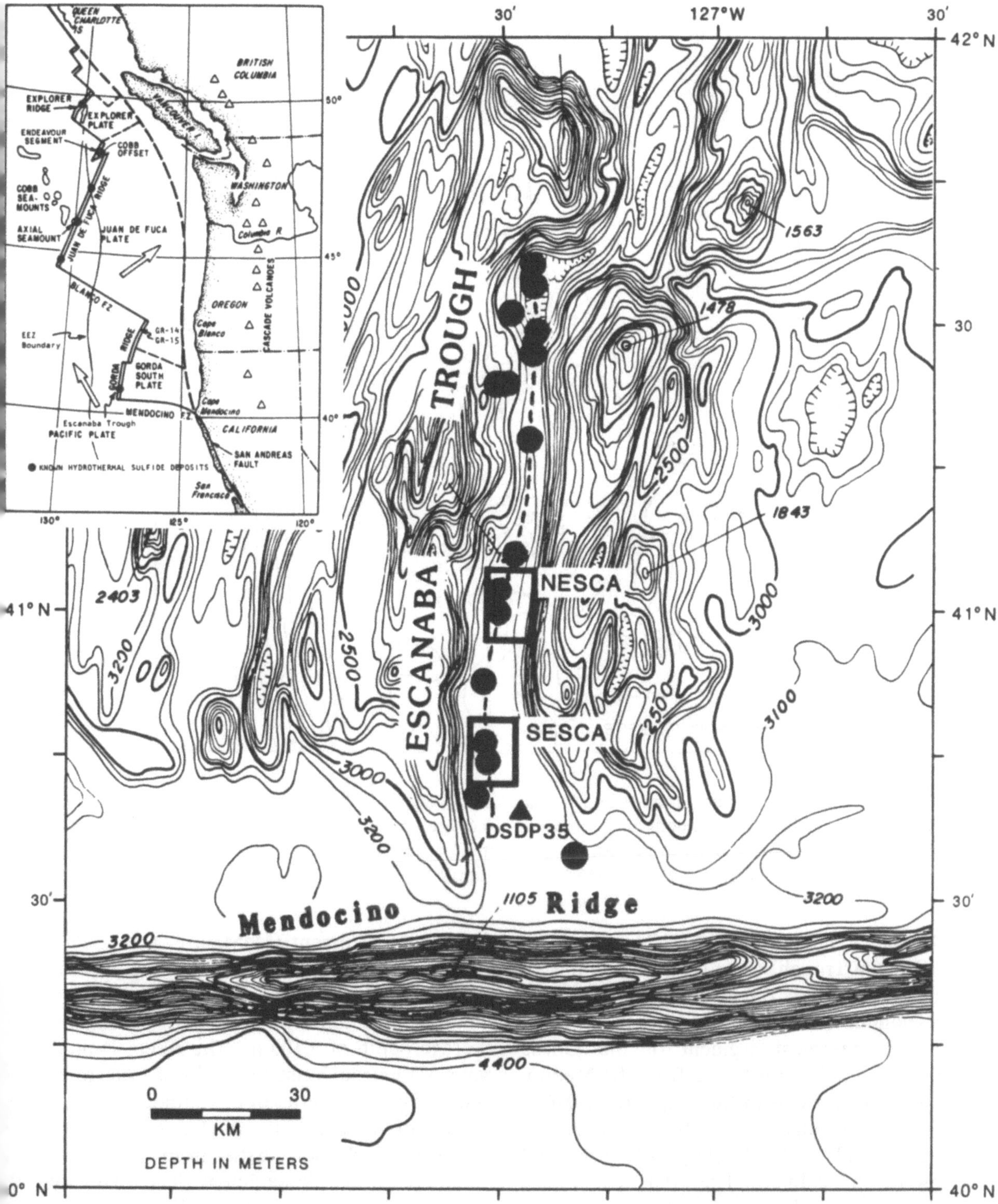

FIGURE 6.1. Bathymetry chart (from Wilde et al, 1978) of southern Gorda Ridge showing locations of volcanic edifices in Escanaba Trough (black circles). Contour interval 100 m. Black triangle indicates location of DSDP Hole 35 (McManus et al, 1970); the NESCA and SESCA study areas (Figs. 6.2 and 6.3, respectively) are shown by boxes. Inset shows regional tectonic setting.

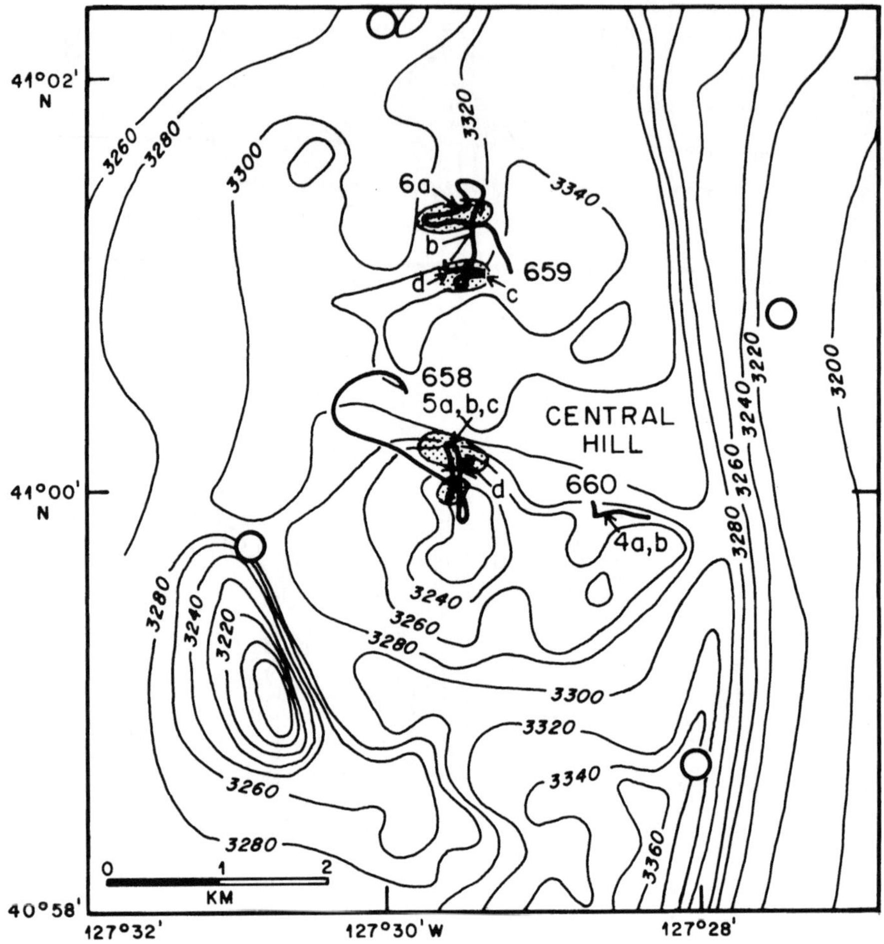

FIGURE 6.2. Bathymetric chart of the NESCA area. Contour interval 20 m. Dark sinuous lines represent *Sea Cliff* dive traverses referred to in text. Bottom transponders used for navigation are denoted by circles. Positions of photographs shown in Figs. 6.4–6.6 are indicated. Approximate limits of sulfide mineralization observed on the dives is shown by the stippled pattern.

ESCAnaba Trough) and SESCA (Southern ESCAnaba Trough) areas (Figs. 6.2, and 6.3). Further details are discussed by Morton et al. (this volume). Escanaba Trough is the 120-km-long southernmost segment of the Gorda Ridge, extending northward from the Mendocino Fracture Zone to the major right-lateral offset on the ridge axis at latitude 41° 38′ N, longitude 127° 24′ W (Fig. 6.1; Fisk and Howard, this volume). The axial valley of the Escanaba Trough segment ranges in width from 5.5 km in the north to more than 20 km in the south, where the axial trend curves eastward. The southern 75 km of the 3,300-m-deep valley is partially filled with more than 500 m of Pleistocene sediment (McManus et al., 1970; Holmes et al., 1986; Morton et al., 1986a), whereas the northern part of the trough is an unsedimented and elevated block having an average depth of 3,100 m. The steep, faulted walls of the trough rise to depths of 1,500 m in places, forming a series of discontinuous ridges parallel to the spreading axis (Fig. 6.1).

Seismic reflection data from Escanaba Trough show that the sediment fill is broken by faults that indicate uplift along the valley margins. Domelike structures, approximately 3 to 6 km in diameter and spaced at 15- to 20-km intervals, occur along the entire length of the trough (Morton et al., 1987b, this volume).

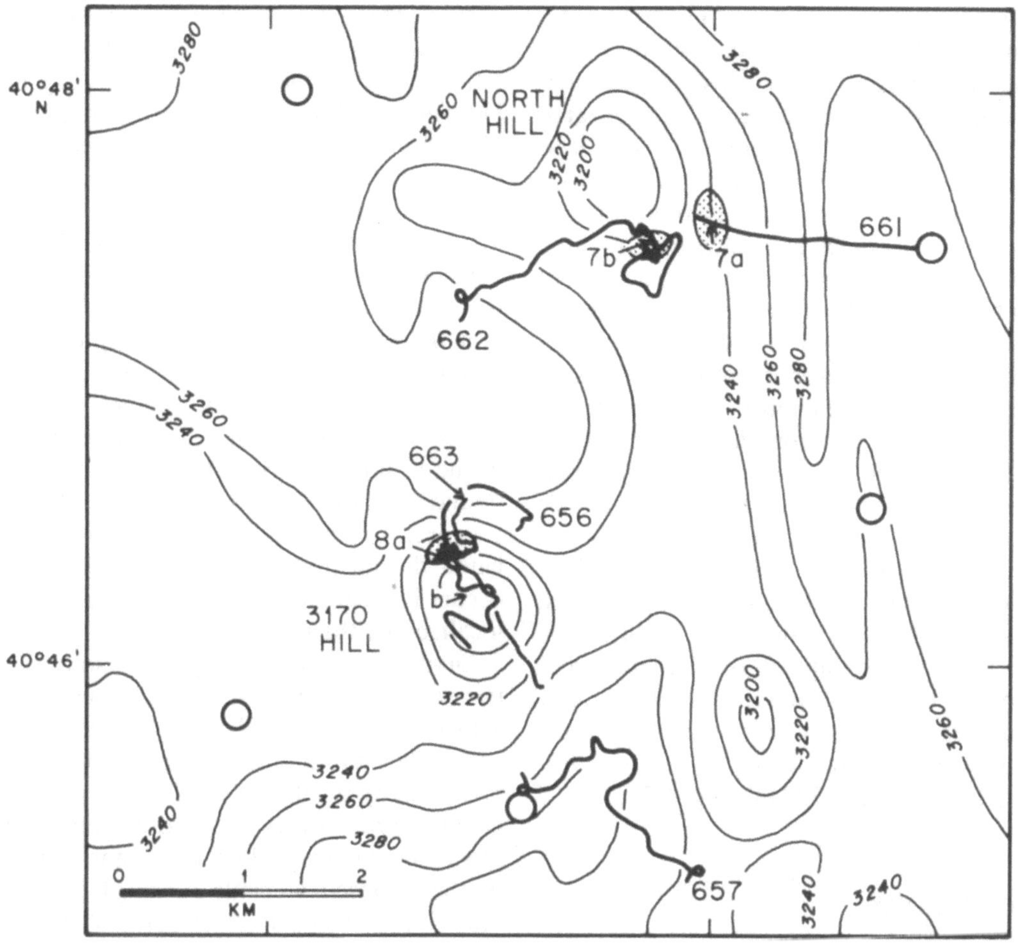

FIGURE 6.3. Bathymetric chart of the SESCA area. Contour interval 20 m. Dark sinuous lines represent *Sea Cliff* dive traverses referred to in text. Bottom transponders used for navigation are denoted by circles. Positions of photographs shown in figs. 6.7–6.8 are indicated. Approximate limits of sulfide mineralization observed on the dives is shown by the stippled pattern.

These volcanic centers are exposed in the northern part of the trough but are completely or partially buried by sediment in the south.

At the NESCA area (Fig. 6.2), a buried domal structure 300 to 400 m high and more than 5 km in diameter is capped by two smaller seafloor hills approximately 100 m high. The central hill rises to a depth of almost 3,200 m and has a steep northern flank. A broad ridge extends eastward from the dome to a north-trending fault scarp formed by the uplift of a turbidite-covered bench near the eastern margin of the valley. Anomalously high heat flow values exceeding 1.2 W/m^2 were measured just north of this central hill (Abbott, 1986; Abbott et al., 1986). About 2 km southwest of this central hill, another seafloor hill rises to a depth of 3,195 m. This feature is strongly asymmetric, with a steep linear northeastern flank (Fig. 6.2). Bottom photographs and samples collected during the 1985 and 1986 cruises showed that the two hills are mostly sediment covered. The crestal area and parts of the southern and eastern flanks of the central hill are covered with fresh basalt pillows and sheet flows. Hydrothermal sulfide deposits were photographed and dredged from both of the hills as well as from three locations north of the central hill (Fig. 6.2).

The SESCA site, 27 km south of NESCA

(Fig. 6.3), is underlain by a much broader basement high (Morton et al., 1986b, 1987a; Holmes and Morton, 1987). This irregular platform is approximately 6 km wide, as delineated by the 3,240-m isobath in Figure 6.3. The eastern part of the SESCA uplift is a north-trending ridge that lies near the axis of Escanaba Trough. The edifice is surmounted by three distinct and almost circular hills, each about 100 m high and 700 m in diameter. Surface ship observations conducted in 1985 and 1986 revealed a sediment-covered bottom over most of the edifice and over all three of the hills. The seismic reflection records showed an abrupt transition from the relatively undeformed sediment cover around the base of the hills to their steep, uplifted flanks (Morton et al., this volume, Fig. 5.5). The sediment on the hills was too thin to be resolved using conventional surface-towed seismic equipment. Sulfide deposits were photographed or dredged from each of the three hills and at one site about 1 km southeast of the western hill (Fig. 6.3). Sulfides were also recovered in a core (9G) about 3 km west of 3170 Hill (Fig. 6.3).

Dive Results

The *Sea Cliff* dive program provided important new information, not available from surface ship observations, which has greatly refined our understanding of the geology of these areas. The highlights of the new information gained during the dive program are presented here; Appendix B contains descriptions of each sample recovered during the 1986 *Sea Cliff* dive program.

NESCA Site

Pillow basalt is exposed over a large area of the volcanic edifice at the NESCA site and was observed on dive 660 (Fig. 6.2, and 6.4). The fresh, glassy pillow rims decorated small buds attest to the recency of volcanism at this site. However, even though these flows are very recent, they are lightly covered with sediment, and sediment has acculated in pockets between pillows. The pillows tend to be large (1 to 2 m) and bulbous rather than elongate. Striations and expansion cracks are common (Fig. 6.4b). No signs of rifting, hydrothermal deposits, or alteration were observed on or near the volcanic rocks.

The area north and west of the exposed pillow basalt was explored on dive 658 (Fig. 6.2). Seismic records showed that the north flank of the volcanic hill is covered by sediment and that the sediment has been disrupted by normal faults, which are down dropped to the north (Morton et al., this volume). Numerous north-block-down fault scarps were observed in the sediment during this dive. Some scarps bound back-tilted sediment terraces. An extensive area of sulfide mineralization was investigated during this dive (Fig. 6.5). The sulfide occurs in a sediment-covered area and may be localized by the normal faults.

Dives 658 and 659 (Fig. 6.2) started their traverses in the plain north of the central hill. This plain is completely sediment covered and generally smooth and is interpreted as an area underlain by sheet flow basalt. Dive 659 investigated an area of very irregular bottom topography and complex geology to the north and west of the exposed volcanic rocks of the central hill. Numerous channels or narrow valleys cut across the area. The channels are generally oriented north-south, vary from a few meters to several tens of meters in width, and are about half as deep as they are wide. They are often asymmetric, with one near-vertical wall and a less steep wall sloping from 45° to 70°. Mass wasting of the channel walls is evident (Fig. 6.6), but no indications of down-channel sediment transport were observed. The steepest channel walls locally expose truncated layers of mudstone and siltstone.

A sediment-covered basalt sheet flow was observed in one vertical channel wall during dive 659. In this case it appears that the roof of a drained lava channel collapsed after burial by sediment. Small outcrops of massive sulfide chimneys and crust were observed throughout this area, especially near the channel margins (Fig. 6.6). One larger area of sulfide talus was observed on a steep scarp, but no extensive sulfide deposits were located.

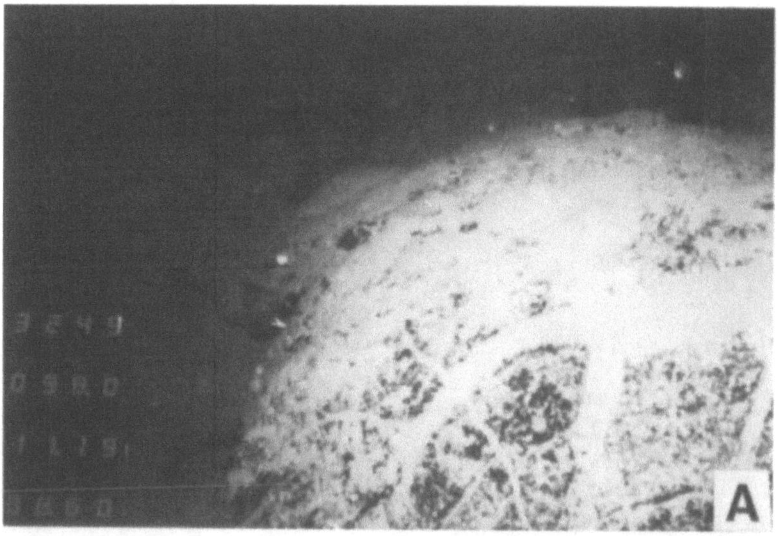

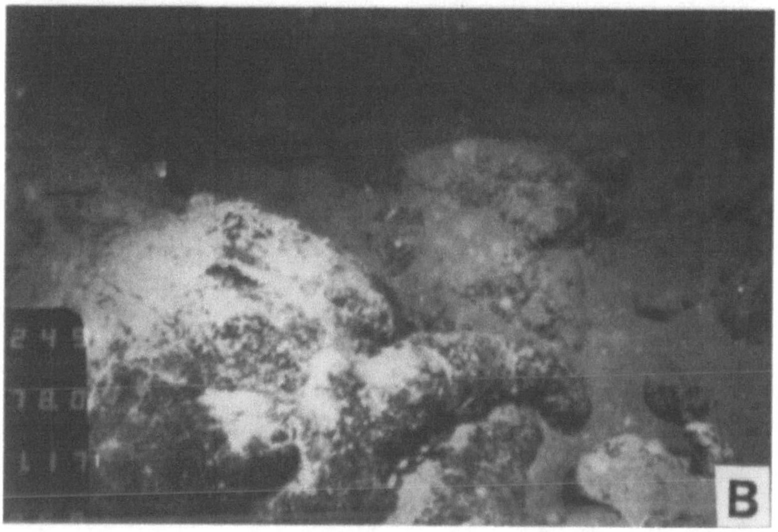

FIGURE 6.4. Photographs from NESCA Dive 660; location shown on Fig. 6.2. (A) Large bulbous basalt pillows with glassy crust and expansion cracks. Sediment fills the cracks and areas between the pillow margins. Water depth 3275 m. (B) Sedimented toothpaste-like basalt pillows on the northeastern flank of the central hill. Water depth 3275 m. Small white globular sponges are common on exposed basalt.

SESCA Site

The volcanic edifice at the SESCA site is exposed at the surface in only a few areas. The most prominent topographic features are the three flat-topped, sediment-covered hills that sit atop the volcanic edifice (Fig. 6.3). Massive sulfide deposits occur around the bases of these hills (Figs. 6.7, 6.8).

The five dives at the SESCA area (Fig. 6.3) concentrated on the northernmost hill (North Hill) and the central hill (3170 Hill; Fig. 6.3). The tops of the hills are generally flat and smooth and are completely covered by bioturbated sediment. A field of ripple marks was seen on the top of 3170 Hill during dives 656 and 663 (Fig. 6.8b). The hills are bounded on all sides by near-vertical erosional scarps. The scarps have a parabolic shape with slopes that flatten as depth increases. In places the scarps are stepped such that the base of one scarp flattens to a smooth plain, which is truncated by the next lower scarp. The sides of the hills are also generally sediment-covered, even on near-vertical slopes. Locally, the steeper scarps expose truncated beds of mudstone and siltstone. The hills are clearly retreating by mass wasting; sediment transport grooves and chan-

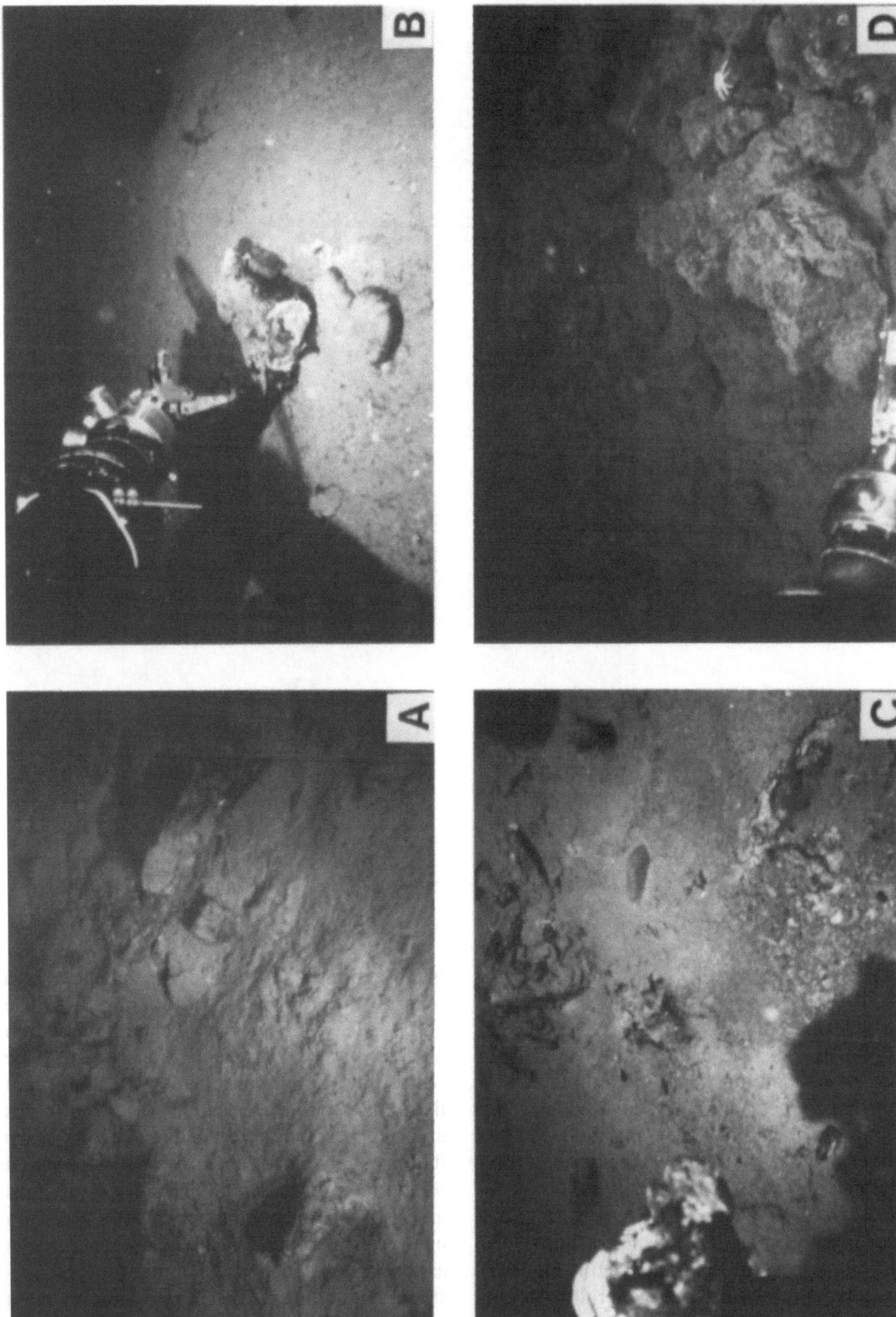
B
D
A
C

Figure 6.5. Photographs from NESCA Dive 658; location shown on Fig. 6.2. (A) Small sulfide growths in sediment near the base of the central hill. The sulfides have broken and lifted the semi-lithified sediments. Water depth 3290 m. (B) *Sea Cliff* manipulator reaching for a freshly broken barite chimney (sample 658 R2, Fig. 6.9A and Appendix II) in a field of massive sulfide deposits. Water depth 3285 m. (C) Chimney sample of massive sulfide, visible in *Sea Cliff* manipulator, has just been taken from beneath the sediment surface (hole at lower right). Water depth 3285 m. (D) A large sulfide mound with talue blocks at the base. Galatheid crab visible at right. Water depth 3255 m.

nels were observed on the flanks. Mudstone talus is commonly observed on the flatter slopes, but there do not appear to be large talus or sediment aprons surrounding the hills. The hills are interpreted to have been formed by intrusion of basaltic bysmaliths into the unconsolidated sediment.

No volcanic rocks were observed or recovered from the slopes of the hills at SESCA during the dive program. However, volcanic rocks older than those observed at the NESCA site are exposed in the flat area southwest of North Hill. They are partially sediment covered, and they lack volcanic glass. The most commonly observed lava form is large (1 to 2 m), nearly spherical pillows. The pillows are expanded and often show trap door breakouts. In some areas the pillows either sit on or were extruded from sheet flow basalt. The low-relief sheet flows are rapidly covered by sediment, so the proportion of surface volcanic rocks exposed as pillows is greater than their true volumetric percent. Near the base of the North Hill, the volcanic rocks are cut by numerous north-trending open fissures a few centimeters to about 1 m wide. The fissure walls generally expose sheet flow basalt. No hydrothermal activity or rock alteration is associated with the fissures. High density of filter feeding organisms in the fissures suggests that, in addition to providing a hard bottom for attachment, the fissures may localize down-flow in zones of hydrothermal recharge.

Sulfide Deposits

Large deposits of massive sulfide were explored on dives 658 (NESCA) and 662 and 663 (SESCA). The sulfide deposit observed at the end of dive 661 (Fig. 6.7a) is probably continuous with the deposits explored on the southeastern flank of SESCA's North Hill during dive 662 (Figs. 6.3, 6.7b). Dive 659 (Fig. 6.3) encountered extensive sulfide mineralization (Fig. 6.6), but no large massive deposits were observed. None of the dives detected any evidence for active discharge of hydrothermal fluid. Relative ages for the sulfide deposits were established during the dive program.

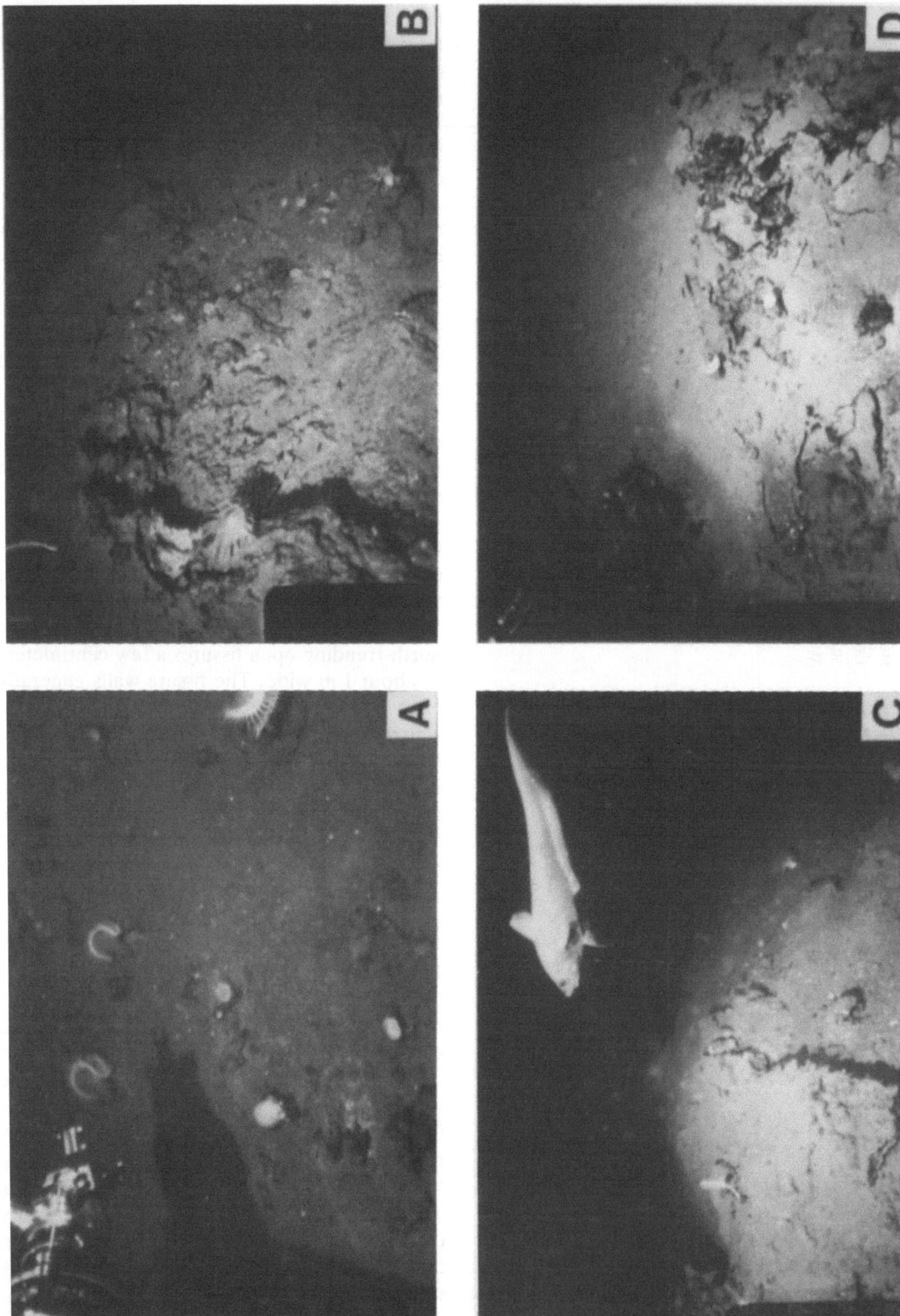
B
D
A
C

Figure 6.6. Photographs from NESCA Dive 659; location shown on Fig. 6.2. (A) Partially buried sulfides, colonized by "venus flytrap" (Actinoschyphid?) anemones. Water depth 3321 m. (B) Hydrothermal barite exposed in the headwall of a slump scarp. Tentacled anemones are visible at lower right and left center; stalked coelenterates at upper left. Water depth 3324 m. (C) Mound of massive sulfide and talus, with white (Spectrunculid?) fish, anemones, and whip coral. Water depth 3322 m. (D) Small projections and broken plates of sulfide and altered sediment, colonized by anemones and small sponges. Water depth 3322 m.

Some of the deposits showed evidence of recent activity (unweathered spires and chimneylike features), while others had been inactive for a sufficient time to be modified by chemical and mechanical weathering.

NESCA Site

The sulfide deposits explored on dive 658 (NESCA, Fig. 6.5) are the youngest and probably the most extensive deposits observed during the dive program. Massive sulfide and sulfide talus were observed discontinuously along a north-south traverse approximately 500 m long between the depths of 3,216 m and 3,270 m (Fig. 6.5). Problems with the positioning system preclude a detailed reconstruction of the true outcrop pattern of the sulfide deposits. However, reconciliation of the depths of the sulfide occurrences with the bathymetry requires at least 500 m of semicontinuous outcrop of sulfide and sulfide talus. Although much of the hydrothermal material observed was talus, constructional mounds and chimneys of sulfide formed in place were observed near both the shallowest and the deepest occurrence of sulfide.

Most of the sulfide occurs as coarse talus blocks ranging in size from 20 cm to a few meters (Fig. 6.5a). Platy slabs of massive sulfide 10 to 20 cm thick and 0.5 to 1 m on a side are common. A notable feature is the fresh appearance and coarse grain size of the sulfide. Reflections from individual crystals could commonly be seen at a distance of a few meters. The sulfide talus occurs as mounds and ridges several meters high and tens of meters long. Areas between mounds are sediment covered, but disturbance of the sediment by the submarine during sampling attempts revealed that fine-grained black sulfide is present at a depth of a few centimeters. Some small, chimneylike projections of sulfide occur within the sediment-covered areas, and one 30-cm-high chimney composed of hydrothermal barite was recovered from a sulfide field (Figs. 6.5c, 6.9a).

Constructional mounds of massive sulfide, tens of meters in diameter and a few meters high, are typically surrounded by sulfide talus

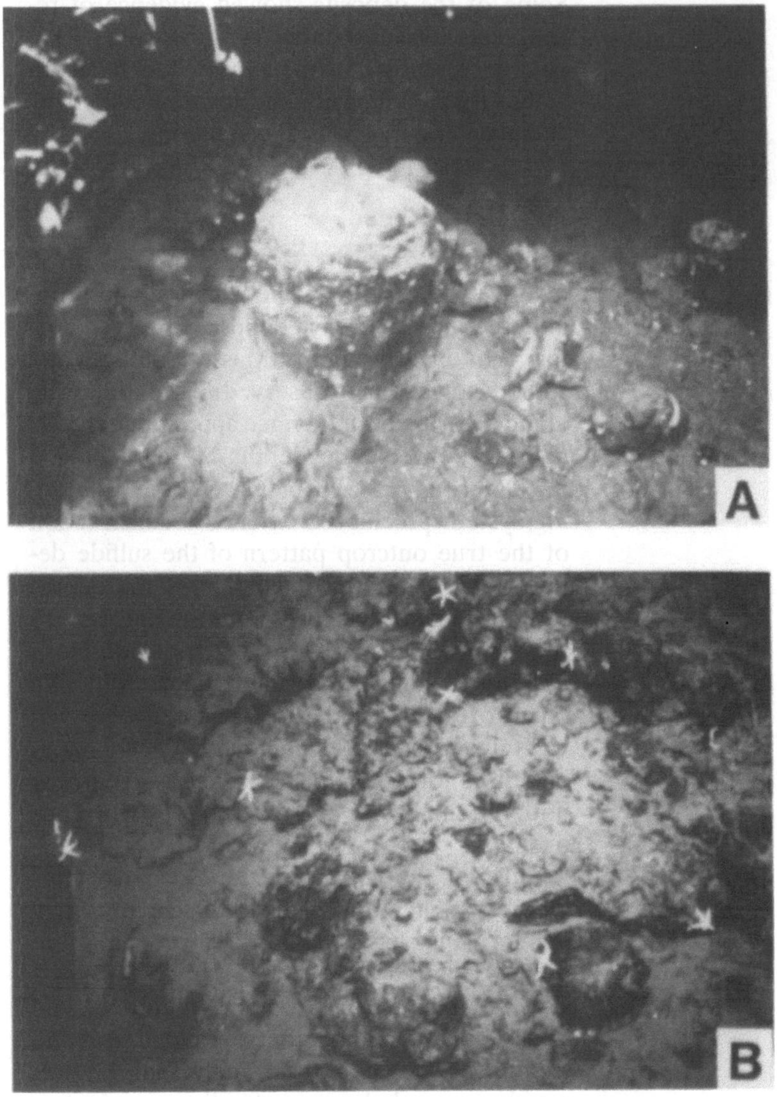

FIGURE 6.7. (A) Photograph from SESCA Dive 661; location shown on Fig. 6.3. Stump of sulfide chimney and some freshly broken sulfide which fragmented during sampling attempt. Water depth 3238 m. (B) Photograph from SESCA Dive 662; location shown on Fig. 6.3. Mound of massive sulfide on southern flank of North Hill, colonized by numerous white starfish. Water depth 3225 m.

and topped by sulfide or barite chimneys about 1 m high. The preservation of the chimneys and the fresh appearance of the pyrrhotite grains imply that the deposits formed very recently. The presence of dead tube worms and the shells of vent-specific clams is further evidence that the deposits were active in the recent past. However, the extensive occurrence of sulfide talus and the coarse grain size of some sulfide suggest that the deeper portions of the deposits have already been disrupted and exposed by faulting.

Dive 659 (NESCA, Fig. 6.2) encountered sulfide mineralization over a wide area north of the central hill. The sulfide occurrences are mostly small, chimneylike projections a few tens of centimeters high, extending above the sediment-covered bottom (Figs. 6.6a, 6.9b). The amount of exposed sulfide is generally small. Thin plates of sulfide formed broken mounds which appeared to have buckled upward as a result sulfide growth below the sediment water interface. Sulfide and hydrothermal barite occur near the edges of collapse channels and are occasionally exposed at the tops of the scarps that form the channel walls (Fig. 6.6b, d). The largest exposure of sulfide observed is a 3-m-high, 60° scarp of massive

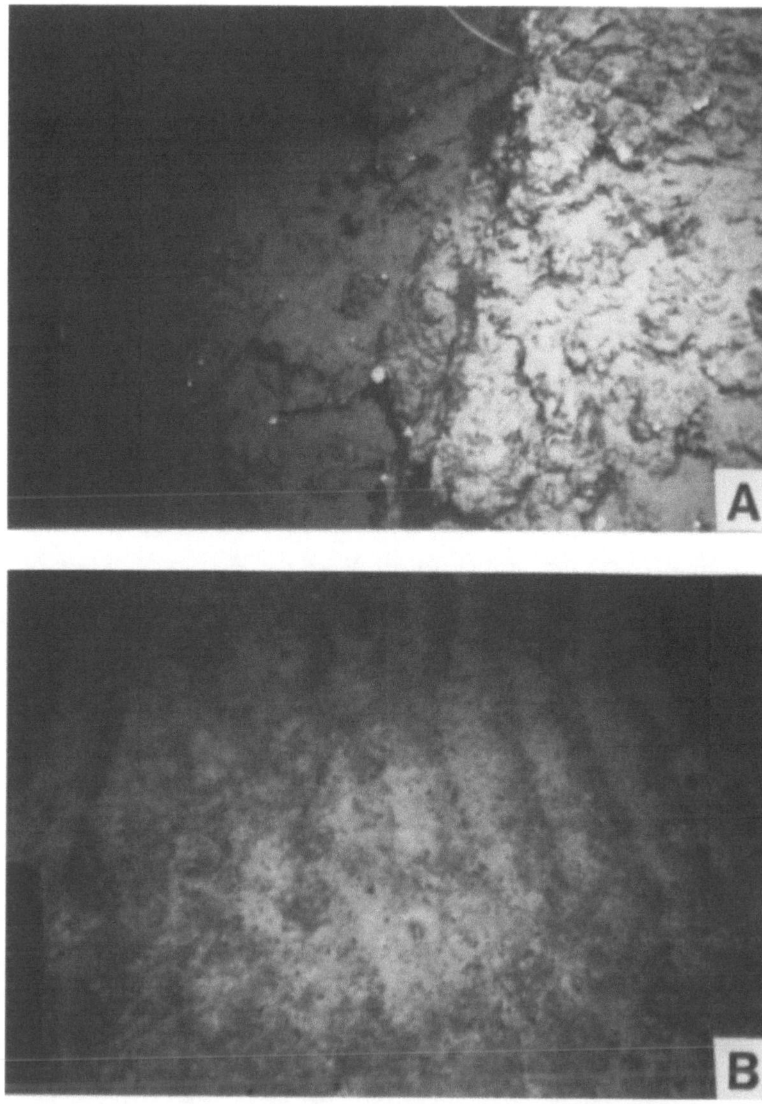

FIGURE 6.8. Photographs from SESCA Dive 663; location shown on Fig. 6.3. (A) Exposure of weathered and sedimented massive sulfide mound colonized by white starfish and small sponges. Water depth 3207 m. (B) Ripple marks on the sediment-covered bottom on a flat bench near the summit of 3170 Hill. Water depth 3175 m.

sulfide talus (Fig. 6.6c). The talus blocks are equant, angular, and 10 to 50 cm on a side. Both the top and bottom of the scarp are flat and sediment covered.

SESCA Site

Sulfide deposits near the base of North Hill at SESCA were encountered near the end of dive 661 and were investigated in greater detail on dive 662. The deposits seen on dive 661, at a depth of 3,230 m, were small mounds of sulfide talus with some chimneylike projections (Fig. 6.7a). An extensive field of mounds and ridges of massive sulfide at the base of North Hill was explored on dive 662 (Figs. 6.3, 6.7b). The sonar display indicated that the hydrothermal deposits are 20 to 30 m wide and more than 100 m long, extending parallel to the slope in the direction of the deposits observed on dive 661. The sulfide occurs as elongate mounds and ridges several meters high. Rounded slabs of sulfide talus are common at the bases of the mounds; a fallen sulfide tower several meters long was observed. The submersible traversed through valleys bounded on both side by steep-sided massive sulfide that extended above the

FIGURE 6.9. Samples collected at NESCA include: (A) barite chimney (658-R2, Appendix II) from the northern flank of the central hill, and (B) a small chimney of massive sulfide (659-R1, Appendix II) from the ridge-and-swale province north of the central hill. The chimney is shown on its side; the basal crust, which was below the sediment surface, is on the left.

view of either observer. The mounds are generally rounded and partially oxidized. Locally, the mounds are topped by 1 to 2-m-high, stubby, chimneylike projections. The rounded morphology of the deposits indicates that they are older than those observed on dive 658 (NESCA) but not as old as the deposits seen on dive 663 (3170 Hill, SESCA).

The sulfide deposits at the base of the 3170 Hill explored on dive 663 (Figs. 6.3, 6.8a) are similar to those at the base of the North Hill. The deposits on North Hill occur on a flat or slightly back-tilted sediment bench at the hill's base, whereas the deposits on the 3170 Hill occur near the base of the slope at a depth of about 3,250 m (Fig. 6.3). The highly weathered deposits are exposed as rough-textured, steep-sided cliffs of massive sulfide up to 6 m thick. Preferential dissolution of sulfide has produced a jagged, porous mass of barite, silica, and secondary Fe-oxide. Degradation of the deposit has removed any original constructional projections, and the top of the deposit is buried by sediment. In some places the deposit is cut by large, mass-wasting canyons that expose several meters of continuous massive sulfide on the walls with sediment and mudstone talus on the canyon floors.

Sulfide Samples

Sea Cliff collected a total of 32 kg of sulfide samples during the dive program (Table 6.1). A general description of the samples is presented in Appendix B; the chemical composition of selected samples is presented in Table 6.2. The samples are generally similar to pyrrhotite-rich massive sulfide samples dredged from the Escanaba Trough (Benninger and Koski, 1987; Koski et al., 1988; Morton et al., this volume).

Pyrrhotite is the predominant mineral in the sulfide samples, generally constituting more than 90% of the samples. Pyrrhotite grain size ranges form very fine to several millimeters, and grains tend to form hexagonal platelets. An interlocking network of pyrrhotite plates leaves a large volume of pore space that is partially filled by other sulfide minerals but is generally open. The coarsest-grained pyrrhotite forms in internal cavities and fluid channelways. Unaltered pyrrhotite is predominantly, if

TABLE 6.1. Summary of observations and samples collected during the dives.

Drive No.	Bottom time (h)	Photographs: External camera	Photographs: Starboard observer	Photographs: Port observer[a]	TV cassettes	Sample sites
656	4.5	432	34	28	—	—
657	6.0	576	30	46	—	2
658	6.5	625	25	—	2	6
659	8.0	610	80	—	2	4
660	1.7	288	—	—	1	—
661	3.5	383	—	—	2	1
662	6.5	822	62	—	3	1
663	4.3	257	10	—	3	2

[a] During the last six dives (658 to 663), the port strobe light was not precisely synchronized to the observer's camera. Because no developing facilities were available on *Transquest*, this problem went undetected, and most of these photographs cannot be recovered.

TABLE 6.2. Chemical composition of selected dive samples.

	Barite chimney (658-R2)	Pyrrhotite-rich talus block (658-R3)	Pyrrhotite-rich massive sulfide (658-R4)	Base of pyrrhotite-rich chimney (659-R1)	Pyrrhotite-rich chimney fragment (662-R1)
wt %					
Fe	1.30	48.10	52.50	17.90	48.20
Cu	0.02	0.55	2.06	7.70	2.22
Zn	0.02	0.61	0.12	1.38	2.05
ppm					
Ag	1.5	15	15	150	150
As	ND	ND	ND	2,000	500
Au	<0.1	<0.1	<0.1	2.2	<0.1
Ba	>20,000	57	53	1,500	22
Bi	5.8	57	53	68	22
Cd	0.8	15.6	6.5	47	67
Hg	106	<2	<2	<2	51
Ni	15	5	5	30	10
Pb	<100	700	800	2,600	3,500
Sb	8.8	7.8	17	44	140
Sn	ND	ND	ND	30	ND

ND: Not determined.

not exclusively, hexagonal. Monoclinic pyrrhotite occurs in altered samples.

Alteration of pyrrhotite takes several forms. Weathering of the samples often leads to formation of goethite and lepidocrosite rims on the pyrrhotite grains. The pyrrhotite is commonly altered to fine-grained aggregates of marcasite, which may include Fe-oxide and monoclinc pyrrhotite. Some of this type of alteration may be associated with the circulation of late-stage, lower-temperature fluids through the sulfide deposits. A common alteration of pyrrhotite, especially in samples formed below the sediment-water interface, is the pseudomorphic replacement of pyrrhotite by monoclinic native sulfur. Fe- and CuFe-sulfate minerals often occur interstitial to the altered pyrrhotite grains.

Isocubanite is the next most abundant primary sulfide mineral and can form up to 5% of the rock. Isocubanite grains are generally subhedral and occur interstitial to pyrrhotite, but nearly monomineralic bands of isocubanite occur in a few samples. Many of the interstitial grains of isocubanite have exsolutionlike fiamme and replacement rims of chalcopyrite, but the volume of chalcopyrite in the majority of samples is small relative to isocubanite.

Sphalerite is nearly as abundant as isocubanite and generally occurs interstitial to pyrrhotite. Sphalerite is dark red to opaque in transmitted light. Sphalerite often shows partial replacement by isocubanite similar to the chalcopyrite disease described by Barton and Bethke (1987).

Marcasite is a common alteration product of pyrrhotite and is the major mineral present in some samples; pyrite is generally absent. Although much of the marcasite is derived from the alteration of pyrrhotite, colloform-banded marcasite is common in some samples. Colloform-banded marcasite occurs as an outer rim overgrowing pyrrhotite and is often itself overgrown by banded Fe-oxide. Botryoidal aggregates of marcasite also occur interstitial to pyrhotite.

Other sulfide phases occur in trace amounts. Arsenopyrite and galena are the most common of the minor phases. Native bismuth has been observed as inclusions in arsenopyrite examined by scanning electron microscope.

The predominant nonsulfide phase is barite, which occurs as crusts of bladed crystals on the exterior surfaces of some samples of massive sulfide. A chimney of nearly monomineralic barite was also recovered on dive 659 from a sediment-covered area next to outcropping massive sulfide (Fig. 6.9; 659-R2, Appendix B). Iron oxide is common in weathered samples, and both goethite and lepidocrosite are present. Some samples contain traces of late-stage amorphous silica. One sample of Fe-stained rubble recovered from the surface of weathered massive sulfide consists predominantly of talc (663-R2; Appendix B). A few of the sulfide samples contain hemipelagic sediment incorporated during sulfide growth. A small number of these samples also contain yellowish hydrocarbon.

The chemical composition of the sulfide samples is consistent with the observed mineralogy and is similar to analyses reported by Benninger and Koski (1987) and Koski et al. (1988). The samples are Fe-rich with up to 2 wt% Cu and Zn. A sample from the base of a small pyrrhotite-rich chimney contains 7.7% Cu, 1.4% Zn, 0.3% Pb, 150 ppm Ag, and 2.2 ppm Au (Table 6.2; 659-R1, Appendix B). This part of the sample formed at and just below the sediment-water interface and contains a small amount of incorporated sediment and some hydrocarbon.

Biology

A majority of the area of seafloor observed on the dives is sediment covered. The most common animals observed on sediment were several varieties of holothurians, brittle stars, basket stars, anemones, pyramid urchins, galatheid crabs, and fish. The sediment was generally bioturbated, and roughly circular arrays of centimeter-size burrows surrounding a central sediment dome a few centimeters high, referred to as "fairy rings", were common. Free-swimming animals other than fish were not common but included an octopus, shrimp, and a few types of jellyfish.

Stalked animals in sediment-covered areas are less abundant but were commonly observed. They included sea pens, corkscrew-like coelenterates, stalked goblet-shaped glass sponges, and a few crinoids. In areas where volcanic rock provided a hard substrate, stalked animals are more abundant, including fan-shaped gorgonians and straight and branch-stalked colonial animals interpreted as corals. Several types of sponges are common in the areas with exposed basalt. A crescent-shaped, bilobate anemone, referred to here as the "Venus flytrap" (Fig. 6.6), was observed on several dives. This anemone is often found on small sulfide or mudstone outcrops exposed to bottom currents. It was also observed to colonize other stalked organisms, in which case it surrounds the stalk of its host with its own stalk. Many stalked animals were host to other animals, including brittle stars and gelatinous, membranelike creatures.

Hydrothermal deposits are generally more heavily colonized than surrounding areas, in part because the deposits present a stable, hard substrate in an area of soft sediment. In addition to stalked animals, the deposits are often heavily colonized by anemones, starfish, and tunicates with less abundant brittle stars and galatheid crabs. Closeup views of the sulfide deposits often revealed a diffuse, fuzzy coating of filamentous material interpreted to be sulfur-oxidizing bacteria. This was especially prevalent on the young deposits seen on dive 658. The abundance of starfish and anemones on the deposits may be in part related to availability of sulfur-oxidizing bacteria as a food source. The only vent-specific fauna observed were the cluster of dead tube worms and a small patch of disarticulated clam shells observed on dive 658.

Summary

The eight dives at the NESCA and SESCA sites in Escanaba Trough provided a total of 41 hours of bottom observation and provided additional data concerning the extent of the metallic sulfide deposits and the structural and geological complexity of the intrusive centers that occur along the axis of southern Gorda Ridge. The emplacement of bysmalithic intrusions into the thick (500 m) sediment that partially fills the trough has resulted in local uplift and the formation of clusters of small hills on the seafloor above larger, deep-seated intrusive/extrusive complexes. This conclusion is supported, in part, by the occurrence of tensional fissures on the flanks of some of the hills. The hills are about 100 m high and 700 m in diameter, although those at NESCA are distinctly more irregular and asymmetrical than those at SESCA. This asymmetry may reflect a greater degree of tectonic control associated with intrusion at the northern site.

The sediment hills appear to have been uplifted approximately 100 m above the surrounding area (Figs. 6.2, 6.3; Morton et al., this volume). The mechanism of uplift is not certain. The hills might be erosional remnants of horst blocks uplifted by intrusion of basalt into the sediment. Basaltic magma intruded at a sediment-covered ridge crest can form sills, unless the restriction to lateral flow exceeds the force necessary to lift the roof rocks. At this point the magma body inflates and acts like a piston to lift the overlying sediment (Lonsdale and Lawver, 1980), producing a small bysmalith intrusion (Morton et al., this volume, Fig. 5.7). This type of intrusion would form a series of faults around the base of the uplifted sediment hill. The intrusion could act to focus hydrothermal circulation; hydrothermal deposits would tend to form over the faults owing to the higher hydrologic continuity of the fracture system compared to the relatively impermeable hemipelagic sediment.

The flanks of the hills are much steeper than conventional bathymetric charts indicate. Where the sediment cover is absent, stratified mudstone and intercalated sheet flows are exposed. Near-vertical slopes are interspersed with benches and gently sloping terraces. Signs of mass wasting abound; extensive accumulations of mudstone and sulfide talus on the terraces and at the hill bases suggest that the slopes have retreated since their initial formation. The tops of the hills are flat, bounded in many cases by steep scarps; bed forms (ripple marks) were observed on the top of 3170 Hill at SESCA.

Extensive deposits of Fe-rich sulfide were

ARRANGEMENT OF SEA CLIFF EQUIPMENT

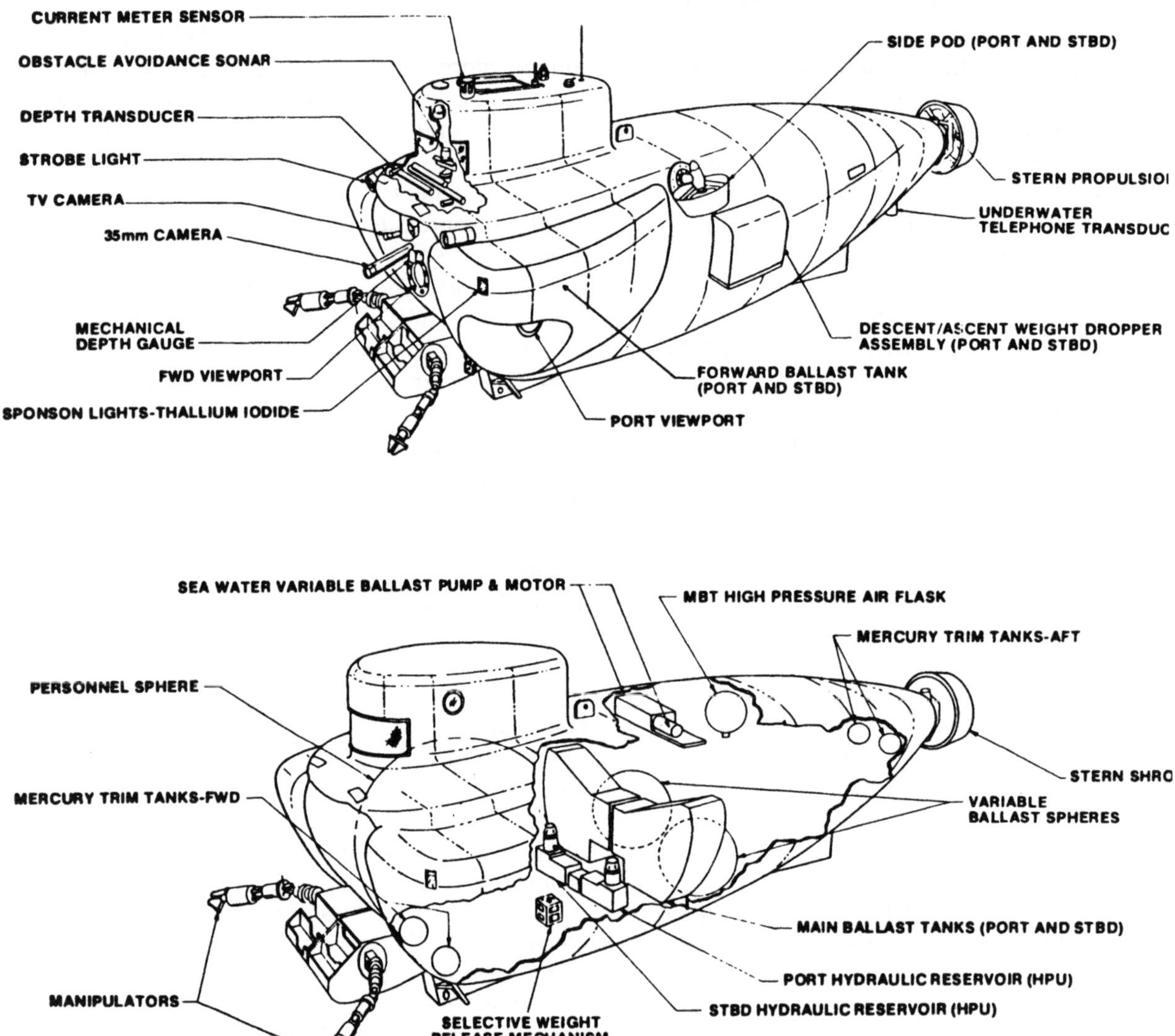

FIGURE 6.10. Schematic external and cut-away drawings of U.S. Navy deep-diving submersible *Sea Cliff* (DSV-4), showing location of major components and instrument systems.

mapped on the mid and lower flanks of the central hill at NESCA and both North Hill and 3170 Hill at SESCA, in water depths that range from 3,180 m to 3,230 m. Several of the deposits are tens of meters wide and more than 100 m long. They are characterized by massive ramparts and ridges of sulfide, chimneylike features up to 10 m high, and extensive taluspiles. The extent of the deposits and composition of the sulfide assemblages suggest that they were formed by sustained high-temperature discharge.

A series of north-trending ridges and valleys about 5 m in amplitude form the margin of the SESCA uplift east of North Hill. Small chimneys and discrete plates of sulfide were photographed just east of the base of North Hill; these deposits may be contiguous with other massive deposits that were mapped on the southeastern flank of North Hill. About 2 km north of the central dome at NESCA the margin of a suspected collapsed lava lake also has ridge-and-swale topography with a relief of up to 10 m. Sulfides have formed along the crests of some of the ridges; broken plates and small chimneys are spaced 1 to 2 m apart. The nature of the deposits at these two sites indicates diffuse hydrothermal circulation and sulfide formation within the upper sediment layers. No signs of active hydrothermal venting were seen during the dives; however, the fresh appearance of sulfide deposits and the presence of dead vent animals at one site suggest recent venting. Other deposits are more extensively weathered and dissected.

Exposed volcanic rocks were seen at only two places. The summit of the central hill at NESCA was the site of extensive accumulation of basalt pillows, some up to 3 m in diameter. The large pillows are thinly covered with sediment in places; glassy rinds were observed on some of the flows. Extensive pillow and sheet flows were also seen just to the southwest of the base of North Hill at SESCA. These deposits are more heavily sedimented than the NESCA volcanics, and in places they are fractured by posteruptive movements.

The sediment-covered areas are populated by holothurians, brittle and basket stars, urchins, anemones, and occasional galatheid crabs. Burrowed areas about 50 cm in diameter surrounding a central dome of castings were commonly observed, although the animal responsible for the burrows was never identified. Stalked animals including sea pens, crinoids, goblet-shaped siliceous sponges, and screwlike coelenterates are more abundant in areas of hard substrate or exposed rock. Straight and branch-stalked corals(?), sea fans, sponges, and the "Venus flytrap" (actinoschyphid?) anemones were also encountered on exposed basalt and sulfides. Nektonic animals such as shrimp, octopods, jellyfish, and spectrunculid fish were encountered near or over the seafloor hills.

The extent and distribution of the massive sulfide deposits encountered during the dive program suggests that such accumulations may cover much of the lower and middle slopes of all of the hills at NESCA and SESCA, and possibly many of the similar features at other locations in Escanaba Trough. Although the sulfide deposits are generally weathered and no signs of active hydrothermal venting were observed, only 0.3 km^2 of sea floor was brought under direct observation, and such a possibility cannot be ruled out.

Note Added in Proof

At the time this paper was submitted for publication in early 1989, active hydrothermal discharge had been discovered in June 1988 at NESCA near the site of dive 658 during a series of dives using the research submersible *Alvin*. These vents were further investigated in September 1988 during a series of *Sea Cliff* dives.

References

Abbott D (1986) Heat Flow Results from the Gorda Ridge. State of Oregon, Department of Geology and Mineral Industries, Open-File Report 0-86-12, 10 pp.

Abbott D, Morton J, Holmes M (1986) Heat flow measurements on a hydrothermally-active, slow-spreading ridge: The Escanaba Trough. Geophys Res Lett 13: 678–680.

Atwater T, Mudie J (1973) Detailed near-bottom

geophysical study of the Gorda Ridge. J Geophys Res 78:8665–8686.
Barton PB, Bethke PM (1987) Chalcopyrite disease in sphalerite: Pathology and epidemiology. Am Mineral 72:451–467.
Benninger LM, Koski RA (1987) Descriptions and Chemical Analyses of Sulfide Samples Dredged in 1986 from Escanaba Trough, Southern Gorda Ridge. U.S. Geological Survey Open-File Report 87-375-B, 18 pp.
Bibee L (1986) Ocean Bottom Seismometer Measurements on the Gorda Ridge. State of Oregon, Department of Geology and Mineral Industries, Open-File Report 0-86-15, 25 pp.
Carey D, Stein D, Taghon G (1986) Analysis of Benthic Epifaunal and Infaunal Community at the Gorda Ridge. State of Oregon, Department of Geology and Mineral Industries, Open-File Report 0-86-11, 34 pp.
EEZ-SCAN 84 Scientific Staff (1986) Atlas of the Exclusive Economic Zone, Western Conterminous United States U.S. Geological Survey Miscellaneous Investigations Seris I-1792, 152 pp.
Holmes M, Morton J (1986) Massive sulfide deposits from Escanaba Trough, Gorda Ridge. Geological Society of America, Program with Abstracts 18:118 (abstract).
Holmes M, Morton J (1987) Regional structure and distribution of hydrothermal activity within Escanaba Trough, southern Gorda Ridge. U.S. Geological Survey Circular 995, USGS Research on Mineral Resources 1987, p 31 (abstract).
Holmes M, Morton J, Clague D, Embley R, Cacchione D, Abbott D (1986) Morphology and mineral resources of Gorda Ridge and Blanco Fracture Zone, northeast Pacific Ocean. Am Assoc Petroleum Geol Bull 70:926–927 (abstract).
Karlin R, Lyle M (1986) Sediment studies on the Gorda Ridge. Oregon State University, College of Oceanography Technical Report 86-14, 43 pp.
Koski RA, Shanks WC III, Bohrson WA, Oscarson RL (1988) The composition of sediment-hosted massive sulfide deposits from Escanaba Trough, Gorda Ridge: Implications for fluid-sediment interaction and depositional processes. Can Mineral 26:655–673.
Kvenvolden K, Rapp JB, Hostettler F, Morton JL, King JD, Claypool GE (1986) Petroleum associated with polymetallic sulfide on the Gorda Ridge, offshore northern California. Science 234:1231–1234.
Lonsdale P, Lawver LA (1980) Immature plate boundary zones studied with a submersible in the Gulf of California. Geol Soc Am Bull 91:555–569.
McManus DA, Burns RE, Von der Borch C, et al. (1970) Site 35. In McManus D et al. (eds): Initial Reports of the Deep Sea Drilling Project, Vol 5. Washington: U.S. Government Printing Office, pps 165–172.
Morton J, Holmes M, Normark W, Ross S, Lyle M (1986a) Sedimentary setting of massive sulfide deposits, Escanaba Trough, offshore northern California. Trans Am Geophys Union 67:1282–1283 (abstract).
Morton J, Normark W, Holmes M, Lyle M (1986b) Geologic setting of two sediment-hosted mineral localities in Escanaba Trough, Gorda Ridge. U.S. Geological Survey Circular 995, USGS Research on Mineral Resources 1987, p 49 (abstract).
Morton J, Normark W, Ross S, et al. (1987a) Preliminary report, cruises L1-86-NC and L2-86-NC, Escanaba Trough, Gorda Ridge. U.S. Geological Survey Open-File Report 87-375A, 20 pp.
Morton J, Holmes M, Koski R (1987b) Volcanism and massive sulfide formation at a sedimented spreading center, Escanaba Trough, Gorda Ridge, northeast Pacific Ocean. Geophys Res Lett 14:769–772.
Wilde P, Chase T, Holmes M, Normark W, Thomas J, McCulloch D, Kulm L (1978) Oceanographic Data Off Northern California—Southern Oregon: Lawrence Berkeley Lab. Pub. 251, 1 Sheet.

Appendix A: *Sea Cliff* Capabilities

Sea Cliff (DSV-4) is a U.S. Navy research submersible capable of carrying three people (pilot, observer/equipment operator, and observer) to a depth of 6,100 m for up to 16 hours (Fig. 6.10). The vessel is 9.6 m long, weighs 26,300 kg, and is capable of speed up to 2.5 knots. Dive personnel and instrumentation are housed in a 2-m-diameter titanium sphere attached to the front of the submersible's frame. Power for propulsion and onboard systems is supplied by special high-capacity silver-zinc batteries. *Sea Cliff* is used primarily for deep-ocean object recovery, search, and scientific research.

The vehicle is equipped with a variety of tools and sensors that provide for operation of the submersible and permit collection of scientific data. The principal systems used during the 1986 Escanaba Trough dives are listed below:

CTFM search sonar	Benthos 35-mm external camera
Mesotech search sonar	Birns incandescent lights (4)
EDO fathometer	Exterior strobe lights (3)
External TV camera (B&W)	Hydraulic manipulators (2)
Current meter	Thallium iodide lights (4)
Sample baskets (4)	UQC underwater telephone
Water bottle	35-mm hand-held cameras (2)
Sonatech transponder	External thermistor probes (3)
Digital depth indicator	Internal video camera (color)

The CTFM obstacle avoidance sonar was used to obtain an indication of major topographic changes along the intended traverse. The CTFM operates on a swept frequency of 72 to 87 kHz and can be adjusted for a forward-looking sector sweep or fixed directional scan. This device displayed the sonar returns on a plan position indicator screen and could be set to scan from 46 m to 1,370 m (50 to 1500 yd) ahead of *Sea Cliff*. The Mesotech sonar, which was installed during the later phases of the study (dives 658 to 663), provides higher resolution and considerably more discrimination of changes in the bottom type and morphology. This device operates at 675 kHz and provides for selection of seven ranges from 5 to 500 m. In addition to the conventional sector scan mode, the Mesotech can be operated as a single-channel side-scan sonar. This capability was used during traverses along some of the steep hill flanks. In some cases, the images could be used to supplement visual observation in mapping the extent of the large sulfide deposits on the Escanaba Trough hills.

The EDO fathometer was used to determine *Sea Cliff*'s altitude above the sea floor. Because of limited visibility downward, the fathometer was of particular importance when conducting bottom approach maneuvers. The display could be adjusted for four depth ranges from 15 to 915 m (50 to 3,000 ft) and was frequently used to provide information on sea floor morphology during times when the bottom was beyond the viewing range of the observers. The submersible's depth below the sea surface was provided continuously by the digital depth indicator.

The manipulators could be equipped with a variety of tools to permit the accomplishment of specific sampling tasks and needs. These included a gripping claw (*Alvin* tool), parallel jaw gripper, a scoop, and a cable cutter. There were no provisions for interchanging tools during a dive, so the two that best fit the requirements of the dive were selected and fitted during predive preparations. Samples that were collected during the dives could be placed in any of the four sample baskets attached to a frame directly below the pilot's viewpoint.

The 35-mm Benthos camera was mounted directly above the pilot's viewport, aimed and focused to permit photographing the seafloor in front of the submersible as well as sampling operations using the manipulators. This camera had a capacity of about 800 frames of color film; a control panel in the personnel sphere permitted photographs to be taken automatically at 4- to 18-sec intervals. This rate selector could also be overridden by a manual shutter control. Two large strobe lights attached to the submersible's Fiberglas fairing above the sphere provided illumination by firing whenever the shutter was activated.

The external TV camera mounted directly above the Benthos camera could be moved with pan-and-tilt controls inside the sphere to provide the two observers with a view of the area in front of the vehicle. This video also allowed the observers to advise the pilot in selection of sampling sites and to monitor sample recovery and storage operations. The camera could be connected to a video recorder whenever the observers wished to obtain a permanent record of these observations.

Hand-held 35-mm cameras and a portable video camera were operated by dive personnel inside the sphere provided additional TV and photographic capabilities. The 35-mm cameras were equipped with wide-angle lenses and were used to obtain photographs through the observer's viewpoints. The field of view of the equipment operator (starboard) was illuminated by incandescent lights, whereas that of the port observer was lit by hull-mounted thallium iodide lights. The observer's camera was connected to an external strobe light which provided the necessary illumination for photography. The small TV camera could be aimed through any of the three view ports to obtain superb color pictures that were displayed on a monitor and recorded on magnetic tape cassettes.

Normal operating procedures call for *Sea Cliff* to be operated by Navy pilots and equipment operators, with space for a single observer on each dive. The U.S. Navy agreement to train and certify scientists as observer/equipment operators provided the maximum scientific return for the limited amount of time available.

Appendix B: Escanaba Trough Sulfide Samples

Sample	*wt(kg)*	*Description*[*]
658-R1	0.6	Fragments of massive sulfide talus up to 10 cm long. Predominantly medium- to coarse-grained Po and Mc inverted from Po. Minor minerals are Ba, Icb, Sl, and rare Gn. Samples had an iron-oxide coating, some of which is Lp.
658-R2	5.4	Barite chimney approximately 25 cm long and 10 cm in diameter. The chimney is very porous and has radial texture defined by dendritic to plumose Ba aggregates growing outward from a small central conduit. The central conduit has a diameter of 2 to 4 mm on the cut section at the base of the sample but does not breach the top. The conduit contains some fine-grained, clear yellow crystals of Sl, but the chimney is nearly monomineralic Ba. The outer surface of the chimney has a 1- to 2-mm-thick crust of sooty Mn-oxide and was heavily colonized by tunicate biota and a few small sponges.
658-R3	9.5	Rounded talus block of massive Po with irregular porous zone of coarse-bladed Po within more compacted zones of fine-bladed to massive Po. Po is highly altered to Mc with minor Fe-oxide. Icb and Sl are fairly abundant, and there are trace amounts of Asp and Gn. The sample is coated with a 1- to 2-mm-thick orange to red Fe-oxide rind composed of Go, Lp, and natrojarosite, and was colonized by tunicates.
658-R4	2.7	Po-rich massive sulfide talus composed of medium- to fine-grained Po and Mc inverted from Po, with some zones of coarser, bladed Po. Po is generally unaltered, Icg is common, and there is trace secondary Go.
659-R1	3.5	Massive sulfide chimney approximately 25 cm high and 10 cm in diameter. The upper part is constructed of chimney walls of fine-grained Po. The irregular walls show that the fluid conduit changed location several times during the growth of the chimney. The fluid channelways are filled with coarse-grained porous aggregates of Po plaes. Icb is common interstitial to Po, and there are traces of Cp and Asp. Parts of the coarse Po are

[*]Abbreviations: Asp, arsenopyrite; Ba, barite; Cp chalcopyrite; Go, goethite; Gn, galena; Icb, isocubanite; Lp, lepidocrosite; Mc, marcasite; Po, pyrrhotite; Sl, sphalerite.

		pseudomorphically replaced by native S, with interstitial Fe and CuFe sulfate. The outer surface of the chimney is an alteration rind of fine-grained Go. The base of the chimney formed at the sediment-water interface. The top layer is predominantly fine-grained Po, with Icb, Sl, Asp, and Gn. Below this is a layer composed predominantly of Icb with minor Asp and Po. The bottom of the sample is mostly coarse Po with Icb, Sl, Asp, and minor Gn and Cp. The bottom layer incorporates pellets of sediment composed of clay, quartz, feldspar, and mica. Yellow hydrocarbon occurs in the pore space.
659-R2	0.7	Fragments of partially oxidized Po-rich massive sulfide composed of medium- to fine-grained Po with variable amounts of Sl, Icb, and Ba. Some samples incorporate sediment pellets composed of clay, quartz, feldspar, mica, and chlorite. Some fragments contain hydrocarbon. Alteration phases include native sulfar, sulfates, and Fe-oxide.
659-R3	1.0	Fragments of partially oxidized Po-rich massive sulfide composed of medium- to fine-grained Po with variable amounts of Sl, Icb, and Ba. Some samples incorporate sediment composed of clay, quartz, feldspar, mica, and chlorite. Some fragments contain hydrocarbon. Alteration phases include marcasite, monoclinic Po, and Fe-oxide, including Lp.
659-R4	0.4	Fragments of Po-rich massive sulfide, generally less oxidized than 659-R2 or 659-R3. Fragments include a few pieces of small chimney tops with partially open fluid channels. Sl is common in some fragments.
662-R1	7	Multiple fragments of a large, stubby, chimneylike projection atop a massive sulfide mound. The sample broke apart when the submersible surfaced. The outer surface is predominantly fine-grained massive Po with Icb, Sl, and Cp. The interior is medium-grained Po that forms porous comblike aggregates, with interstitial Icb. Late-stage botryoidal Mc is common. Some fragments contain Ba. Outer surface contains secondary Fe-oxide and native S.
663-R1	1.2	Fragments of porous and weathered, originally Po-rich massive sulfide. Much of the material is altered to Mc and Fe-oxide, including Go and poorly crystallized hematite. Minor amorphous silica, natrojarosite, Fe-sulfate, and barite.
663-R2	0.3	Weathered, porous, Fe-stained, white rubble, predominantly talc with some Go and secondary sulfates. One fragment was darker and smelled petroliferous.

7—A Comparison of Sulfide Deposits from Modern Sediment-Covered Spreading Axes with Besshi-Type Deposits of Japan

Randolph A. Koski

Abstract

A comparison of the occurrence, associated lithologies, compositions, and tectonic settings of massive sulfide deposits from the sediment-covered spreading axes at Escanaba Trough (ET), Middle Valley (MV), and Guaymas Basin (GB) in the northeast Pacific with massive sulfide deposits of the Besshi type in the Sanbagawa metamorphic belt of southwestern Japan and at the Shimokawa mine in northern Japan reveals numerous similarities but also some important differences. Major differences are greater thickness of sediment and the absence of an igneous basement complex in the Sanbagawa terrane. At the Shimokawa deposit, an ophiolitic assemblage is interleaved with turbidites and massive sulfides in a tectonic melange. In addition, the mafic lavas associated with Besshi-type deposits have both alkalic and midocean ridge basalt (MORB) compositions whereas the basalts associated with ET, MV, and GB are restricted to MORB compositions. The principal compositional differences are the lower Cu/Zn ratios; higher Pb and Ba contents; more abundant sulfate and carbonate gangue mineralization in the sediment-hosted seafloor deposits; and the greater pyrite content and Co enrichment in the Besshi-type massive sulfides. The massive polymetallic (Zn-Cu-Pb-Ag-As-Sb) sulfide present at ET is apparently absent in the Besshi deposits. Pb isotope ratios for Besshi deposits are somewhat similar to those for GB sulfides and hydrothermal fluids; they indicate a contribution of Pb from both basalt and sediment. The similarities in deposits formed on sediment-covered ridge crests in different tectonic settings (open-ocean spreading axis and rifted continental margin) indicate that deposits with Besshi-type characteristics, both ancient and modern, are not unique to one tectonic environment. Deposits with similar characteristics may be forming in other modern oceanic rift environments (e.g., narrow back-arc or interarc basins, small ocean basins surrounded by continental highlands, and subduction zones in which an oceanic ridge is underthrusting the continental margin) where the sedimentation rate exceeds the spreading rate.

Introduction

Massive sulfide deposits formed at sediment-starved oceanic ridges share numerous attributes with deposits preserved in ophiolite sequences (Oudin et al., 1981; Koski, 1987). It is reasonable to expect that massive sulfides on sediment-covered spreading axes, such as Escanaba Trough (ET) on the southern Gorda Ridge, Middle Valley (MV) on the northern Juan de Fuca Ridge, and Guaymas Basin (GB) in the Gulf of California (Fig. 7.1a), might also have analogs preserved in ancient assemblages of oceanic crust and marine sediment.

The late Paleozoic Sanbagawa metamorphic belt of Shikoku and the Mesozoic Hidaka terrane of Hokkaido, Japan (Fig. 7.1b), contain more than 200 massive Fe-Cu sulfide deposits that form concordant tabular bodies in thick sequences of metasedimentary and metavolcanic rocks (Kanehira and Tatsumi, 1970). Recent work in Japan led to the conclusion that these deposits, collectively referred to as the Besshi type after the largest and best-known example in the Sanbagawa terrane, represent

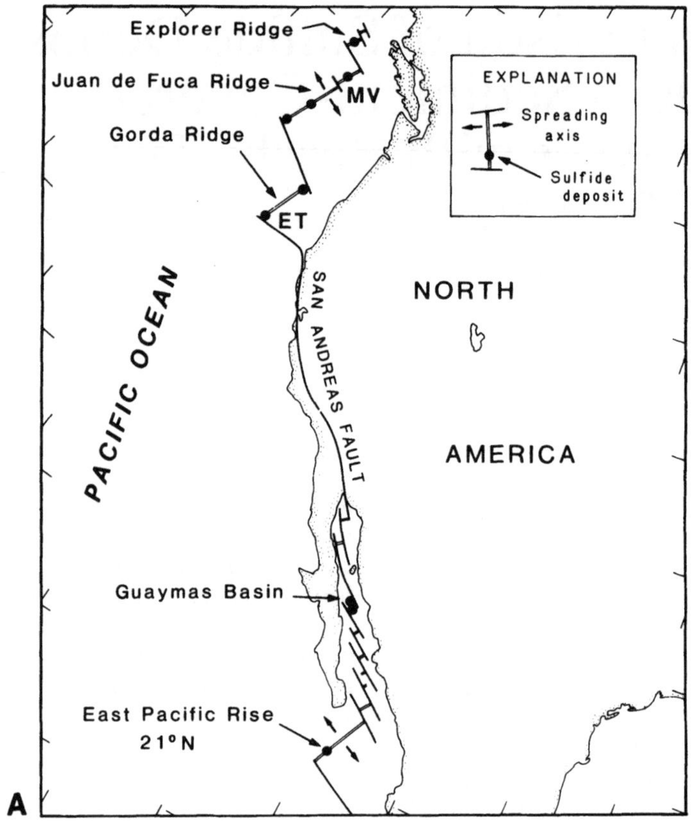

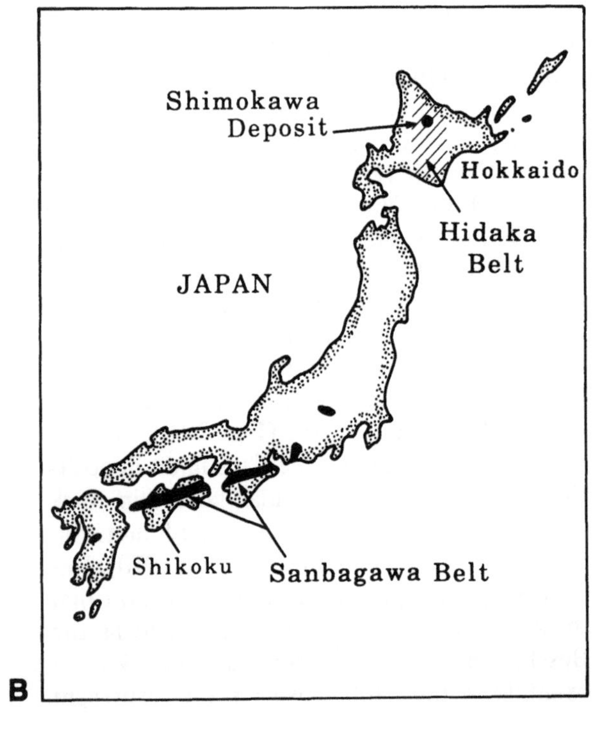

FIGURE 7.1. (A) Location of sulfide deposits on spreading axes of the northeast Pacific including Escanaba Trough (ET) on the southern Gorda Ridge, Middle Valley (MV) on the northern Juan de Fuca Ridge, and Guaymas Basin (GB) in the Gulf of California. (B) Location of the Sanbagawa metamorphic belt and the Shimokawa deposit in the Hidaka belt of Japan.

synvolcanic mineralization in ocean ridge environments (Mariko, 1984; Miyake, 1988). Fox (1984) and Scott (1985) proposed that Besshi-type deposits are ancient analogs of the hydrothermal sulfide mounds on the sediment-covered floor of Guaymas Basin in the Gulf of California.

Although the specific tectonic setting of the Besshi-type massive sulfide deposits of Japan is uncertain (Kase and Yamamoto, 1988), they share a number of common attributes with sediment-hosted massive sulfide deposits in other Phanerozoic terranes (Sawkins, 1976); on a worldwide basis, these deposits comprise an important subclass of the volcanogenic-exhalative type of base-metal sulfide deposits. The geology of Besshi-type deposits in Japan and elsewhere has been described in detail by several authors (Kanehira and Tatsumi, 1970; Sawkins, 1976; Klau and Large, 1980; Franklin et al., 1981; Fox, 1984; Kase and Yamamoto, 1988), and a detailed recapitulation of their descriptions is beyond the scope of this paper. The purpose of this paper is to compare selected characteristics of Besshi-type deposits in Japan and the deposits in ET, MV, and GB in an attempt to evaluate the similarities and differences between these deposits. The comparison also shows that deposits with characteristics like those in ET, MV, and GB are not unique to one tectonic setting and are probably forming at other contemporary oceanic rifts where the sedimentation rate exceeds the spreading rate.

Comparative Geology of Sea-Floor and Besshi-Type Sulfide Deposits

Table 7.1 summarizes and compares the occurrences, lithological associations, compositions, and tectonic settings of the sediment-hosted deposits in ET, MV, and GB with the Besshi-type deposits in Japan. The Besshi-type deposits include (1) massive sulfides of the late Paleozoic Sanbagawa metamorphic belt of Shikoku in southwestern Japan, and (2) the Shimokawa massive sulfide in the Mesozoic Hidaka terrane of Hokkaido in northern Japan.

Occurrence

Massive sulfide deposits at ET, MV, and GB have formed on the sediment-covered floors of narrow troughs. These deposits generally occur in clusters of individual mounds and chimneys that extend for tens of kilometers along the spreading axis. At GB many of the deposits are clustered in linear arrays parallel to faults (Lonsdale and Becker, 1985), whereas the bulk of the sulfide at ET forms semicontinuous outcrops on the flanks of steep-sided sediment hills (Morton et al., 1987). Sulfide deposits at MV occur within flat-topped mound structures, as much as 60 m high and 1,000 m across, that are covered by a thin veneer of sediment (Goodfellow and Blaise, 1988).

The sulfide mounds and chimneys are upright constructional deposits but are inherently unstable features subject to decomposition and degradation. The abundant sulfide talus surrounding sulfide mounds indicates that the ET deposits are undergoing rapid mass wasting (Holmes and Zierenberg, this volume). Clastic sulfide debris recovered in cores form the MV mounds (Davis et al., 1987) indicates that at least some of these deposits were eroded prior to burial. Similarly, it is probable that the seafloor weathering and collapse of fragile mounds and chimneys in GB, composed largely of sulfate and carbonate phases (Koski et al., 1985; Peter and Scott, 1988), will result in lateral redistribution of hydrothermal material on the seafloor.

Besshi-type deposits in Japan typically are concordant sheetlike bodies of massive and banded sulfide with total thicknesses of only a few meters and lateral dimensions of a few kilometers (Table 7.1). Deposits within several ore districts on Shikoku appear to be confined to a narrow stratigraphic interval. The remarkable ratio of lateral extent to thickness (>100) in many Besshi deposits may be the result of primary deposition or erosional processes on the paleoseafloor; however, the original shape and texture of sulfide deposits in the Sanbagawa belt have been modified by folding and high-pressure metamorphism (Banno, 1986). Although lobe- and pod-shaped sulfide bodies occur at the Shimokawa deposit, these forms

TABLE 7.1. Comparison of massive sulfide deposits at three sediment-covered ridge axes in the northeast Pacific with Besshi-type deposits in Japan.

	Escanaba Trough, Southern Gorda Ridge	Middle Valley, Northern Juan de Fuca Ridge
Occurrence	The ridge axis is covered by 0 to 500 m of terrigenous and hemipelagic sediment. Discontinuous ledges and mounds of massive sulfide, as much as 100 m long and 30 m thick, occur along the base of 1-km-wide, sediment-capped hills and on the step-faulted, sediment-covered flanks of a volcanic hill (Morton et al., 1987). Hydrothermal venting at 220°C was measured at the NESCA site in 1988 (Campbell et al., 1988a). Sulfides occur above two domelike volcanic edifices that locally breach and uplift the sediment. Small chimneys and thin sheets of sulfide and sulfate are associated with collapse features in thinly sediment-covered lava flows. Interlayered massive sulfide and sulfide-cemented sediment occur in subsurface. Area of each site with sulfide deposits exceeds 15 km^2. Considerable sulfide rubble near mounds is the result of mass wasting.	Eight moundlike structures of massive and clastic sulfide, averaging 400 m in diameter and 50 m in height, overlain by a thin (<10 m) layer of pelagic sediment and underlain by 300 m of terrigenous and hemipelagic sediment near east side of 15-km-wide axial valley. Active venting through sulfide chimneys occurs near the mounds. Mounds have sublinear trend, 11 km in length, oblique to the eastern valley wall (Davis et al., 1987; Goodfellow and Blaise, 1988).
Associated lithologies	Silt, sand, and mud turbidite, predominantly arkosic in composition (Vallier et al., 1973) overlying pillow basalt. Samples of pillow lava from Gorda Ridge north of Escanaba Trough have low-potassium MORB compositions (Davis and Clague, 1987).	Fine-grained turbidite of terrigenous origin overlying igneous basement, presumably of MORB composition. Basalt elsewhere on the Juan de Fuca Ridge to the south and the Explorer Ridge segment to the north have Fe-enriched MORB compositions (Eaby et al., 1984; Cousens et al., 1984). In areas of high heat flow, sediment is hydrothermally altered with carbonate cement (Goodfellow and Blaise, 1988).
Mineralogy	*Pyrrhotite-rich sulfide:* pyrrhotite, marcasite > isocubanite, chalcopyrite, sphalerite >> löllingite, galena > native bismuth. Also barite, talc. *Polymetallic sulfide:* sphalerite, pyrrhotite, galena > isocubanite, arsenopyrite > löllingite, tetrahedrite, boulangerite, argentite. Also barite.	Pyrrhotite, pyrite, marcasite, sphalerite, barite, amorphous silica, talc > isocubanite, chalcopyrite > covellite, galena.
Important metals and other elements: major, (minor), [trace]	*Pyrrhotite-rich sulfide:* Fe, S, (Cu), (Zn), (Ba), [Mn], [As], [Au] *Polymetallic sulfide:* Zn, Fe, Pb, S, (Ba), (As), (Ag), (Cu), [Sb], [Sn] Mineral and chemical data from Benninger and Koski (1987) and Koski et al. (1988).	Fe, S, (Zn), (Ba), (Cu), [Mn], [As], [Sb], [Ag], [Se] Mineral and chemical data from Davis et al. (1987) and Goodfellow and Blaise (1988).
Tectonic setting	Open-ocean ridge near a continental margin; slow-rate spreading axis.	Mid-ocean ridge near continental margin, medium-rate spreading axis.

TABLE 7.1. (*continued*)

	Guaymas Basin, Gulf of California	Besshi-type Deposits, Sanbagawa Metamorphic Belt (Shikoku)
Occurrence	Two short spreading axes separated by a transform fault result in two sediment-covered basins (Lonsdale and Lawver, 1980). Sediment cover approximately 500 m thick over igneous basement. Approximately 130 individual and coalesced massive sulfide-sulfate-carbonate mounds and chimneys are located around high-temperature vents on the sediment pond of the Southern Trough in an area 8.5 km long and 1.5 km wide (Lonsdale and Becker, 1985). Deposits occur in elongate clusters or sublinear arrays associated with fractures and small scarps displacing sediment. Ledges of hydrothermal material (mostly talc and pyrrhotite) also occur in the sediment-covered Northern Trough of Guaymas Basin (Lonsdale et al., 1980).	Thin, semicontinuous, sheetlike or lenticular bodies of massive and banded sulfide generally confined to single horizons in folded sequences of pelitic, quartzose, and mafic schists. Spatially associated with mafic volcanic rocks (Kanehira and Tatsumi, 1970). Lateral dimensions much greater than vertical: thickness <1 to 5 m, width to 1 km, length to 7 km. The largest deposit, Besshi, contained 33 million tons of massive sulfide averaging 2.6% Cu.
Associated lithologies	Mud turbidite and interbedded diatomaceous ooze intruded by diabasic sills of MORB composition (Saunders et al., 1982; Perfit et al., 1982).	Interbedded with mafic schist, originally mafic lavas, tuffs, and gabbroic rocks of both MORB and alkalic basalt composition (Sugisaki and others, 1972; Fox, 1984). Protoliths of the metasedimentary rocks include shale, graywacke, chert, and limestone (Sugisak et al., 1972; Banno, 1986). Less common are Mn-silicate-bearing schists (footwall) and hematite or magnetite iron formation (hanging wall).
Mineralogy	*Mounds:* Calcite, barite, amorphous silica, pyrrhotite, marcasite, talc, stevensite, anhydrite > pyrite, sphalerite, chalcopyrite, isocubanite > galena *Chimney:* Calcite, barite, anhydrite, amorphous silica, talc > pyrrhotite, sphalerite, wurtzite, isocubanite > galena, chalcopyrite	Pyrite, pyrrhotite, chalcopyrite > sphalerite, bornite, magnetite >> valeriite, galena, tetrahedrite, cobaltite, cubanite, stannite, molybdenite, native gold, native silver, electrum. Gangue minerals, generally minor in abundance, include quartz, calcite, barite, anhydrite, muscovite, albite, chlorite, tourmaline. Pyrrhotite is more abundant in deposits of lower metamorphic grade (Kanehira and Tatsumi, 1970).
Important metals and other elements: major, (minor), [trace]	Fe, Si, Ca, Ba, S, (Zn), (Cu), (Pb), [Mn], [Ag] Mineral and chemical data from Peter (1986), Peter and Scott (1988), and Koski et al. (1985).	Fe, Cu, S, (Zn), [Pb], [Ag], [Au], [Co], [Mo], [Mn] Mineral and chemical data from Kanehira and Tatsumi (1970) and Kase and Yamamoto (1988).
Tectonic setting	Rifted continental margin; medium-rate spreading axis.	Uncertain. Sugisaki et al. (1972) suggested a late Paleozoic geosyncline, possibly a deep intracontinental rift filled with several kilometers of terrigenous sediment. Partial melting in the upper mantle resulted in basaltic volcanism. Horikoshi (quoted by Ohmoto and Takahashi, 1983) suggested a trench setting.

TABLE 7.1. (*continued*)

	Besshi-type deposits, Shimokawa, Hidaka Belt (Hokkaido)
Occurrence	Tabular and pod-shaped deposits of massive and banded sulfide in a zone approximately 1,800 m in length, 600 m in width, and averaging 7.6 m in thickness. Total mine production: 6 million tons at 2.5% Cu. Deposits commonly occur along faulted contacts between diabase or basalt and black slate within melange of the Hidaka terrane (Miyake, 1988).
Associated lithologies	Ophiolitic assemblage of serpentinite, pyroxenite, gabbro, diabase, massive and pillow basalt with interlayered sandstone, argillite, black slate, chert, and minor limestone (Mariko, 1984, 1988). Mafic rocks are metamorphosed to greenschist and amphibolite facies.
Mineralogy	Pyrite, pyrrhotite, chalcopyrite > sphalerite, magnetite > cobalt pentlandite, cobaltite, cobaltian mackinawite, cubanite, galena. Gangue minerals include quartz, calcite, siderite, chlorite, amphibole, and apatite. In addition, carbonaceous materials are reported in the massive sulfide.
Important metals and other elements: major, (minor), [trace]	Fe, Cu, S, (Zn), (Co), [Ag] Mineral and chemical data from Mariko (1988) and Miyake (1988).
Tectonic setting	Ocean ridge near a continental margin and subduction zone.

are attributed to deformation of a single sulfide layer (Miyake, 1988). Sulfide mound and chimney structures in the Besshi deposits have not been reported.

At the sites of sulfide deposition in ET, MV, and GB, the sediment thickness above igneous basement is generally between 0 and 500 m. At ET, massive sulfide deposits occur on the sediment-covered flanks of broad volcanic edifices spaced at 15- to 20-km intervals (Morton et al., 1987). GB deposits are forming on sediment that is also intruded by a complex of diabase dikes and sills (Einsele, 1985).

Besshi-type deposits also have a close spatial association with interlayered basaltic and sedimentary rocks in the Sanbagawa belt (Kanehira and Tatsumi, 1970). The total thickness of the Sanbagawa terrane in central Shikoku exceeds 10 km in thickness, and for the entire section mafic volcanic and pelitic sedimentary units are present in subequal amounts (Aoki et al., 1970; Ernst et al., 1970). However, basaltic rocks and massive sulfide deposits are relatively more abundant in the middle member of the Minawa Formation, which in turn lies within the middle part of the Sanbagawa section.

In the Shimokawa mining district, massive sulfide occurs within a thick (>1 km) section of interlayered sandstone, slate, pillow basalt, and diabase. Although the larger-scale stratigraphy and structural relations in the Shimokawa area are obscured by deformation within a melange terrane (Miyake, 1988), the pyroxenite, gabbro, diabase, and pillow basalt units within the melange appear to comprise an ophiolitic assemblage (Mariko, 1984).

Associated Lithologies

The sediments hosting sulfide deposits at ET, MV, and GB are unconsolidated to semi-indurated mud, silt, and sand turbidites (Table 7.1). The turbidite section at GB contains nearly pure layers of diatom ooze as much as 0.1 m in thickness (Einsele and Kelts, 1982). The sediments also contain an average of approximately 2% organic carbon (Simoneit and Bode, 1982), and both sediment and sulfide deposits locally are saturated with hydrothermal petroleum (Simoneit, 1985). Furthermore, surface sediments near hydrothermal mounds in GB contain as much as 1.2% Mn derived from fallout from particulate clouds above active vents (Campbell et al., 1988b).

The basal unit within the Sanbagawa belt, the Oboke Formation, contains conglomerate schists with pebbles of grantic rocks, andesite-basalt, quartzite, and slate (Ernst et al., 1970). Metasedimentary units interlayered with mafic schists and massive sulfide deposits on Shikoku include pelitic, siliceous, and calcareous schists, metasandstones, and "black schists" containing abundant graphite. The protoliths for the Sanbagawa schists are sandstone, chert, shale, limestone, and basalt (Kawachi et al., 1982). At the Okuki mine in western Shikoku, the massive sulfide horizon is overlain by a thin layer of "red chert" composed of alternating hematite- and aluminosilicate-rich layers (Watanabe et al., 1970). Schistose rocks composed of magnetite, garnet, and quartz occur in the hanging wall and footwall of the Besshi deposit (Doi, 1961). Manganiferous garnet and epidote are minor phases in metashales and metasandstones near the Shirataki sulfide deposit (Ernst et al., 1970).

The sedimentary rocks in the Shimokawa mining area are predominantly black slate and fine-grained arkosic sandstone (Mariko, 1984). The massive sulfide deposits are concentrated along contacts between black slate and diabase or pillow lava.

Pillow lavas from ET and other sites on Gorda Ridge, basalt from Juan de Fuca Ridge to the south and Explorer Ridge to the north of MV, and diabasic sills intruding the sediment at GB all have normal midocean ridge basalt (MORB) compositions. Diabasic rocks associated with the Shimokawa deposit also have normal MORB characteristics (Miyake, 1980); however, the basaltic metavolcanic rocks in the Sanbagawa belt have major and trace-element compositions of both alkalic basalt and MORB (Sugisaki et al., 1972; Fox, 1984).

Composition of Deposits

Massive sulfide deposits at ET, MV, and GB have lower Cu/Zn ratios and higher Pb and Ba contents than Besshi-type deposits (Fig. 7.2). Notably absent from Besshi are the polymetallic (Zn-Cu-Pb-Ag-As-Sb) massive sulfide assemblages present at ET (Koski et al., 1988) and the abundant sulfate gangue mineralization present at all of the modern sites (Table 7.1). The formation of polymetallic Sn-As-Zn-Ag veins in massive sulfide and siliceous schist at the Besshi mine has been attributed to a contact metamorphic event associated with a hidden granitic pluton of Miocene age (Kase, 1988). The principal Fe-sulfide phase in Besshi-type deposits is pyrite, whereas deposits in ET, MV, and GB are dominated by pyrrhotite. The Co enrichment observed in pyrite from Besshi-type deposits (Kase and Yamamoto, 1988) and the presence of Co-rich pentlandite, Co-bearing mackinawite, and cobaltite at the Shimokawa deposit (Bamba and Motoyoshi, 1985; Mariko, 1988) have not been recognized in the seafloor deposits on sediment-covered ridge crests.

Massive sulfide and sulfide-cemented sediment from the GB and ET deposits are impregnated with petroleum derived from the rapid hydrothermal maturation of organic matter in the sediment (Simoneit and Lonsdale, 1982; Kvenvolden et al., 1986). Carbonaceous mate-

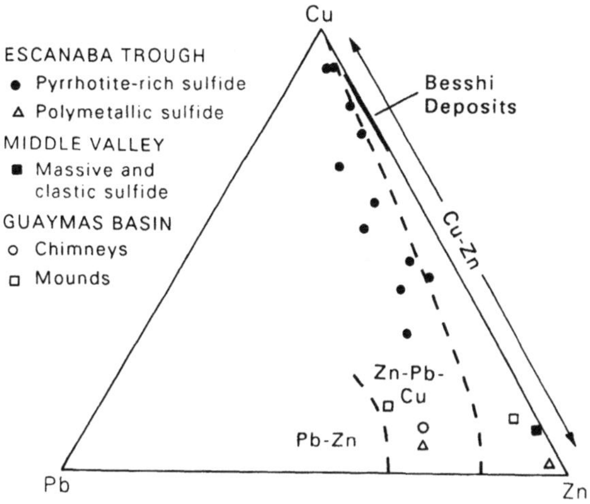

FIGURE 7.2. Cu-Zn-Pb ratios for massive sulfide deposits from Escanaba Trough (Koski et al., 1988), Middle Valley (Davis et al., 1987), Guaymas Basin (Koski et al., 1985; Peter, 1986), and the Sanbagawa metamorphic belt (Doi, 1961). Besshi deposits are restricted to the Cu-rich part of the Cu-Zn field, whereas most of the sea floor deposits plot in the Zn-rich part of the Cu-Zn field or in the Zn-Pb-Cu field. The compositional fields are from Franklin et al. (1981).

rial has also been observed in the massive ores of the Shimokawa deposit (Mariko, 1988).

Tectonic Setting

Sugisaki et al. (1972) concluded that the geosynclinal sequence represented by the Sanbagawa terrane was deposited during the late Paleozoic rifting of Precambrian continental crust. If their interpretation is correct, the "best-fit" modern analog may be a rifted continental margin setting (Gulf of California) or an intracratonic rift setting (Red Sea). In contrast, Kunugiza et al. (1986) suggest that ultramafic and gabbro bodies in the Sanbagawa terrane are syntectonic blocks that originated earlier in the lower crust–upper mantle region of an island arc. It is apparent that despite extensive recent investigations of the geology of the Sanbagawa metamorphic belt, the tectonic history of this region remains uncertain.

Mariko (1984) proposed that the Shimokawa deposit formed at a midocean spreading axis near a land mass that was shedding clastic sediment. This tectonostratigraphic setting might be comparable to ET and MV, both of which are located within 300 km of the continental margin of North America. In a somewhat different interpretation, Miyake (1988) suggested that the Shimokawa sulfide deposit formed at a sediment-free spreading axis near a west-dipping subduction zone. Following mineralization, the sulfide horizon and associated upper crustal rocks were juxtaposed and interleaved with turbidites in a subduction melange that resulted from underthrusting of the oceanic plate beneath the Eurasian continental margin.

Pb Isotope Data

Pb isotope ratios of sulfide minerals can help identify the source of Pb and associated metals in ore deposits and may be a useful criterion for comparison of modern and ancient analogs. The Pb isotope ratios of sulfide ores from Besshi-type deposits of Japan and Pb isotope data for massive sulfide, sediment, basalt, and hydrothermal fluid from GB and ET (Fig. 7.3) indicate that radiogenic Pb derived from sediment is a variable but significant component of the total Pb budget in both the ancient and modern deposits.

Pb isotope data for sulfides in several Paleozoic Besshi-type deposits, mostly from the Sanbagawa terrane, and sulfides from the Shimokawa deposit (Sato and Sasaki, 1980) are compared to data for MORB and deep-sea sediment in Figure 7.3a. The $^{208}Pb/^{204}Pb$ and $^{206}Pb/^{204}Pb$ ratios for Besshi sulfides have a wide range in values. Some discordance between Besshi-type sulfide (including samples from Shimokawa) and sea floor basalt values is indicated, and a few Besshi samples have ratios that are similar to or more radiogenic than deep-sea sediment values (Fig. 7.3a). Additional Pb isotope data on the spatially associated sedimentary and volcanic rocks are required for a more rigorous interpretation of the data, but it is probable that a significant fraction of the Pb in at least some of the Besshi deposits was derived from a sedimentary source.

The Pb isotope ratios for samples of GB sulfide and GB hydrothermal fluid plot near the convergence of trends for Gulf of California sediment and East Pacific Rise basalt (Fig. 7.3b). This relationship indicates a combined sediment-basalt source for Pb in GB sulfide deposits (LeHuray et al., 1988). This conclusion is consistent with fluid circulation models for GB based on other criteria, such as silica geobarometry and sulfur isotope ratios, that require penetration of fluid into the igneous basement at depth prior to further interaction with the sediment pile and discharge onto the sea floor (Von Damm et al., 1985; Koski et al., 1985).

Lead isotope ratios for ET sulfides, both pyrrhotite-rich and polymetallic types, are nearly identical to values for ET sediment (Fig. 7.3c); no contribution from Gorda Ridge basalt is indicated (LeHuray et al., 1988). This is surprising, considering the relatively thin blanket of sediment between the deposits and underlying volcanic edifices.

FIGURE 7.3. (A) Lead isotope diagrams (from Sato and Sasaki, 1980) comparing ratios of Besshi-type deposits in the Sanbagawa terrane (open circles) and the Shimokawa mine (solid circles) to ratios of ocean ridge basalt (Church and Tatsumoto, 1975; Tatsumoto, 1978) and deep-sea sediment (Reynolds and Dasch, 1971; Church, 1976; Meijer, 1976). (B) Lead isotope ratios for massive sulfide (LeHuray et al., 1988) and hydrothermal fluid (Chen et al., 1986) from Guaymas Basin, sediment from the Gulf of California (Brevart et al., 1981), and basalt from the East Pacific Rise near 21° N lat. (White et al., 1987). (C) Lead isotope ratios for massive sulfide, sediment, and one basalt samples from the Escanaba Trough, Gorda Ridge (LeHuray et al., 1988) and basalt from the northern Gorda Ridge (Church and Tatsumoto, 1975).

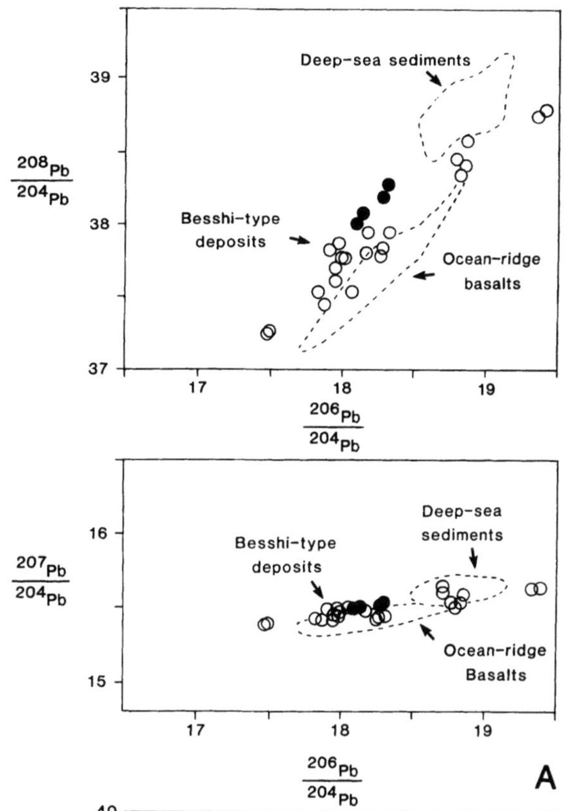

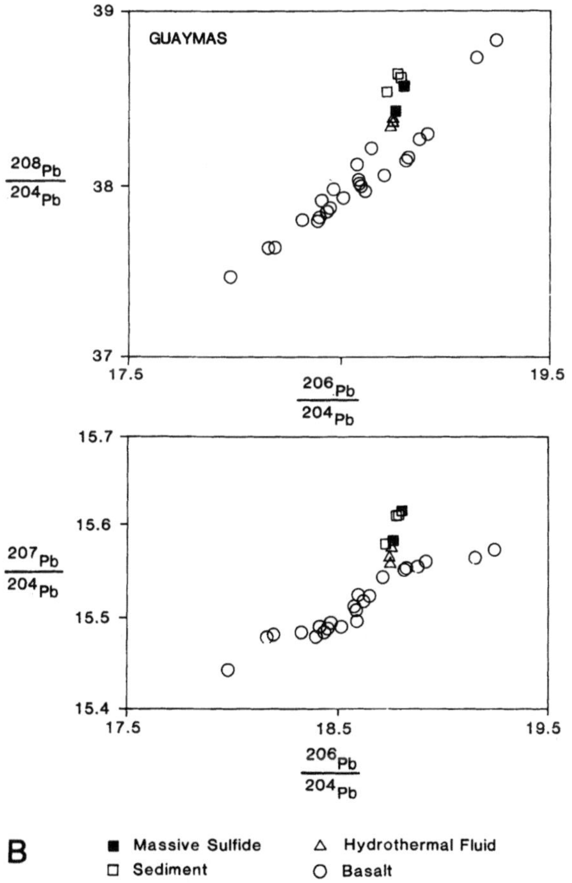

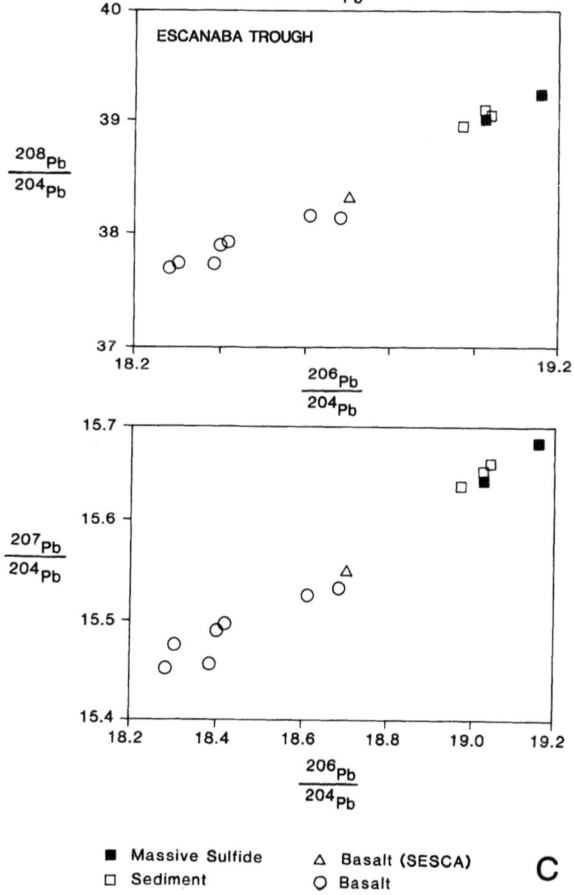

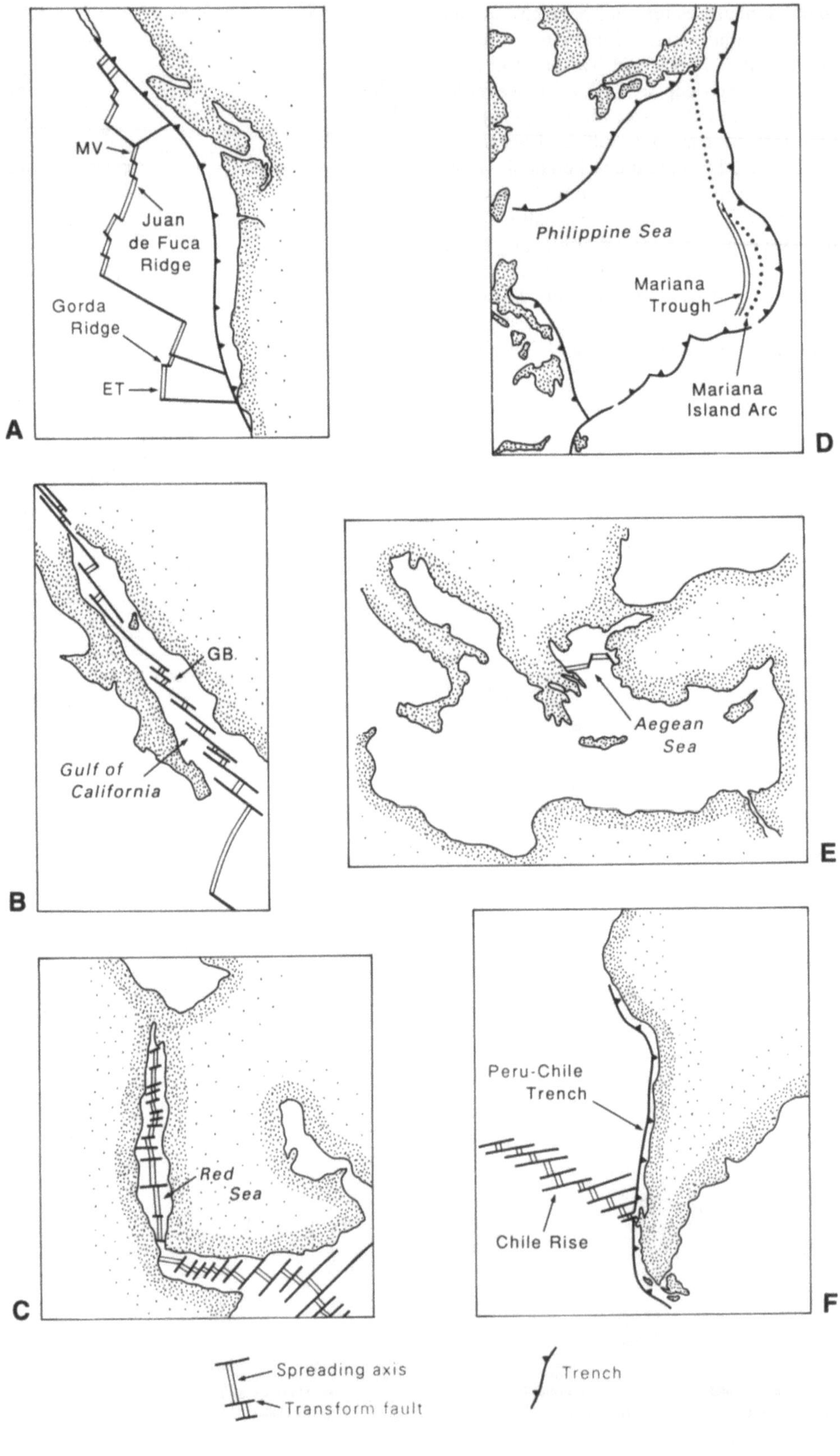
MV
Juan de Fuca Ridge
Gorda Ridge
ET
A
Philippine Sea
Mariana Trough
Mariana Island Arc
D
GB
Gulf of California
B
Aegean Sea
E
Red Sea
C
Peru-Chile Trench
Chile Rise
F
Spreading axis
Transform fault
Trench

Conclusions and Implications for Besshi-Type Deposits in Other Settings

Close similarities in the occurrence (wide distribution within a sediment-covered sea floor environment), associated lithologies (fine-grained turbidites and basalts with MORB-type chemistry), and composition (Fe-Cu-Zn sulfide assemblage, presence of carbonaceous material, Pb isotope ratios) of Besshi-type deposits in Japan and the massive sulfide deposits at ET, MV, and GB generally support the proposal that these ancient and modern deposits are metallogenic analogs. Major differences include the greater thickness of sediment, the presence of both alkalic and MORB-type basalt, and the absence of an underlying mafic igneous complex in the Sanbagawa terrane; the sheetlike nature of Besshi deposits compared to the moundlike forms at ET, MV, and GB; the lower Cu/Zn ratio and higher Pb and Ba contents in the modern deposits; and the greater amount of pyrite relative to pyrrhotite and the Co enrichment in the Besshi-type deposits. The massive polymetallic (Zn-Cu-Pb-Ag-As-Sb) sulfide assemblage at ET and the abundant sulfate and carbonate gangue mineralization in the modern sites are apparently absent in the Besshi deposits.

The differences can be attributed, at least in part, to the effects of metamorphism and deformation on the ancient deposits, differences in the proportion and composition of volcanic and sedimentary source rocks, and variations in the tectonic setting. Although the ET, MV, and GB deposits share numerous morphological and compositional characteristics, they formed in different tectonic settings (open-ocean, slow-rate spreading axis; open-ocean, medium-rate spreading axis; and rifted continental margin, respectively). Although the specific environment of formation of Besshi-type deposits in Japan remains uncertain, there is evidence that deposits may have formed in both open-ocean (Shimokawa) and rifted continental margin (deposits in the Sanbagawa belt) environments. Thus, it is clear that deposits with Besshi-type characteristics, ancient and modern, are not unique to one tectonic setting.

These observations imply that deposits with characteristics similar to ET, MV, and GB may be forming on other active oceanic rift zones buried under relatively thin mantles of sediment (Fig. 4). In addition to the known sulfide occurrences at MV, ET, and GB (Fig. 7.4a, b) and in the Red Sea, a young ocean basin within an intracratonic rift (Fig. 7.4c; also see Zierenberg, this volume), other deposits might be forming elsewhere at spreading axes in narrow back-arc or interac basins such as the Mariana Trough that are covered in part by detritus shed from the active island arc (Fig. 7.4d), back-arc spreading axes developed on old continental crust in small ocean basins such as the Aegean Sea, which is surrounded by continental highlands (Fig. 7.4e), and active spreading axes such as the Chile Rise that are subducted at continental margins (Fig. 7.4f).

Acknowledgments. I thank A. Davis, J. Hein, and G. McMurray for reviewing the manuscript, and L. Benninger and P. Swenson for drafting the figures.

References

Aoki H, Oida T, Sasaki Y, et al. (1970) Crustal structure in central Japan derived from explosion seismic observations along line from Atumi to Noto. Rep Res Group Explosion Seismol 28:30–49.

FIGURE 7.4. Spreading axes, partly or completely covered by sediment, that may contain Besshi-type mineralization. (A) Open-ocean: Escanaba Trough (ET) on southern Gorda Ridge and Middle Valley (MV) on northern Juan de Fuca Ridge. (B) Rifted continental margin: Guaymas Basin (GB) in the Gulf of California. (C) Intracratonic rift: Red Sea. (D) Back-arc basin: Mariana Trough. (E) Back-arc basin on continental crust: Aegean Sea. (F) Ocean ridge underthrusting continental margin along subduction zone: Chile Rise. Sulfide deposits are known at ET, MV, GB, the Red Sea, and the Mariana Trough (sediment-free segment).

Bamba T, Motoyoshi Y (1985) Study on massive sulfide ores from the Shimokawa mine, Hokkaido, Japan. Marine Geol 35:211–225.
Banno S (1986) The high-pressure metamorphic belts of Japan. A review. Geol Soc Am Mem 164:365–374.
Benninger LM, Koski RA (1987) Descriptions and chemical analyses of sulfide samples dredged in 1986 from Escanaba Trough, southern Gorda Ridge. U.S. Geological Survey Open-File Report 87-375-B:26 p.
Brevart O, Dupre B, Allegre CJ (1981) Metallogenesis at spreading centers: Lead isotope systematics for sulfides, manganese-rich crusts, basalts, and sediments from the Cyamex and Alvin areas (East Pacific Rise). Econ Geol 76:1711–1720.
Campbell AC, German C, Palmer MR, Edmond JM (1988a) Preliminary report on the chemistry of hydrothermal fluids from the Escanaba Trough. EOS, Trans Am Geophy Union 69:1271.
Campbell AC, Gieskes JM, Lupton JE, Lonsdale PF (1988b) Manganese geochemistry in the Guaymas Basin, Gulf of California. Geochim Cosmochim Acta 52:345–357.
Chen JH, Wasserburg GJ, Von Damm KL, Edmond JM (1986) The U-Th-Pb systematics in hot springs on the East Pacific Rise at 21° N and Guaymas Basin. Geochim Cosmochim Acta 50:2467–2479.
Church SE (1976) The Cascade Mountains revisited: A re-evaluation in light of new lead isotope data. Earth Planetary Sci Lett 29:175–188.
Church SE, Tatsumoto M (1975) Lead isotope relations in oceanic ridge basalt from the Juan de Fuca–Gorda Ridge Area, NE Pacific Ocean. Contrib Mineral Petrol 53:253–279.
Cousens BL, Chase RL, Schilling JG (1984) Basalt geochemistry of the Explorer Ridge area, northeast Pacific Ocean. Can J Earth Sci 21:157–170.
Davis AS, Clague DA (1987) Geochemistry, mineralogy, and petrogenesis of basalt from the Gorda Ridge. J Geophys Res 92:10,467–10,483.
Davis EE, Goodfellow WD, Bornhold BD, et al. (1987) Massive sulfides in a sedimented rift valley, northern Juan de Fuca Ridge. Earth Planetary Sci Lett 82:49–61.
Doi M (1961) Geology and cupriferous pyrite deposits (Besshi type) of the Sambagawa metamorphic zone including the Besshi and the Sazare mines. Mining Geol 11:610–626.
Eaby J, Clague DA, Delaney JR (1984) Sr isotopic variations along the Juan de Fuca Ridge. J Geophys Res 89:7883–7890.
Einsele G (1985) Basaltic sill-sediment complexes in young spreading centers: Genesis and significance. Geology 13:249–252.
Einsele G, Kelts K (1982) Pliocene and Quaternary mud turbidites in the Gulf of California: Sedimentology, mass physical properties and significance. Deep Sea Drilling Project Leg 64. In Curray JR, Moore DG, et al. (eds): Initial Reports of the Deep Sea Drilling Project Leg 64. Washington: U.S. Government Printing Office 64:511–528.
Ernst WG, Seki Y, Onuki H, Gilbert MC (1970) Comparative study of low-grade metamorphism in the California Coast Ranges and the outer metamorphic belt of Japan. Geol Soc Am Mem 124:276.
Fox JS (1984) Besshi-type volcanogenic sulphide deposits—a review. Can Inst Mining Bull 77:57–68.
Franklin JM, Lydon JW, Sangster DF (1981) Volcanic-associated massive sulfide deposits. Econ Geol 75th Anniv Vol 485–627.
Goodfellow WD, Blaise B (1988) Sulfide formation and hydrothermal alteration of hemipelagic sediment in Middle Valley, northern Juan de Fuca Ridge. In Barrett TJ, Jambor JL (eds): Seafloor Hydrothermal Mineralization. Can Mineral 26:675–696.
Kanehira K, Tatsumi T (1970) Bedded cupriferous iron sulphide deposits in Japan, a review. In Tatsumi T (eds): Volcanism and Ore Genesis. Tokyo: University of Tokyo Press, pp 51–76.
Kase K (1988) Tin, arsenic, zinc and silver vein mineralization in the Besshi mine, central Shikoku, Japan. Mining Geol 38:407–418.
Kase K, Yamamoto M (1988) Minerals and geochemical characteristics of ores from the Besshi-type deposits in the Sambagawa belt, Japan. Mining Geol 38:203–214.
Kawachi Y, Watanabe T, Landis CA (1982) Origin of mafic volcanogenic schists and related rocks in the Sanbagawa belt, Japan. J Geol Soc Jpn 88:797–817.
Klau W, Large DE (1980) Submarine exhalative Cu-Pb-Zn deposits—a discussion of their classification and metallogenesis. Geol Jahr 40:13–58.
Koski RA (1987) Sulfide deposits of the seafloor: Geological models and resource perspectives based on studies in ophiolite sequences. In Teleki PG, Dobson MR, Moore JR, Von Stackelberg U (eds): Marine Minerals, Advances in Research and Resource Assessment. Dordrecht, D Reidel, pp 301–316.
Koski RA, Lonsdale PF, Shanks WC, Berndt ME, Howe SS (1985) Mineralogy and geochemistry of a sediment-hosted hydrothermal sulfide deposit

from the Southern Trough of Guaymas Basin, Gulf of California. J Geophy Res 90:6695–6707.

Koski, RA, Shanks WC III, Bohrson WA, Oscarson RL (1988) The composition of massive sulfide deposits from the sediment-covered floor of Escanaba Trough, Gorda Ridge: Implications for depositional processes. Can Mineral 26:655–673.

Kunugiza K, Takasu A, Banno S (1986) The origin and metamorphic history of the ultramafic and metagabbro bodies in the Sanbagawa metamorphic belt. Geol Soc Am Mem 164:375–385.

Kvenvolden KA, Rapp JB, Hostettler FD, Morton JL, King JD, Claypool GE (1986) Petroleum associated with polymetallic sulfide in sediment from Gorda Ridge. Science 234:1231–1234.

LeHuray AP, Church SE, Koski RA, Bouse RM (1988) Pb isotopes in sulfides from mid-ocean ridge hydrothermal sites. Geology 16:362–365.

Lonsdale P, Becker K (1985) Hydrothermal plumes, hot springs, and conductive heat flow in the Southern Trough of Guaymas Basin. Earth Planetary Sci Lett 73:211–225.

Lonsdale P, Bischoff JL, Burns VM, Kastner M, Sweeney RE (1980) A high-temperature hydrothermal deposit on the seabed at a Gulf of California spreading center. Earth Planetary Sci Lett 49:8–20.

Lonsdale P, Lawver LA (1980) Immature plate boundary zones studied with a submersible in the Gulf of California. Geol Soc Am Bull 91:555–569.

Mariko T (1984) Sub-sea hydrothermal alteration of basalt, diabase and sedimentary rocks in the Shimokawa copper mining area, Hokkaido, Japan. Mining Geol 34:307–321.

Mariko T (1988) Ores and ore minerals from the volcanogenic massive sulfide deposits of the Shimokawa mine, Hokkaido, Japan. Mining Geol 38:233–246.

Meijer A (1976) Pb and Sr isotopic data bearing on the origin of volcanic rocks from the Mariana island-arc system. Geol Soc Am Bull 87:1358–1369.

Miyake T (1980) Genesis of Shimokawa ore deposit, Hokkaido. Subsidized Science Research Report for Ministry of Education, pp 1–20.

Miyake T (1988) Geology and mineralization of the Shimokawa mine: An allochthonous ridge-type massive sulfide ore deposit. Mining Geol 38:215–231.

Morton JL, Holmes ML, Koski RA (1987) Volcanism and massive sulfide formation at a sedimented spreading center, Escanaba Trough, Gorda Ridge, northeast Pacific Ocean. Geophys Res Lett 14:769–772.

Ohmoto H, Takahashi T (1983) Geologic setting of the Kuroko deposits, Japan. III. Submarine calderas and Kuroko Genesis. In Ohmoto H, Skinner BJ (eds): The Kuroko and Related Volcanogenic Massive Sulfide Deposits. Econ Geol Monogr 5:39–54.

Oudin E, Picot P, Pouit G (1981) Comparison of sulphide deposits from the East Pacific Rise and Cyprus. Nature 291:404–407.

Perfit MR, Saunders AD, Fornari DJ (1982) Phase chemistry, fractional crystallization, and magma mixing in basalts from the Gulf of California, Deep Sea Drilling Project Leg 64. In Curray JR, Moore DG, et al. (eds): Initial Reports of the Deep Sea Drilling Project Leg 64. Washington: U.S. Government Printing Office 64:649–666.

Peter JM (1986) Genesis of Hydrothermal Vent Deposits in the Southern Trough of Guaymas Basin, Gulf of California: A Mineralogical and Geochemical Study. Toronto: Univesity of Toronto, 164 pp.

Peter JM, Scott SD (1988) Mineralogy, composition, and fluid-inclusion microthermometry of seafloor hydrothermal deposits in the Southern Trough of Guaymas Basin, Gulf of California. In Barrett TJ, Jambor JL (eds): Seafloor Hydrothermal Mineralization. Can Mineral 26:567–587.

Reynolds PH, Dasch EJ (1971) Lead isotopes in marine manganese nodules and the ore-lead growth curve. J Geophys Res 76:3402–3413.

Sato K, Sasaki A (1980) Lead isotope features of the Besshi-type deposits and its bearing on the ore lead evolution. Geochem J 14:303–315.

Saunders AD, Fornari DJ, Joron JL, Tarney J, Treuil M (1982) Geochemistry of basic igneous rocks, Gulf of California, Deep Sea Drilling Project Leg 64. In Curray JR, Moore DG, et al. (eds): Initial Reports of the Deep Sea Drilling Project Leg 64. Washington: U.S. Government Printing Office 64:595–642.

Sawkins FJ (1976) Massive sulphide deposits in relation to geotectonics. Geol Assoc Can Spec Paper 14:221–240.

Scott SD (1985) Seafloor polymetallic sulfide deposits: Modern and ancient. Marine Mining 5:191–212.

Simoneit BRT (1985) Hydrothermal petroleum: Genesis, migration, and deposition in Guaymas Basin, Gulf of California. Can J Earth Sci 22:1919–1929.

Simoneit BRT, Bode GW (1982) Appendix II. Carbon/carbonate and nitrogen analyses, Leg 64, Gulf of California. In Curray JR, Moore DG, et al. (eds): Initial Reports of the Deep Sea Drilling Project. Washington: U.S. Government Printing

Office 64:1303–1305.

Simoneit BRT, Lonsdale PF (1982) Hydrothermal petroleum in mineralized mounds at the seabed of Guaymas Basin. Nature 295:198–202.

Sugisaki R, Mizutani S, Hattori H, Adachi M, Tanaka T (1972) Late Paleozoic geosynclinal basalt and tectonism in the Japanese Islands. Tectonophysics 14:35–56.

Tatsumoto M (1978) Isotopic composition of lead in oceanic basalt and its implication to mantle evolution. Earth Planetary Sci Lett 38:63–87.

Vallier TL, Harrold PJ, Girdley WA (1973) Provenances and dispersal patterns of turbidite sand in Escanaba Trough, northeastern Pacific Ocean. Marine Geol 15:67–87.

Von Damm KL, Edmond JM, Measures CI, Grant B (1985) Chemistry of submarine hydrothermal solutions at Guaymas Basin, Gulf of California. Geochim Cosmochim Acta 49:2221–2237.

Watanabe T, Iwao S, Tatsumi T, Kanehira K (1970) Folded ore bodies of the Okuki mine. In Tatsumi T (ed): Volcanism and Ore Genesis. Tokyo: University of Tokyo Press, pp 105–117.

White WM, Hoffman AW, Puchelt H (1987) Isotope geochemistry of Pacific midocean ridge basalts. J Geophys Res 92:4881–4893.

8—Deposition of Metalliferous Sediment Beneath a Brine Pool in the Atlantis II Deep, Red Sea

ROBERT A. ZIERENBERG

Abstract

The Atlantis II Deep in the central Red Sea is the site of formation of a large submarine hydrothermal sulfide deposit. It is the largest deposit discovered to date on the seafloor and the only site given serious consideration for economic exploitation in the near future. The metalliferous deposit in the Atlantis II Deep is very different from other known submarine massive sulfide deposits forming on seafloor spreading centers. The Atlantis II deposit is a thin, but laterally extensive, blanket of interlayered metal sulfide, oxide, and silicate minerals. The Atlantis II deposit is genetically and geochemically similar to the other oceanic ridge crest sulfide deposits in that it forms from a seawater-derived hydrothermal fluid which is modified by high-temperature interaction with oceanic crust. However, the presence of evaporite strata adjacent to the rift and a submarine brine pool in the Atlantis II Deep controls the fluid composition and depositional environment and determines the form of the deposit. The interplay of the hydrotermal system with the depositional environment is one of the critical variables that controls the form, composition, concentration, preservation, and economic potential of submarine exhalative mineral deposits.

Introduction

The discovery of metalliferous sediment in the Atlantis II Deep, Red Sea (Fig. 8.1), enabled the first observations of the formation of a submarine exhalative ore deposit. The subsequent discovery of active hydrothermal vents on the Galapagos Ridge, the East Pacific Rise–Juan de Fuca Ridge system, and the Mid-Atlantic Ridge provided additional opportunities for direct observation of the formation of massive sulfide from an ore fluid venting onto the seafloor. The results of numerous investigations of the Atantis II Deep and ridge crest hydrothermal sites indicate that the geologic and geochemical processes responsible for their origin are generally similar. However, the two types of deposits are distinctly different in their tonnage, geometry, mineralogy, and probability of preservation. These differences relate to the different depositional environments of the two types of deposits. The Atlantis II Deep metalliferous sediments are deposited under a density-stratified anoxic brine pool, but the sulfide mineralization at the ridge crest hydrothermal sites is the result of venting of a buoyant plume of hydrothermal fluid directly onto the seafloor (Fig. 8.2). Differences in the depositional environments of the two deposit types are the major factors controlling the different forms of the deposits.

Geologic Setting

The Red Sea occupies an intercontinental rift separating the Precambrian shield areas of the Arabian and Nubian plates. Extension began during the late Oligocene or early Miocene. Thin sequences of Oligocene continental sedimentary rocks overlie basement rocks in some areas, but Miocene rocks generally lie directly on the basement. The Red Sea was connected to the Mediterranean Sea in the Miocene, and both contain thick (~4 km)

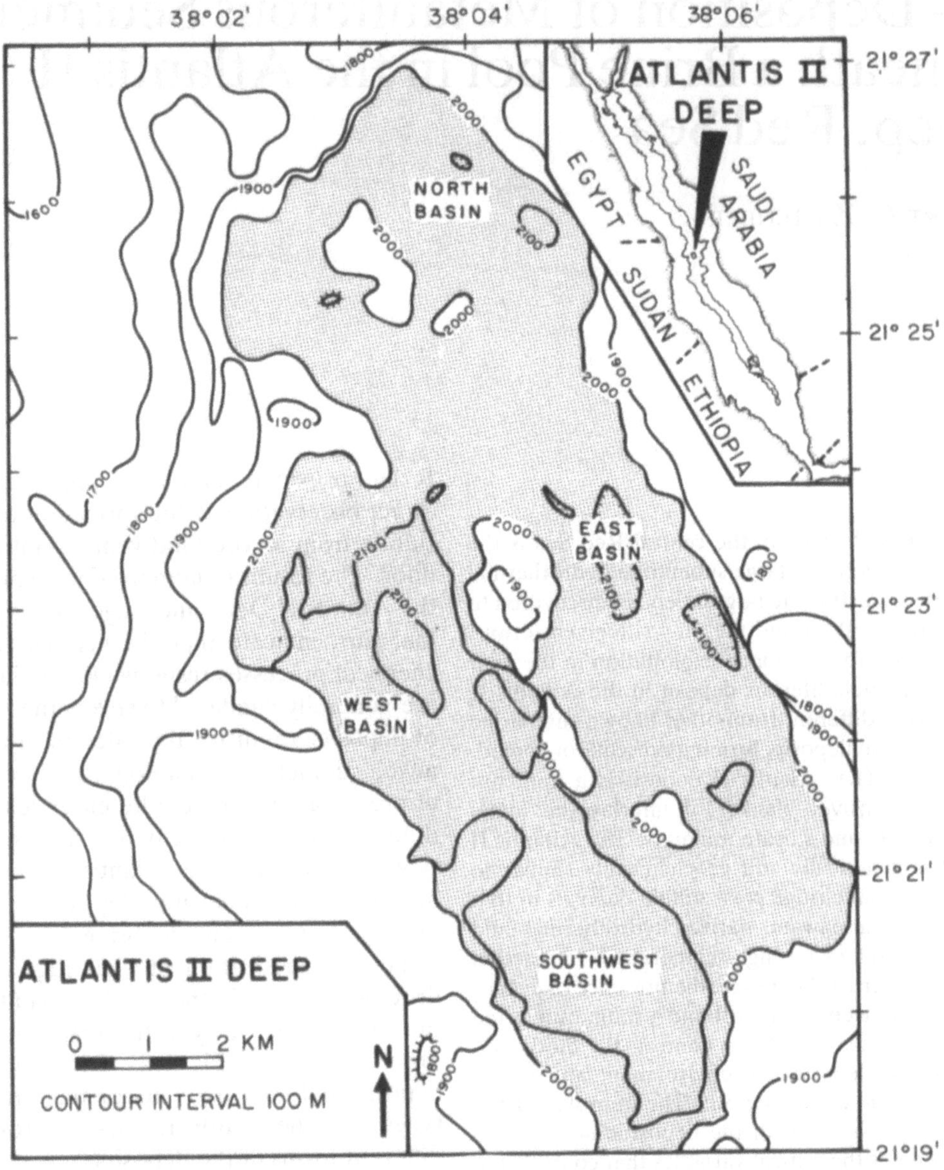

FIGURE 8.1. Inset map showing the location of the Atlantis II Deep in the central Red Sea. The 500-m contour line defines the main trough of the Red Sea. The bathymetric map of the Atlantis II Deep graben shows the location of the different basins. The patterned area below approximately 2,000 m is covered by brine. (From Bäcker and Richter, 1973).

sequences of Miocene evaporites. The upper 500 m of the evaporites in the Red Sea is predominantly anhydrite and halite (Stoffers and Kühn, 1974). Petrographic evidence of shallow-water origin for the upper evaporites includes oncolites and hopper halite crystals (Stoffers and Kühn, 1974) but does not preclude a deep-water origin (Schmalz, 1969) for the bulk of the lower evaporite sequence. Along the flanks of the Red Sea, the evaporites grade into continental clastic rocks of Miocene age. A permanent connection to the Indian Ocean, established approximately 5 million years ago, ended evaporite deposition. The sill between the southern Red Sea and the Indian Ocean is less than 140 m deep and is located in

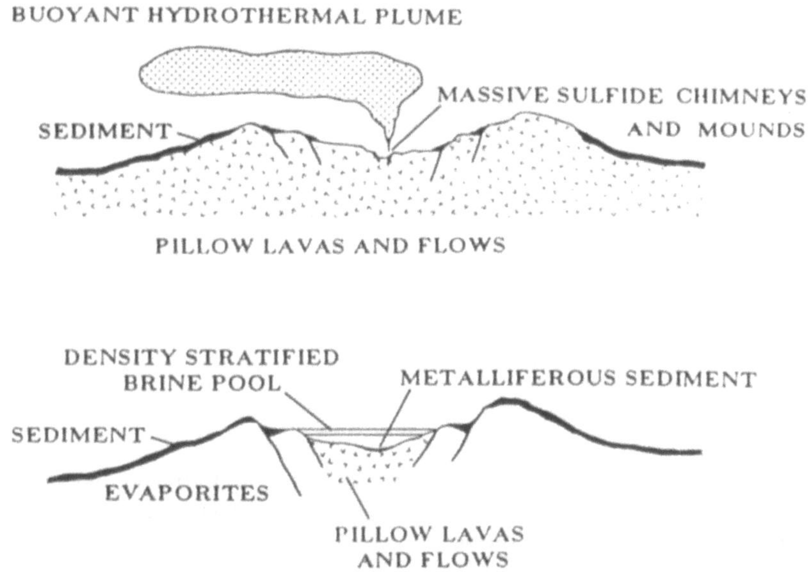

FIGURE 8.2. Schematic diagram contrasting the depositional setting of ridge crest hydrothermal deposits (e.g., 21° N, East Pacific Rise) which form buoyant plumes with the deposition of metalliferous sediment beneath a brine pool, as exemplified by the Atlantis II Deep.

a volcanically and tectonically active area. Because of the high evaporation rate and lack of fluvial discharge into the Red Sea, closure of the connection to the Indian Ocean would lead to a geologically rapid ($\sim 10^3$ years) change from deep water (2000 m) deposition to evaporite deposition and complete desiccation.

True seafloor spreading appears to have been initiated in the southern Red Sea about 3 million to 5 million years ago (Cochran, 1983). The total opening rate is approximately 1.5 cm/year, and the ridge crest has a slow spreading, Mid-Altantic Ridge-type morphology. Extension in the northern Red Sea is not localized along a discrete spreading center and probably occurs by intrusion of dikes and sills into extended continental crust (Cochran, 1983; Pautot et al., 1984; Bicknell et al., 1986). The Atlantis II Deep is near the transition from normal seafloor spreading to more diffuse extension. The graben that contains the brine pool is actively spreading and is floored by oceanic tholeiite (Bäcker and Richter, 1973).

Carbonate reef and shelf sediment extends approximately 60 km from both flanks of the Red Sea to the main trough at approximately 500 m water depth. In the southern Red Sea, an axial trough approximately 20 km wide is developed by garben faulting associated with seafloor spreading (Bäcker et al., 1975). Sea floor relief can exceed 1,000 m along steep bounding faults; water depths are greater than 2,000 m within the axial trough. Sediment in the axial trough is predominantly biogenic pelagic carbonate and eolian detritus derived from the flanking cratons. Sedimentation rates in the central Red Sea are 5 to 10 cm per 10^3 years.

Brine Pools

Hydrographic investigation in 1964 confirmed the presence of submarine brine pools in the area of the Atlantis II Deep and provide data on the temperature and salinity of the brines (Swallow, 1969; Dietrich and Krause, 1969). Brines with a range of temperature, salinity, and chemistry are found within grabens along the central axis of the Red Sea, and diverse origins have been proposed for the different brine pools (Bäcker and Schoell, 1972; Schoell and Faber, 1978; Zierenberg and Shanks, 1986).

The Atlantis II Deep contains a multi-

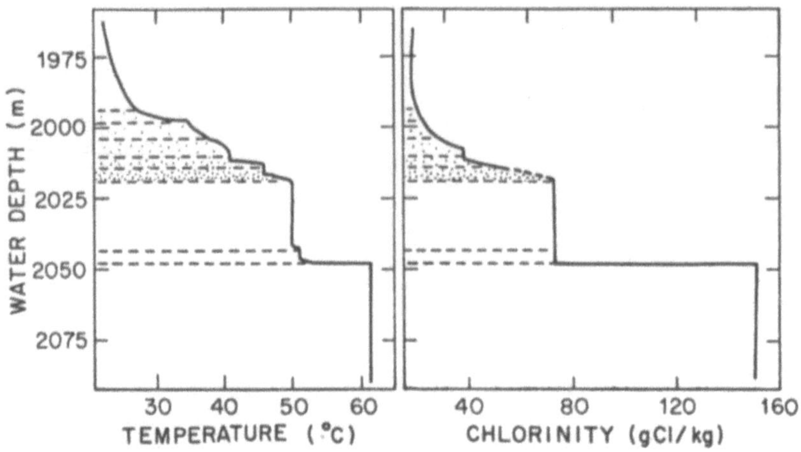

FIGURE 8.3. The density stratified nature of the brine pool is illustrated by the depth vs. temperature and salinity plots. Dashed lines identify acoustic reflectors. (From Hartmann, 1980. Reprinted with permission from Pergamon Press PLC.)

TABLE 8.1. Chemical composition of Atlantis II Lower Brine (A II), 21° N, East Pacific Rise (21° N), and Southern Juan de Fuca Ridge (SJFR) Hydrothermal Fluid, and seawater (SW).

		A II[a]	21° N[b]	SJFR[b]	SW[b]
Li	μmol	37.7	891–1,322	1,110–1,810	26
Na	mmol	4,038	432–510	700–800	464
K	mmol	47.8	23.2–25.8	37.3–51.6	9.79
Rb	μmol	—	27–33	28–37	1.3
Be	nmol	—	10–150	95–150	0.02
Mg	mmol	30.86	0	0	52.7
Ca	mmol	118.7	11.7–20.8	77.3–96.4	10.2
Sr	μmol	480	65–97	230–312	87
Al	μmol	—	4.0–5.2	1.9	0.005
Cl	mmol	4,406	489–579	896–1,090	541
SiO_2	mmol	0.92	15.6–19.5	22.7–23.3	0.16
SO_4	mmol	7.11	0	0	27.9
H_2S	mmol	—	6.6	8.4	0
Mn	mmol	1.5	0.70–1.00	2.61–4.48	<0.001
Fe	mmol	1.5	0.75–2.42	10.3–18.7	<0.001
Cu	μmol	0.33	<2–44	<2	0.007
Zn	μmol	46	40–106	≤900	0.01
Pb	nmol	1,100	183–359	—	0.01
pH		5.5	3.3–3.8	3.2	7.8

[a]Data from Hartmann (1985), Danielsson et al. (1980), Craig (1969).
[b]Data from Von Damm and Bischoff (1987), Von Damm et al. (1985).

layered, density-stratified brine with sharp temperature and salinity interfaces between the brine layers (Fig. 8.3). The chemical composition of the brine has been determined by Brewer and Spencer (1969), Hartmann (1980, 1985), and Danielsson et al. (1980). The composition of the lower brine layer in the Atlantis II Deep is compared with the 21° N, East Pacific Rise (EPR) and southern Juan de Fuca Ridge (SJFR) hydrothermal fluids, and normal seawater in Table 8.1. The lower brine is nearly saturated with NaCl and is strongly enriched in alkali and alkaline earth elements, but it is depleted in magnesium and sulfate compared with seawater. The low pH and high salinity contribute to the extreme enrichment of trace metals in the brine. Isotopic analyses by Craig (1969) indicate that the brine forms from Red Sea water by dissolution of Miocene evaporites. The brine is heated by interaction with rift-zone basalt (Lupton et al., 1977; Shanks and Bischoff, 1977); the basalt also affects the chemistry of the brine by contributing dissolved metals and reduced sulfur (Delevaux and Doe, 1974; Shanks and Bischoff, 1977; Zierenberg and Shanks, 1986).

Detailed hydrographic observations (Schoell, 1976; Schoell and Hartmann, 1978; Hartmann, 1980) indicate that the brine pool is a dynamic system with significant temporal changes in temperature caused by influx of hot brine into

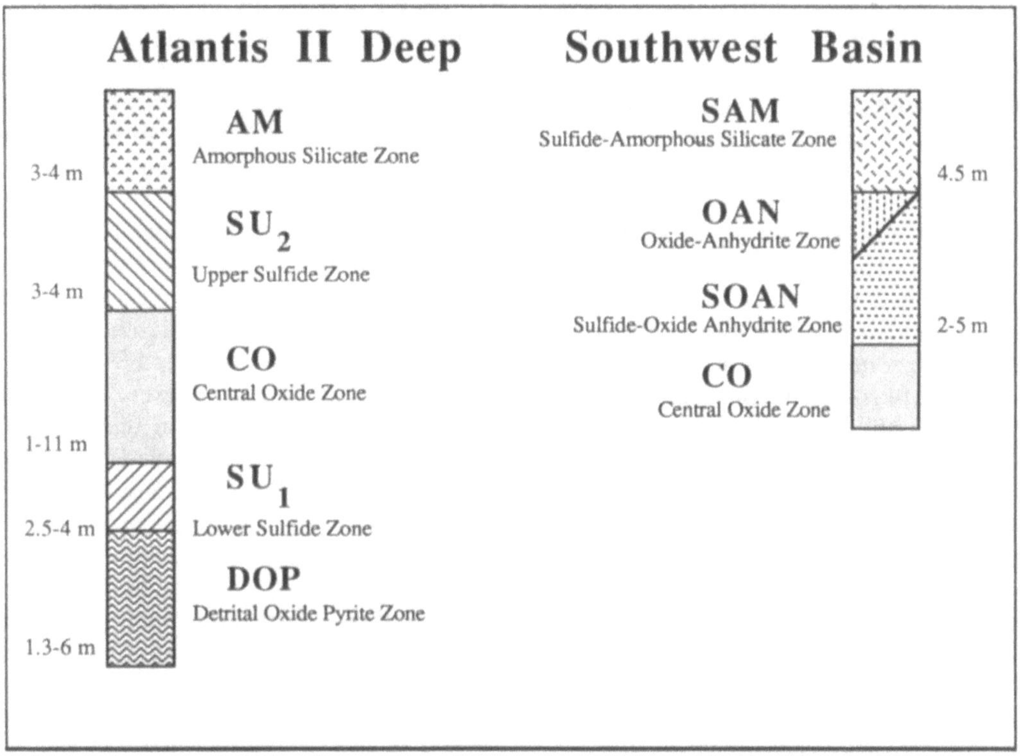

FIGURE 8.4. Generalized lithostratigraphic sequence of Atlantis II Deep metalliferous sediments. (From Bäcker and Richter, 1973).

the deep. The temperature of the lower layer increased from 55.9°C, measured in 1965, to 61.5°C in November 1977 (Schoell and Hartmann, 1978). Hydrographic profiles obtained by bathysonde (Schoell, 1976) indicate that hot brine is currently venting in the Southwest Basin, the deepest part of the Atlantis II Deep. New brine rises vertically to the interface between the lower and upper brines and then spreads laterally below this interface through the West Basin to the East and North Basins (Schoell, 1976). Correspondingly, the highest concentration of suspended base-metal sulfide is present in the lower brine in the Southwest Basin (Hartmann, 1973), due to precipitation from the rising plume of inflowing hotter brine. Long-term variations in the inflow of hot, saline brine from a number of vent localities have led to the formation of distinctive facies of metalliferous sediment, which can be correlated throughout the Deep (Bäcker and Richter, 1973; Hackett and Bischoff, 1973).

Metalliferous Sediments

A general lithostratigraphy of the metalliferous sediments (Fig. 8.4) has been established by Bäcker and Richter (1973). The basal sediment of the detrital-oxide-pyrite (DOP) zone may be as old as 28,000 years BP (Shanks and Bischoff, 1980), and it consists predominantly of biogenic-detrital marl that directly overlies basalt. Zones enriched in goethite and manganese oxides occur intermixed and interlayered with the marls, especially toward the top of this zone. In addition, there are pyritic sulfide layers, some of which have high copper contents. Sulfur isotope ratios of sulfides from the DOP zone suggest that some of the pyritic sulfide layers have formed by biogenic sulfate reduction, but others contain a hydrothermal sulfide component (Shanks and Bischoff, 1980). On the basis of the distribution of limontic material in the DOP zone, Bäcker and

Richter (1973) suggested that hydrothermal input originated in the northern part of the Deep.

The lower sulfide zone (SU_1) overlies the DOP zone. This sedimentary unit represents the establishment of a permanent brine pool (Bäcker and Richter, 1973), which occurred approximately 15,000 years ago (Shanks and Bischoff 1980). The metalliferous sediment is very fine grained (generally $<2\ \mu$), is finely laminated, and has high interstitial water contents. The SU_1 zone is composed predominantly of dark red-brown, Fe-rich clays that are interlayered with black to violet, very fine grained, sulfide-rich sediments. The sulfide-rich layers contain pyrite and sphalerite with lesser amounts of chalcopyrite, manganosiderite, and anhydrite. Bäcker and Richter (1973) suggest that the vent area for this zone may have been in the West Basin.

The SU_1 zone grades into the central oxide (CO) zone, which is composed of very fine grained to amorphous Fe oxyhydroxide and lesser amounts of Fe-rich phyllosilicate. Like the SU_1 zone, it is thickest in the West Basin. In one area north of the Southwest Basin, the CO zone is locally recrystallized to coarse-grained hematite and magnetite (Hackett and Bischoff, 1973; Zierenberg and Shanks, 1983). Concentrations of finely layered sedimentary manganite and hematite are also observed in various places. This zone contains some discontinuous layers rich in anhydrite, especially near the top, where it grades into the upper sulfide zone.

The upper sulfide zone (SU_2) is similar in most characteristics to the lower sulfide zone (SU_1). It consists of interlayered sulfides and Fe-rich phyllosilicates and contains as much as 20% Zn (dry, salt-free weight basis). The silicates of the SU_2 zone, however, are typically green, in contrast to the red to brown colors observed in the SU_1 zone and the overlying AM zone. Local concentrations of massive anhydrite are present, especially at the base of this zone.

The amorphous silicate zone (AM) is the youngest zone and is still being formed today. It is most similar to the two sulfide zones, containing admixed Fe-rich clay minerals and base-metal sulfides. In addition, the AM zone contains disseminated crystalline anhydrite and thin layers of Fe oxyhydroxides. These very young sediments average ~95 wt% interstitial brine (Bäcker and Richter, 1973) and contain abundant X-ray amorphous material, including Fe oxides and silica.

Base-metal sulfides from the AM zone and the two sulfide zones have similar sulfur-isotope distribution and an average $\delta^{34}S$ value of +5.4‰ (Kaplan et al., 1969; Shanks and Bischoff, 1980; Zierenberg and Shanks, 1988). Distribution of facies (Bäcker and Richter, 1973), suspended material in the lower brine (Hartmann, 1973), and detailed hydrographic profiles (Schoell, 1976) indicate that the vent for the AM zone is the Southern Basin.

The lithostratigraphy in the Southwest Basin is anomalous owing to the presence of hydrothermal fluid vents and disruption by rift zone volcanism. In contrast to CO zone mineralogy in other basins, the Southwest Basin CO zone contains a greater proportion of hematite.

Bäcker and Richter (1973) describe three special zones with distribution limited to the Southwest Basin. The sulfide-oxide-anhydrite zone (SOAN) is probably correlative with the SU_2 zone. It is characterized by disrupted sequences of hematite, Fe-rich phyllosilicates, sulfides, and anhydrite. Slumping and resedimentation are common features.

Above the SOAN zone is the oxide-anhydrite zone (OAN), which consists of hematite and anhydrite with admixed Fe-rich clays and sulfides. This zone is characterized by breccias and intrabasinal turbidites. Breccias range from heterolithic to homolithic. Breccia fragments range in size from less than 1 cm to at least 15 cm, are angular to rounded, and display sharp to diffuse boundaries with the matrix. Fe silicate and Fe oxide mud fragments predominate, accompanied by occasional sulfide fragments. Varying proportions of hematite, Fe silicate, and sulfide form the matrix and are generally accompanied by anhydrite. In comparing these breccias to similar features in ancient ore deposits, it is important to note that the metalliferous sediment is unlithified and uncompacted and contains more than 60% interstitial brine. Breccia

TABLE 8.2. Chemical composition of sulfide, oxide, and silicate facies of Atlantis II Deep metalliferous sediment (wt% dry salt-free sediment) and massive sulfide from 21° N, East Pacific Rise, and Southern Juan de Fuca Ridge.

			Atlantis II deep		
	21° N[a]	SJFR[b]	Sulfide[c]	Oxide[d]	Silicate[e]
wt%					
SiO_2	7.76	2.0	23.3	8.7[f]	25.7
Al_2O_3	0.4	0.1	4.4	1.1[f]	1.8
CaO	1.8	0.1	3.1	0.9	3.0
MgO	0.03	0.03	1.46	0.6	1.3
Fe	19.2	8.9	18.6	51.3	28.2
Zn	32.3	54.0	10.0	0.2	1.8
Cu	0.81	0.22	1.35	0.2	0.5
Pb	0.32	0.25	0.15[g]	—	—
ppm					
Ag	160	260	160[g]	—	—
Au	0.17	0.13	2[g]	—	—
Cd	600	780	320	—	8
Mn	380	400	1,200	1,100	1,200

[a] Bischoff et al. (1983), average of three basal mound samples.
[b] Bischoff et al. (1983), average of two samples.
[c] Bäcker (1976), average of six samples from sulfidic horizons.
[d] Bäcker (1976), average of 100 samples of the Limonite Facies.
[e] Bäcker (1976), average of five samples of the Silicate Facies.
[f] Bischoff (1969).
[g] Oudin (1987).

formation may result from disruption of sediment by inflowing brine and seismic and/or volcanic disturbance of sediment deposited on steep, graben-bounding fault scarps. Above the OAN zone in the Southwest Basin, the sulfide-amorphous silicate zone (SAM) is still being deposited. It is distinguished from the AM zone by higher base-metal sulfide content.

The metalliferous sediment below the SAM zone in the Southwest Basin is commonly cut by open fissures containing minerals precipitated from hydrothermal fluid up-welling into the brine pool. Hydrothermally precipitated anhydrite forms the vein walls and is the predominant vein-filling mineral. The mineralogy and mineral chemistry of the veins are described in detail by Zierenberg and Shanks (1983) and Pottorf and Barnes (1983).

Precipitation of minerals in the veins is interpreted to result from mixing of a high-temperature, reduced hydrothermal fluid with a cooler, sulfate-bearing fluid similar to the lower brine pool (Pottorf and Barnes, 1983; Zierenberg and Shanks, 1988). Pottorf and Barnes (1983) report the presence of isocubanite and pyrrhotite in vein samples from the Southwest Basin. Their interpretation of coexisting pyrite-pyrrhotite-isocubanite as an equilibrium assemblage, based on textual observations, implies depositional temperatures in excess of 334°C. Fluid inclusion measurements on vein anhydrite (Pottorf and Barnes, 1983; Oudin et al., 1984; Ramboz et al., 1988) support high temperatures for the incoming fluid.

The lithostratigraphic units defined by Hackett and Bischoff (1973) and Bäcker and Richter (1973) can be considered to be mixtures of several different chemical facies of metalliferous sediment (Bischoff, 1969; Bäcker, 1976). Average chemical composition of sulfide, silicate, and oxide facies from the Atlantis II Deep are compared with the composition of hydrothermal sulfides from 21° N, EPR, and southern Juan de Fuca Ridge in Table 8.2. The major difference in the chemical compositions

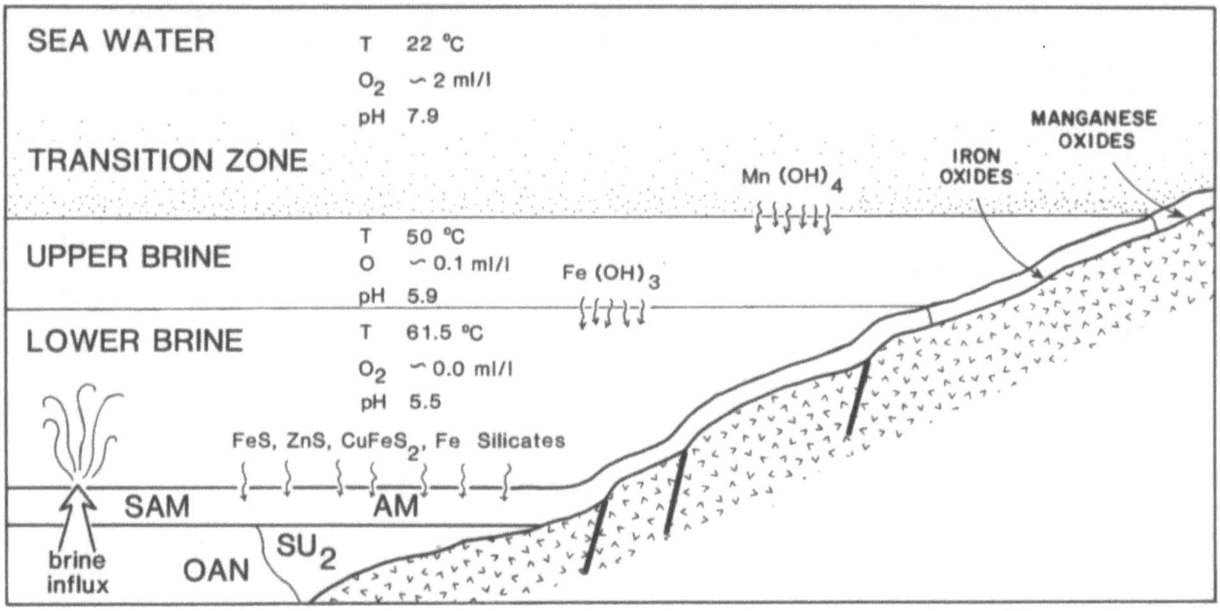

FIGURE 8.5. Schematic representation of precipitation reactions within the density-stratified Atlantis II Deep brine.

results from the large amounts of Fe oxide and Fe silicate that are deposited with the sulfide in the Atlantis II Deep.

Venting of high-salinity brine, with a density higher than that of seawater, into a graben leads to the formation of a submarine brine pool. The unique depositional environment of the brine pool affects the chemistry, sedimentology, and geometric form of the metalliferous deposit. Following formation of the brine pool, there is no direct mixing of the lower, metal-bearing brine with overlying seawater. Both the lower brine layer and the upper brine layer are convectively mixed and homogeneous. However, the density interface between the two brines is sharp, and mixing between the two brines appears to be minor. Heat and dissolved species are transported across the boundary by diffusion. Above the upper boundary of the upper brine is a series of thin, convectively mixed brine layers of increasingly lower temperature, salinity, and density (Fig. 8.3). These graded upward through a diffusive profile into the overlying seawater.

The lower brine is anoxic, and the redox potential is low enough to enable the precipitation of sulfide. Pyrite, sphalerite, and chalcopyrite precipitate in the lower brine layer and are distributed throughout the Deep (Fig. 8.5). In addition to sulfide, the uppermost layers of metalliferous sediment contain Fe-rich clays, amorphous silica, and Fe oxides. Much of the uppermost sediment is X-ray amorphous; more deeply buried sediment layers have well-crystallized, Fe-rich clays. This suggests that much of the clay is formed by diagenesis of amorphous silica and Fe oxides (Weber-Diefenbach, 1977).

The diffsion of Fe from the lower brine to the more oxidizing upper brine results in the precipitation of Fe oxyhydroxides (Fig. 8.5). Particulate Fe oxyhydroxides entering the lower brine should dissolve because the lower brine has a higher temperature and lower fO_2 and pH and is undersaturated with respect to Fe oxide. Resorption is not complete, however, as Fe oxide does occur in the uppermost sediment. Fe oxide that settles out of the upper brine around the edges of the brine pool does not fall into the lower brine; therefore it forms layers enriched in Fe oxides (Fig. 8.5). Similarly, managanese oxides are undersaturated in both the lower and upper brine layers but are precipitated in the overlying mixing zone, where dissolved manganese diffusing upward from the brine encounters dissolved oxygen from overlying seawater. This process of fractional precipitation (Fig. 8.5) of metals is very well illustrated by the sediments in the Chain Deeps (Bäcker and Richter, 1973). The upper brine of the Atlantis II Deep is at an elevation such that it can overflow into the Chain Deeps. The deep closest to the Atlantis II (Chain Deep A) contains a thick sequence of Fe oxide–rich sediments, while Chain Deep B contains layers of nearly pure manganese oxides. Bignell et al. (1976) suggest that the dispersion of manganese might serve as a useful guide for exploration because it forms a broad geochemical halo outside of the area covered by the brine pool.

The presence of the brine pool in the Atlantis II Deep has other effects that distinguish it from ridge crest hydrothermal sites. The high temperature, salinity, and heavy metal content form a sterile environment (Watson and Waterbury, 1969). Not only does this ensure that the finely laminated metalliferous sediments are not disturbed by bioturbation, it also precludes bacterially mediated process such as sulfate reduction or oxidation of previously deposited sulfide. More importantly, the anoxic brine protects the sulfide from seafloor oxidation by overlying seawater.

Deposition under a brine pool thus leads to the formation of a massive sulfide deposit characterized by deposition of a thin (~20 m), blanketlike metal deposit over a large area (~60 km^2). The deposit is finely laminated, very fine grained sediment composed of admixed sulfides, silicates, and oxides. Most of the metal transported by the fluid is deposited within the graben containing the brine pool (Fig. 8.2). This leads to the formation of a large, but lower-grade, deposit due to the dilution of the economically valuable sulfides by silicates and oxides. The anoxic brine pool and the rapid sedimentation rate of metalliferous sediment (Shanks and Bischoff, 1980) greatly increase the chances for preservation of this type of deposit.

In contrast, hot, evolved seawater that vents onto the seafloor at ridge crest deposits, such as at 21° N, EPR, and southern Juan de Fuca Ridge, has a lower density than cold seawater, so it forms a buoyant plume (Fig. 8.2). Massive sulfide deposits form near the vent sites owing to rapid decrease in solubility of base metal with falling temperature as the hydrothermal fluid mixes with cold seawater. The massive sulfide deposits have high metal grades, coarse grain sizes, and high-temperature mineral assemblages. The dissolved constituents of the fluid, including essentially all the silica and Mn and most of the Fe and base metals, are lost from the immediate depositional site and dispersed into the water column by the hydrothermal plume. Because most of the dissolved metal load is dispersed by the buoyant hydrothermal plume, the deposits tend to be small.

On the ridge crests, the sulfides are chemically unstable in their depositional environment and will be quickly oxidized by exposure to seawater when hydrothermal venting ceases. Unless the deposits form in deep graben structures or are rapidly buried by sediment or lava flows, their presence may be marked only by a residual gossan of Fe oxides. The features we see on the ridge crest today are thus ephemeral and are not likely to be preserved as ore deposits.

Sediment-covered ridge crests, such as Guaymas Basin (EPR), Escanaba Trough (Gorda Ridge), and Middle Valley (Juan de Fuca Ridge) are potential sites of formation of large metal deposits (Koski et al., 1985, 1988; Davis et al., 1987). Capping of a hydrothermal system by a sediment blanket can result in dispersion of the hydrothermal fluid through the sediment with increased subsurface mixing, cooling, and precipitation of sulfide. Chemical interaction of the hydrothermal fluid with the sediment can also affect the composition of the resultant deposit (Koski et al., 1988).

In summary, the strikingly different base metal deposits represented by ridge crest hydrothermal systems and the Atlantis II Deep are genetically related owing to the similarity of their tectonic settings and geochemical systems. The Atlantis II and the ridgecrest hydrothermal deposits both occur in extensional environments associated with seafloor spreading and tholeiitic volcanism. Both deposit types form from a hydrothermal fluid that is derived from seawater, and fluid chemistry in both systems is controlled by water-rock interaction with basalts at temperatures around 350°C. However, the presence of evaporites in the hydrothermal circulation system of the Atlantis II Deep leads to formation of a high-salinity brine. The higher density of the brine facilitates formation of a submarine brine pool, which significantly influences the size, grade, and style of mineralization. Most of the differences in the deposits can be related to their different depositional settings.

Economic Geology

The Atlantis II Deep is the only seafloor massive sulfide deposit that has received serious consideration for economic exploitation, as reviewed by Shanks (1983). Extensive core sampling of the deposit by Preussag AG and the Saudi-Sudanese Red Sea Joint Commission provides a data base for evaluation of the size of the deposit (Guney et al., 1988). The total estimated reserves of dry, salt-free metalliferous sediment are approximately 92 million tons (Mt), including 1.9 Mt Zn (average grade 2.1%), 0.4 Mt Cu (0.46%), 3,750 t Ag (41 g/t), 47 t Au (0.51 g/t), and 5,400 t Co (59 g/t) (Guney et al., 1988). The richest part of the deposit is in the Southwest Basin, the site of hydrothermal venting, which contains metalliferous sediment enriched in both base metals and precious metals, relative to the deposit average.

The high interstital water content and lack of lithification of the metalliferous sediments enable them to be pumped from the seafloor as a slurry. Prepilot mining tests have successfully recovered metalliferous sediment, and high-grade ore concentrates have been produced by sulfide floatation at sea. Exploitation of indurated deposits, such as those found on the Juan de Fuca and Gorda Ridges, will require a different technology, and this technology has yet to be developed.

References

Bäcker H (1976) Fazies und chemische Zusammensetzung rezenter Ausfällungen aus Mineralquellen im Roten Meer. Geolog Jahrb 17:151–172.

Bäcker H, Richter H (1973) Die rezente hydrothermalsedimentäre Lagerstätte Atlantis-II-Tief in Roten Meer. Geol Rundsch 62:697–740.

Bäcker H, Schoell M (1972) New deeps with brines and metalliferous sediments in the Red Sea. Nature 240:153–158.

Bäcker H, Lange K, Richter H (1975) Morphology of the Red Sea central graben between Subair Islands and Abul Kizaan. Geol Jahrb 13:79–123.

Bicknell JD, Macdonald KC, Miller SP, Lonsdale PF, Becker K (1986) Tectonics of the Nereus Deep, Red Sea: A deep-tow investigation of a site of initial rifting. Marine Geophys Res 8:131–148.

Bignell R, Cronan D, Tooms J (1976) Metal dispersion in the Red Sea as an aid to marine geochemical exploration. Inst Mining Metallurgy Trans 8:B274–B278.

Bischoff JL (1969) Red Sea geothermal brine deposits: Their mineralogy, chemistry, and genesis. In Degens ET, Ross DA (eds): Hot Brines and Recent Heavy Metal Deposits in the Red Sea. New York: Springer-Verlag, pp 368–401.

Bischoff JL, Rosenbauer RJ, Aruscavage PJ, Baedecker PA, Crock JC (1983) Sea-floor massive sulfide deposits from 21° N, East Pacific Rise; Juan de Fuca Ridge; and Galapagos Rift: Bulk chemical composition and economic implications. Econ Geol 78:1711–1720.

Brewer GB, Spencer DW (1969) A note on the chemical composition of the Red Sea Brines. In Degens ET, Ross DA (eds): Hot Brines and Recent Heavy Metal Deposits in the Red Sea. New York: Springer-Verlag, pp 174–179.

Cochran JR (1983) A model for development of Red Sea. Am Assoc Petroleum Geol Bull 67:41–69.

Craig H (1969) Geochemistry and origin of the Red Sea brines. In Degens ET, Ross DA (ed): Hot Brines and Recent Heavy Metal Deposits in the Red Sea. New York: Springer-Verlag, pp 208–242.

Danielsson L, Dyrssen D, Granéli A (1980) Chemical investigations of Atlantis II and Discovery brines in the Red Sea. Geochim Cosmochim Acta 44:2051–2065.

Davis EE, Goodfellow WD, Bornhold BD, et al. (1987) Massive sulfides in a sedimented rift valley, northern Juan de Fuca Ridge. Earth Planetary Sci Lett 86:49–61.

Delevaux MH, Doe BR (1974) Preliminary report on uranium, thorium and lead contents and lead isotopic composition in sediment samples from the Red Sea. Initial Reports of the Deep Sea Drilling Project, Vol 23, pp. 943–946.

Dietrich G, Krause G (1969) The observations of

the vertical structure of hot salty water by *R.V. Meteor*. In Degens ET, and Ross DA (eds): Hot Brines and Recent Heavy Metal Deposits in the Red Sea. New York: Springer-Verlag, pp 10–14.
Guney M, Al-Marhoun MA, Nawab ZA (1988) Metalliferous submarine sediments of the Atlantis-II-Deep, Red Sea. Can Inst Mining Bull 81:33–39.
Hackett J, Bischoff JL (1973) New data on the stratigraphy, extent, and geological history of the Red Sea geothermal deposits. Econ Geol 68:553–564.
Hartmann M (1973) Untersuchung von suspendiertem Material in den Hydrothermallaugen des Atlantis II-Tiefs. Geol Rundsch 62:742–754.
Hartmann M (1980) Atlantis II Deep geothermal brine system. Hydrographic situation in 1977 and changes since 1965. Deep-Sea Res 27A:161–171.
Hartmann M (1985) Atlantis II Deep geothermal brine system. Chemical processes between hydrothermal brines and Red Sea Deep Water. Marine Geol 64:157–177.
Kaplan IR, Sweeney RE, Nissenbaum A (1969) Sulfur isotope studies on Red Sea geothermal brines and sediments. In Degens ET, Ross DA (eds): Hot Brines and Recent Heavy Metal Deposits in the Red Sea. New York: Springer-Verlag, pp 474–498.
Koski RA, Lonsdale PF, Shanks WC, Berndt ME, Howe SS (1985) Mineralogy and geochemistry of a sediment-hosted hydrothermal sulfide deposit from the southern trough of Guaymas Basin, Gulf of California. J Geophys Res 90:6695–6707.
Koski RA, Shanks WC III, Bohrson WA, Oscarson RL (1988) The composition of massive sulfide deposits from the sediment-covered floor of the Escanaba Trough, Gorda Ridge: Implications for depositional processes. Can Mineral 26:655–673.
Lupton JE, Weiss RF, Craig H (1977) Mantle helium in the Red Sea brines. Nature 266: 244–246.
Nawab ZA (1984) Red Sea mining: A new era. Deep Sea Res 31:813–822.
Oudin E (1987) Trace elements and precious metal concentrations in East Pacific Rise, Cyprus and Red Sea submarine sulfide deposits. In Teleki PG et al. (eds): Marine Minerals. Nordrecht: D. Reidel Publishing Company, pp. 349–362.
Oudin E, Thisse Y, Ramboz C (1984) Fluid inclusion and mineralogical evidence for high-temperature saline hydrothermal circulation in the Red Sea metalliferous sediments: Preliminary results. Marine Mining 5:3–31.
Pautot G, Guennoc P, Contelle A, Lyberis N (1984) Discovery of a large brine deep in the northern Red Sea. Nature 310:133–136.
Pottorf R, Barnes HL (1983) Mineralogy, geochemistry, and ore genesis of hydrothermal sediments from the Atlantis II Deep, Red Sea. Econ Geol Monogr 5:198–223.
Ramboz C, Oudin E, Thisse Y (1988) Geyser-type discharge in Atlantis II Deep, Red Sea: Evidence of boiling from fluid inclusions in epigenetic anhydrite. Can Mineral 26:765–786.
Schmalz RF (1969) Deep-water evaporite deposition: A genetic model. Am Assoc Petroleum Geol Bull 53:798–823.
Schoell M (1976) Heating and convection within the Atlantis II Deep geothermal system of the Red Sea. In Proceedings, Second United Nations Symposium on the Development and Use of Geothermal Resources, Vol 1, pp 583–590.
Schoell M, Faber E (1978) New isotopic evidence for the origin of Red Sea brines. Nature 275:436–438.
Schoell M, Hartmann M (1978) Changing hydrothermal activity in the Atlantis II Deep geothermal system. Nature 274:784–785.
Shanks WC III (1983) Economic and exploration significance of Red Sea metalliferous brine deposits. In Shanks WC III (ed): Unconventional Mineral Deposits. New York: American Institute of Mining Engineers, pp 157–171.
Shanks WC III, Bischoff JL (1977) Ore transport and deposition in the Red Sea geothermal system: A geochemical model. Geochim Cosmochim Acta 41:1507–1519.
Shanks WC III, Bischoff JL (1980) Geochemistry, sulfur isotope composition and accumulation rates of the Red Sea geothermal deposits. Econ Geol 75:445–459.
Stoffers P, Kühn R (1974) Red Sea evaporites: A petrographic and geochemical study. Initial Reports of the Deep Sea Drilling Project, Vol 23, pp. 821–847.
Swallow JC (1969) History of exploration of the hot brine area of the Red Sea: *Discovery* account. In Degens ET, Ross DA Degens (eds): Hot Brines and Recent Heavy Metal Deposits in the Red Sea. New York: Springer-Verlag, pp 3–9.
Von Damm KL, Edmond JM Grant B, Measures CI, Walden B, Weiss RF (1985) Chemistry of submarine hydrothermal solutions at 21°N, East Pacific Rise. Geochim Cosmochim Acta 49:2197–2221.
Von Damm KL, Bischoff JL (1987) Chemistry of hydrothermal solutions from the southern Juan de Fuca Ridge. Geophys Res 92:11,334–11,346.
Watson SW, Waterbury JB (1969) The sterile hot brines of the Red Sea. In Degens ET, Ross DA (eds): Hot Brines and Recent Heavy Metal De-

posits in the Red Sea. New York: Springer-Verlag, pp 272–281.

Weber-Diefenbach K (1977) Geochemistry and diagenesis of recent heavy metal ore deposits at the Atlantis II Deep (Red Sea). In Klemm DD, Schneider HJ (eds): Time and Stratabound Ore Deposits. New York: Springer-Verlag, pp 419–436.

Zierenberg RA, Shanks WC III (1983) Mineralogy and geochemistry of epigenetic features in metalliferous sediment, Atlantis II Deep, Red Sea. Econ Geol 78:57–72.

Zierenberg RA, Shanks WC III (1986) Isotopic constraints on the origin of the Atlantis II, Suakin and Valdivia brines, Red Sea. Geochim Cosmochim Acta 50:2205–2214.

Zierenberg RA, Shanks WC III (1988) Isotopic studies of epigenetic features in metalliferous sediment, Atlantis II Deep, Red Sea. Can Mineral 26:737–753.

9—Economic Perspective on Development of Potential Gorda Ridge Sulfide Deposits

JOSEPH L. RITCHEY, NICHOLAS WETZEL, SCOTT A. STEBBINS, and BENJAMIN W. HAYNES

Abstract

The Turner-Albright massive sulfide deposit is a well-studied example of a polymetallic ore body formed at a site of seafloor spreading in the geologic past. Construction and evaluation of a hypothetical mining and processing scenario based on this deposit, including cost estimates and financial analyses, identified key economic parameters that provide an economic perspective for future development of potential massive sulfide deposits on Gorda Ridge. Sensitivity analyses of cost parameters indicate that profitability is most sensitive to increases in operating costs. Both capital and operating costs are expected to be substantially higher for seafloor mining operations, thus requiring increased revenues to reach economic viability. Consequently, near-term development of massive sulfide deposits on or beneath the seafloor appears unlikely. New developments in exploration and mining technology are needed to locate, delineate, and develop seafloor deposits before an accurate quantitative assessment of economic viability can be made.

Introduction

The Gorda Ridge Technical Task Force requested the U.S. Bureau of Mines to provide an economic perspective on future development of potential massive sulfide deposits associated with Gorda Ridge spreading center. An analysis of the economic viability of a proposed mining operation requires detailed knowledge of the ore deposit to be mined, as well as the mining and processing methods to be utilized. At the present time, no ore deposit has been delineated on Gorda Ridge, and techniques for mining consolidated deposits on the seafloor are in the hypothetical stage of development. Some economic perspective on future seafloor mineral production can be gained by identifying key parameters affecting the economic viability of a well-known, dryland, massive sulfide deposit, which formed under conditions similar to those associated with Gorda Ridge.

Massive sulfide deposits associated with ocean spreading centers are products of the interaction between hydrothermal fluids and the rocks underlying the seafloor. Depending on the tectonic setting, the rock types involved, and the locus of sulfide deposition, a broad range of deposit types is possible (Scott, 1985). Ancient analogs spanning this range of deposit types have been discovered and mined on the present-day continental land masses. Several volcanogenic sulfide deposits in the Klamath Mountains of southern Oregon are representative of hydrothermal processes active on the seafloor during the Mesozoic Period. Of these, the Turner-Albright deposit was selected as a case study of economic parameters for the following reasons: (1) the deposit was formed within mafic volcanic rocks at a former site of seafloor spreading (Koski and Derkey, 1981), analogous to Gorda Ridge: (2) extensive data are available on the size and character of the deposit; and (3) the deposit is enriched in metals of strategic and economic significance.

Past attempts to economically exploit onshore polymetallic deposits of the Turner-

Albright type using traditional processing methods have proved unsuccessful, owing to the complexity of the ore mineralogy. The recently developed oxidative pressure leaching process (Engineering and Mining Journal, 1981) has permitted a fundamentally different approach to metal recovery. This new process shows promise for solving metallurgical problems associated with onshore polymetallic sulfide deposits, thereby enhancing the potential for economic recovery of metals from these deposits, and may be applicable to recovery of metals from present-day seafloor analogs in the Exclusive Economic Zone (EEZ).

The engineering and economic analysis of a hypothetical mining and processing scenario for the Turner-Albright deposit is based on a very large resource data base and was performed following the guidelines of the Minerals Availability Program's (MAP) Cost Estimating System (CES) (U.S. Bureau of Mines, 1987a,b). The CES has been used as a generic analytical tool on hundreds of mineral deposit evaluations, and the results have been published in numerous Bureau reports and articles. The accuracy of the economic and financial analyses by CES is about ±25%.

TABLE 9.1. Comparative assay data for polymetallic sulfide deposits in the Klamath Mountains Province.

	Assay, wt %				
Deposit	Cu	Zn	Pb	Co	Au, tr oz/st
Turner Albright					
(UHZ)	2.24	1.58	ND	0.09	0.25
(MUZ)	0.32	0.87	ND	0.07	0.05
(MLZ)	0.80	2.13	ND	0.03	0.05
Grey Eagle	2.34	0.17	ND	0.10	0.03
	9.90	0.13	Tr	0.10	0.21
	1.48	0.18	ND	0.05	0.04
Blue Ledge	8.43	15.80	0.30	Tr	0.21
	1.88	2.55	0.15	Tr	0.02
	0.59	12.54	0.40	Tr	0.09
Queen of Bronze	0.32	0.46	Tr	0.02	Tr
	12.70	0.14	Tr	0.15	0.58
	0.23	0.02	Tr	0.01	0.09
Copper Queen	3.20	1.00	0.03	0.02	Tr
	3.20	0.80	Tr	0.01	Tr
Other Takilma deposits	2.00	0.20	0.03	0.03	0.02
	2.00	0.10	0.07	0.15	0.02
	2.00	0.07	Tr	0.02	0.04

ND, not determined. Tr, trace.
Source: R.A. Koski, U.S. Geological Survey, Menlo Park, CA, written communication, and Bureau of Mines assay data.

Turner-Albright Deposit

The Turner-Albright deposit is located astride the Oregon-California border, about 64 km southwest of Grant's Pass, Oregon, and is one of several massive sulfide deposits distributed throughout the Klamath Mountains province (Koski and Derkey, 1981). Average metal concentrations reported for deposits in the Klamath Mountains are shown in Table 9.1. All these deposits appear to be similar in tonnage and grade, based on present estimates; as exploration continues, the total resources should be better defined.

Extensive drilling, in excess of 75,000 total feet in 80 separate drill holes, has yielded a large body of data on the resource size and grade, as well as the geological and structural characteristics, of the Turner-Albright deposit (Strickler, 1986). The sulfide ore occurs near the base of extrusive pillow lavas and flows of the Josephine Ophiolite. The deposit comprises three distinct zones: (1) the Upper High Grade Zone (UHZ), (2) the Main Upper Zone (MUZ), and (3) the Main Lower Zone (MLZ). The UHZ is enriched in contained metals and is small in size relative to the other zones (Table 9.2) but appears to be laterally extensive and to possess textures suggestive of collapsed chimney structure constructed on the paleo sea floor (Strickler, 1986).

Average grades for the Turner-Albright ore zones (Table 9.2) are based on more than 18,000 analyses of drill core, and the reported tonnages are drill-indicated reserves. These data indicate total recoverable resources of about 450,000 troy ounces (tr oz) gold, 2 million tr oz silver, 8 million lb cobalt, 90 million lb copper, 200 million lb zinc, and 3 million metric tons (mt) sulfur. Geophysical surveys indicate that additional resources may exist at greater depth, and lower-grade sulfides (pri-

TABLE 9.2. Average grades and reserves of mineral-rich zones in the Turner-Albright deposit, southern Oregon. (See text for explanation of zones.)

	UHZ	MUZ	MLZ	Total
Grade				
Cu, wt %	2.24	0.32	0.80	0.51
Zn, wt %	1.58	0.87	2.13	1.45
Co, wt %	0.09	0.07	0.03	0.05
Ag, tr oz/st	1.06	0.10	0.33	0.22
Au, tr oz/st	0.25	0.05	0.05	0.05
Tonnage, 10^3 mt	119	4,142	3,544	7,805

marily cobaltiferous pyrite) are known to be present north of the principal ore body.

Figure 9.1 is a possible reconstruction of the geological environment in which the Turner-Albright deposit was formed. The exhalative zone, representing seafloor deposition, is small compared with the zone of massive sulfides deposited by circulating waters in basaltic rubble below the seafloor. If this exhalative zone is equivalent to UHZ in the present deposit, then sulfide deposits exposed on the seafloor at slow-spreading ridges, such as Gorda Ridge, may represent as little as 2% of an underlying deposit. By analogy with the Turner-Albright deposit, basalt-hosted massive sulfide deposits on the order of 8 million to 10 million mt, with average grades of about 1.5% zinc, 0.5% copper, 0.05% cobalt, 0.2 tr oz/short ton (st) silver, and 0.05 tr oz/st gold, may underlie Gorda Ridge.

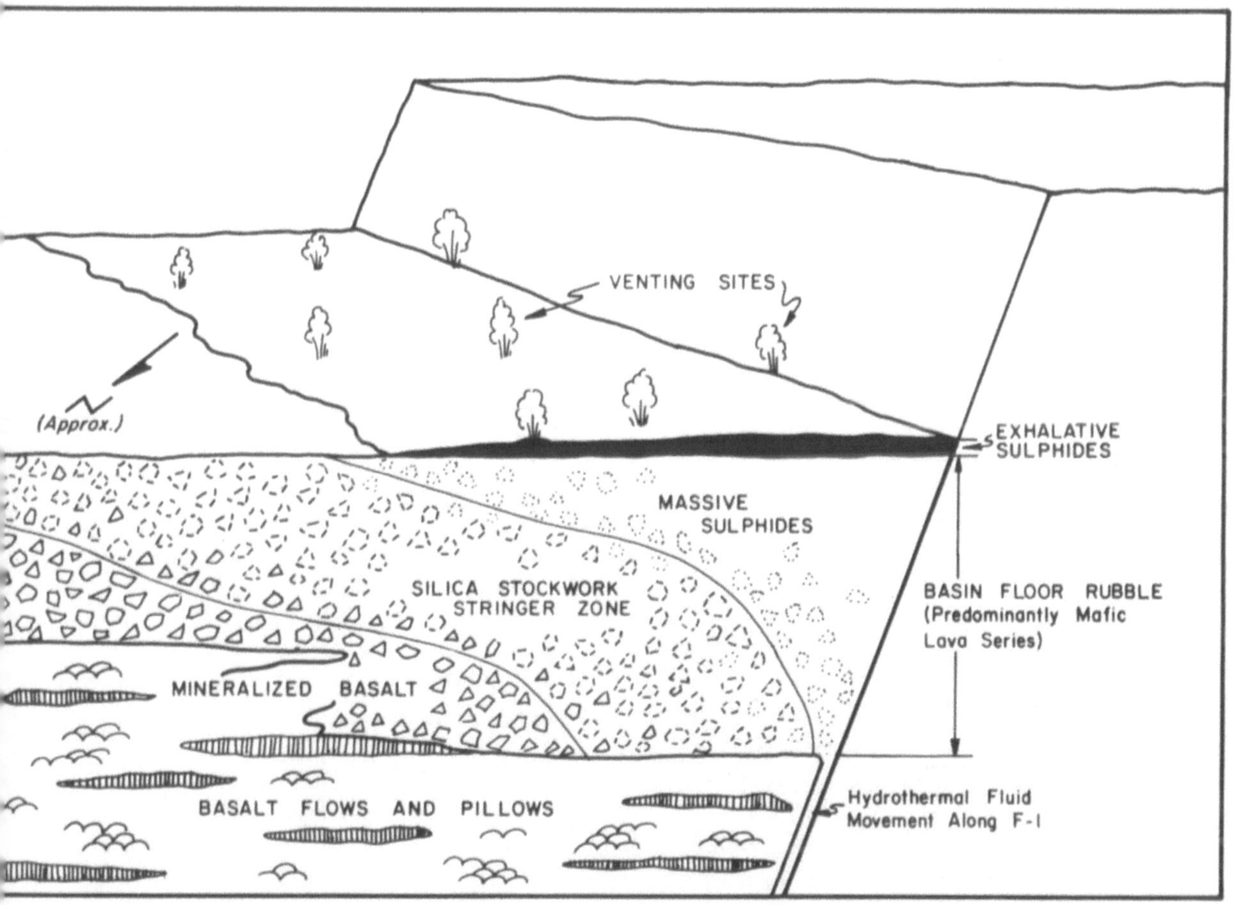

FIGURE 9.1. Stratigraphic cross section during the late stages of the development of the Turner-Albright deposit (after Strickler, 1986).

Hypothetical Development Plan

To identify economic parameters significant to future development of seafloor massive sulfide deposits that may underlie Gorda Ridge, a hypothetical mining and processing scenario has been prepared for an analogous onshore deposit, based on the Turner-Albright resource data. The scenario assumes that about two thirds of the onshore deposit will be mined from the surface and the remaining one third by underground methods. Grades and tonnages were adjusted to reflect anticipated mine recovery and ore dilution. A schematic diagram of the integrated operation is shown in Figure 9.2.

Mining

Assuming that the surface and underground mines are developed concurrently, demonstrated reserves would be sufficient for a 13-year life for both operations. Each mine would be operated on a two-shift/day, 260-day/year schedule. The open-pit mine would yield 1,380 mt/day ore (plus 2,910 mt/day waste at a 3:1 stripping ratio); the ore would be hauled an average of 3,250 m to the mill site stockpile. The underground mine, using a combination of end-slice and vertical crater retreat (VCR) mining methods, would yield 639 mt/day, which would be hauled by truck through a 1,525-m decline to the surface and the mill stockpile.

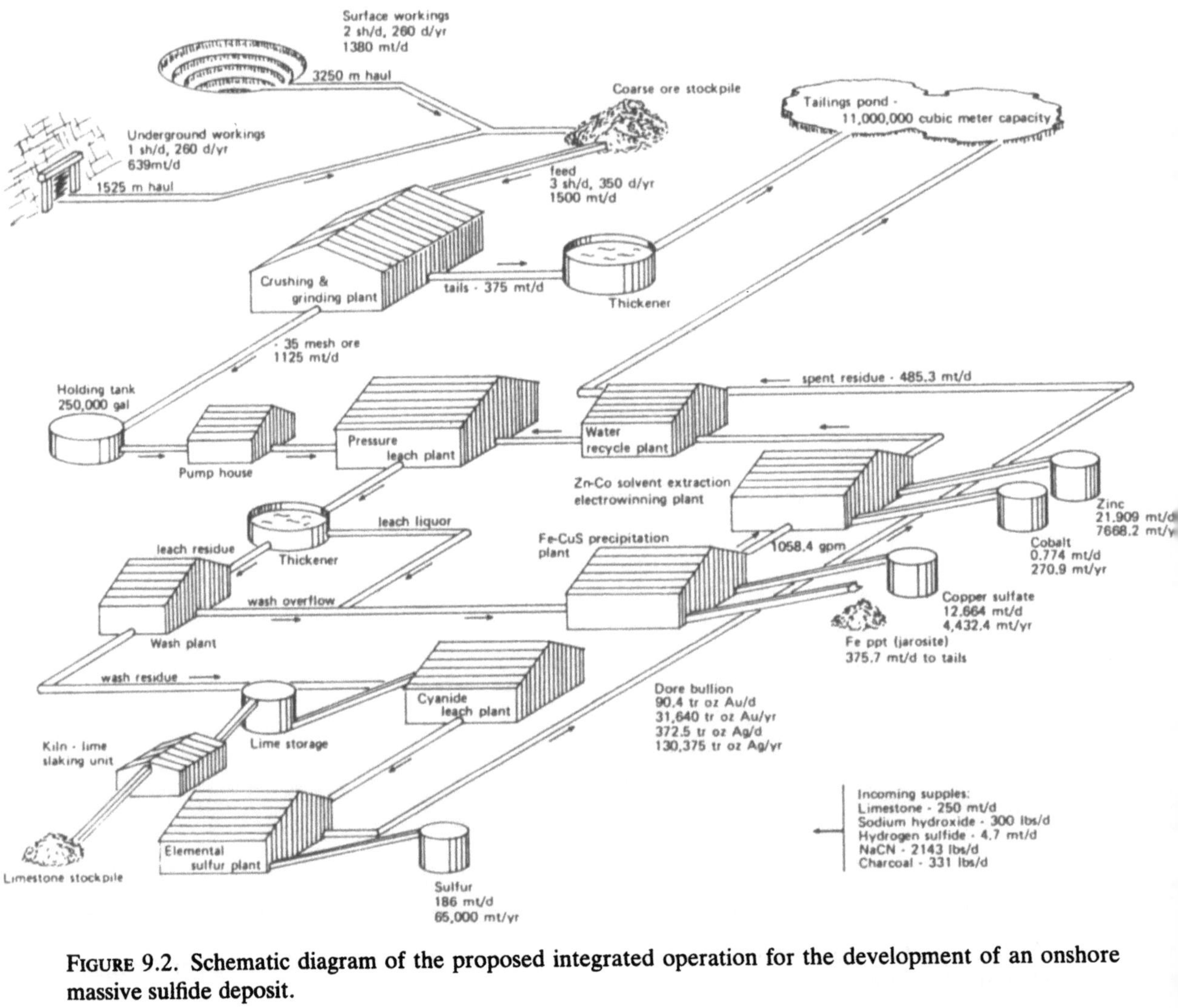

FIGURE 9.2. Schematic diagram of the proposed integrated operation for the development of an onshore massive sulfide deposit.

Processing

The complex intergrowth of fine sulfide grains in the ore does not allow separation of the sulfides into suitable copper and zinc concentrates using standard flotation techniques. However, pressure leaching tests of ore samples from the Turner-Albright deposit recovered more than 92% of the cobalt, copper, and zinc contained in the ore. Therefore, hydrometallurgical processes are assumed to be most suitable for the development of these types of deposits.

Oxidative pressure leaching of polymetallic sulfides requires crushing and grinding of the ore to a minus-35-mesh grain size. The scenario employs a 1,500 mt/day crushing, grinding, and classifying plant to prepare feed to the leach plant, while eliminating approximately 25% of the mine run as waste rock. The ground ore concentrate would be slurried at 35% solids and delivered to the autoclaves. Leaching conditions would include a temperature of 200°C, an oxygen overpressure of 200 psi, and a duration of approximately 2 hours. The leach process is expected to dissolve about 95% of the cobalt, copper, and zinc and about 30% of the iron contained in the ore concentrate. Cobalt, copper, and zinc sulfides react with oxygen and sulfuric acid in the autoclave to form dissolved metal sulfates, elemental sulfur, and water. The leachate solution leaving the autoclave is expected to have a pH of about 0.2 and would be physically separated from the solid residue.

Elemental sulfur would be recovered from the leach residue, and the leachate would be treated to recover cobalt, copper, and zinc through solvent extraction in separate plants. Gold and silver would be recovered from the leach residue in a carbon-in-pulp cyanide leach plant. The proposed processing scenario also calls for an on-site limestone calcining and lime slaking plant to provide the large volume of lime required in the pressure leach and cyanide leach plants. Over the proposed 13-year life of operations, the deposit could produce each year a total of 600,000 lb cobalt, 6 million lb copper, 130,000 tr oz silver, 31,600 tr oz gold, 69,000 mt sulfur, and 16.9 million lb zinc.

Mining, Processing, Infrastructure Development Costs*

The total capital investment for these facilities is estimated to be $72 million in July 1986 dollars. Annual operating cost estimates are $7.6 million for surface and underground mining ($11.79/mt and $20.40/mt, respectively) and $12.7 million for processing ($24.15/mt). With revenues computed using market prices of $390/tr oz gold, $5.50/tr oz silver, $0.60/lb copper, $0.45/lb zinc, $11.50/lb cobalt, and $140/mt sulfur, it is estimated that a positive discounted cash flow rate of return (ROR) on investment could be achieved.

Economic Parameters

Several assumptions concerning mining and processing procedures are inherent in an evaluation of this type. For example, some of the important assumptions affecting costs are (1) the ability to physically concentrate, solely by size classification, the feed to the pressure leach circuit; (2) the ability to leach at 30% solids; and (3) the rates of reagent consumption. With variation in these parameters, significant changes would be expected in capital and operating costs.

Sensitivity analyses were performed to investigate the effects on expected ROR of variations in key cost parameters. Expected ROR is relatively insensitive to changes in capital costs (Fig. 9.3). If capital expenses were increased by 100% over the base case (100% anticipated cost in Fig. 9.3), the expected ROR would be reduced by a little more than one half. On the other hand, variations in operating costs will significantly affect the ROR, reducing it to zero with an increase in costs of 76% (Fig. 9.4). At a zero ROR, operating costs would be met, and revenues and invested capital would be recovered over the life of the mining operation, but no profit would be realized.

*The economic data presented in this report have been developed for analytic and illustrative purposes only and should not be considered for investment decisions.

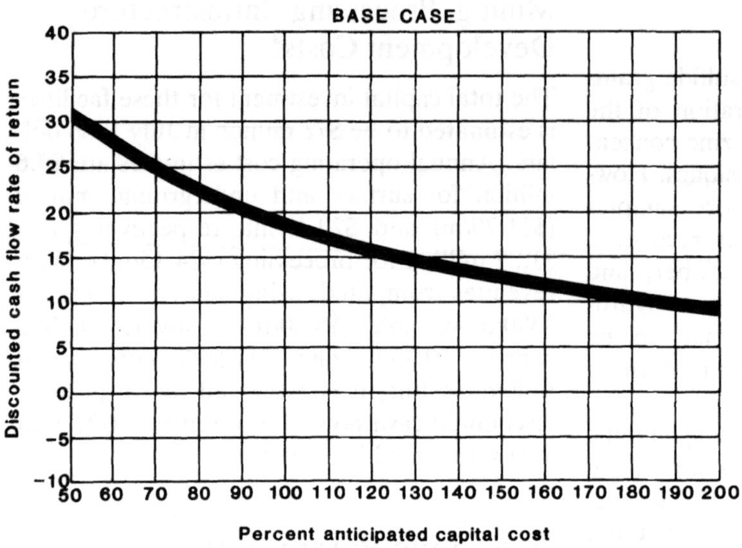

FIGURE 9.3. Diagram illustrating the effect of capital costs on DCFROR.

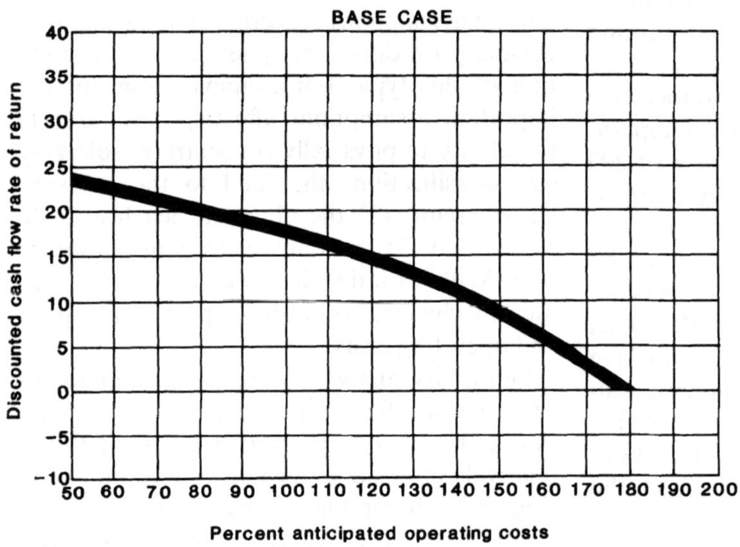

FIGURE 9.4. Diagram illustrating the effect of operating costs on DCFROR.

The anticipated metal recoveries, based on a single leach test, are 92% recovery of cobalt, copper, gold, and zinc. If the gold recovery is lowered to 50%, the ROR decreases by 22 relative percent (Fig. 9.5). Lowering the assumed cobalt, copper, and zinc recoveries to 70% leads to a similar decrease in ROR (Fig. 9.6). If actual revenues received are 63% of those anticipated (e.g., falling metal prices), the expected ROR would decrease to zero (Fig. 9.7).

Discussion

Using the Turner-Albright case study as a basis for comparison, the significant cost parameters affecting economic viability of future seafloor mining operations can be identified and qualitatively evaluated. Sensitivity analyses of cost parameters for the Turner-Albright development scenario indicate that variations in operating costs have a significantly greater impact on profitability than increases in capital costs.

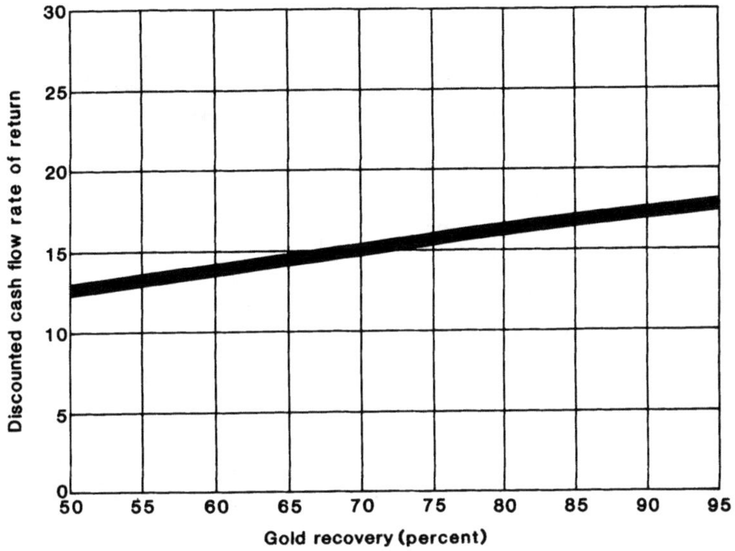

FIGURE 9.5. Diagram illustrating the effect of gold recoveries on DCFROR.

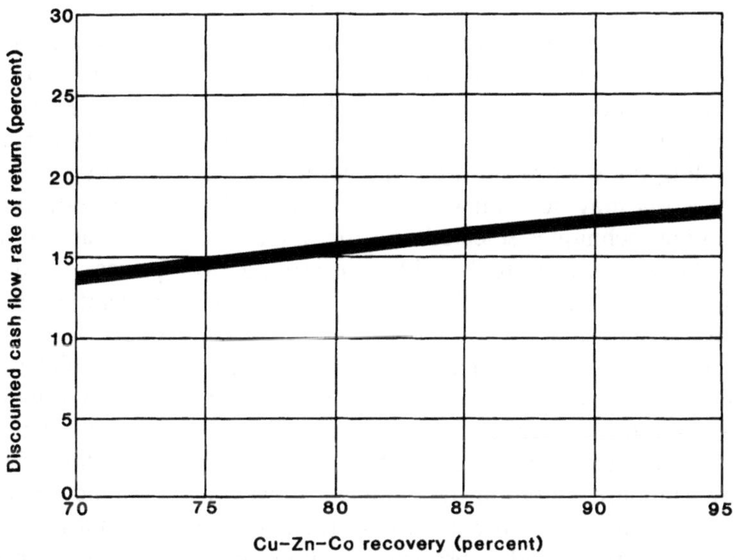

FIGURE 9.6. Diagram illustrating the effect of Cu-Zn-Co recoveries on DCFOR.

Capital expenditures for seafloor mining operations are expected to exceed, at least initially, the range of sensitivity analyses for the hypothetical onshore scenario. Preproduction expenditures will include construction or acquisition of oceangoing ships and the development of new mining technology. The discovery of several ore bodies similar to the Turner-Albright deposit, or the development of a much larger seafloor deposit, would allow the initial capital costs to be prorated over a longer period of time than the 13-year mine life used in the scenario, and thereby lessen the negative impact on profitability.

Operating costs for seafloor mining are also expected to be significantly higher than those for a comparable onshore mine. These higher costs are largely attributable to ore extraction at a water depth on the order of 3,000 m, reduced operating days because of inclement weather and high sea states, and the longer haulage distances to onshore processing plants.

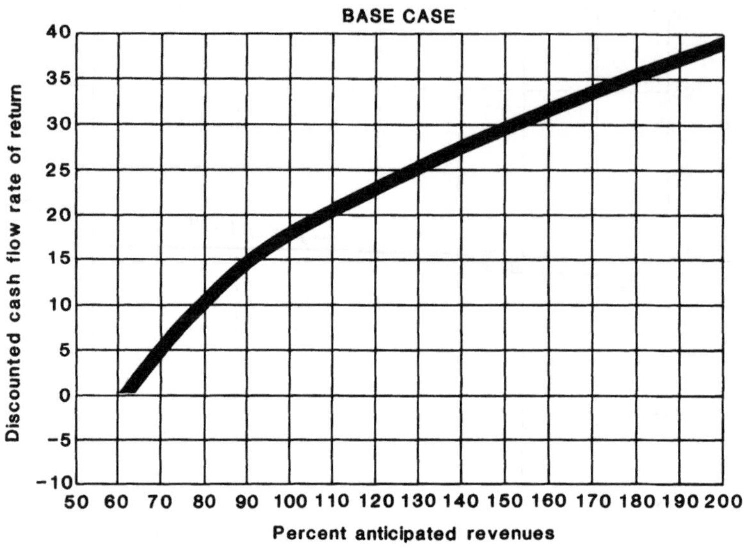

FIGURE 9.7. Diagram illustrating the effect of metal revenues on DCFROR.

On the other hand, if the ore can be recovered from the surface of the seafloor without the need for stripping of overburden, high-energy disaggregation of host rock, or extensive underground extraction, the increased operating costs may be somewhat ameliorated. The mining techniques selected and the associated costs will undoubtedly differ according to deposit type (e.g., basalt-hosted vs. sediment-hosted deposits).

Costs in the scenario associated with processing are not expected to differ greatly for ore recovered from seafloor deposits. The Turner-Albright ore is not amenable to concentration by standard flotation techniques; however, more than 92% of the principal metals in the ore were recovered in pressure leaching tests. Ore recovered from offshore basalt-hosted deposits should also be susceptible to pressure leaching. In addition, there is evidence that metals from sediment-hosted deposits may be recovered most efficiently by the pressure leaching process. Sulfide samples recovered in 1986 from Escanaba Trough at the southern end of Gorda Ridge (Table 9.3) are associated with turbiditic sediments that have interacted with the hydrothermal fluids and provided loci for sulfide deposition (Koski et al., 1988). Initial bench tests performed by the Bureau's Salt Lake City Research Center on two dredged sulfide samples provided by the U.S. Geological Survey were unable to concentrate metal contents by standard flotation techniques (W.C. Hirt, personal communication). These results suggest that a leaching process may be the preferred processing method.

In addition to capital and operating costs, the grade and composition of the ore and the prices of marketable commodities will significantly affect the profitability of seafloor operations. Assuming available markets, ore grades higher than the base case or an increase in the number of recoverable commodities will have a positive impact on profitability. Future escalation of market prices, owing to increased demand or supply shortages, would increase revenues and serve to offset at least some of the higher costs of seafloor mineral production.

Conclusions

The Bureau's analysis of a hypothetical mining scenario for an onshore massive sulfide deposit indicates that cobalt, copper, gold, silver, sulfur, and zinc could be recovered under certain conditions at 1986 commodity prices with a positive ROR. Pressure leaching technologies appear to be suitable for recovery of metals from polymetallic sulfide deposits like the Turner-Albright in Klamath Mountain

TABLE 9.3. Bulk compositions of sulfide samples[a] recovered in 1986 from Escanaba Trough, Gorda Ridge.

	658-R3-A (Pyrrhotite)	659-R2-5 (Pyrrhotite)	662-R1-B (Pyrrhotite)	663-R1-1 (Marcasite)	L286NC14D-1 (Sphalerite, pyrrhotite)	L186NC24D (Pyrrhotite)
Element, wt %						
As	0.003	0.30	0.04	0.04	1.7	ND
Co	0.02	0.01	0.15	0.01	ND	ND
Cu	1.4	1.6	2.4	2.1	1.8	1.2
Fe	53.2	53.7	62.0	46.2	33.1	42.0
Pb	0.09	0.60	0.36	0.16	7.6	0.2
Zn	3.1	3.7	1.8	0.1	20.3	0.34
Element, ppm						
Ag	24	103	81	18	681	13.7
Cd	100	190	40	<5	630	4.4
Mn	110	160	<50	100	ND	ND
Se	50	70	30	80	ND	ND

ND, not determined. ( ), principal sulfide mineral.
[a]Samples 658-R3-A through 663-R1-1 were collected by DSV *Sea Cliff* and analyzed by the Bureau of Mines; samples L286 and L186 were dredged by RV *S.P. Lee* and analyzed by the U.S. Geological Survey, with splits provided to the Bureau for metallurgical testing, courtesy of R.A. Koski.

ophiolite settings, and are likely to be the preferred processing method for potential sulfide deposits on Gorda Ridge. Reconstruction of the Turner-Albright depositional environment suggests that deposits of similar size (8 million to 10 million mt) may be found associated with spreading centers like Gorda Ridge as exploration offshore continues. Since technology for offshore mining of these sulfide deposits has yet to be developed, estimates of the economic viability for future mineral production cannot be determined at this time. Sensitivity analyses performed in this study indicate that increases in capital expense would have less economic impact on viability than higher operating costs. Clearly, both operating and capital costs for seafloor mining would be higher than those onshore. Increased revenues resulting from higher ore grades, better economies of scale, and/or escalated metal prices would have the effect of offsetting these higher costs. Considering the present state of technology and world economic conditions, development of seafloor sulfide deposits would appear to be a relatively long-term endeavor. The greatest need at the present time is for the development and application of new exploration and mining technology to locate, delineate, and develop mineral deposits on and beneath the seafloor.

Acknowledgments. The authors extend thanks to William R. Normark and Randolph A. Koski, U.S. Geological Survey, Menlo Park, California, for providing sulfide samples from Escanaba Trough. William C. Hirt, Bureau of Mines Salt Lake Research Center, provided the results of beneficiation tests on the Escanaba Trough samples. The paper benefited greatly from critical reviews by Greg McMurray and two anonymous reviewers.

References

Engineering and Mining Journal (1981) Sherritt commercializes zinc pressure leaching. Eng Mining J 182:76–79.

Koski RA, Derkey RE (1981) Massive sulfide deposits in oceanic-crust and island-arc terranes of southwestern Oregon. Ore Geol 43:119–125.

Koski RA, Shanks WC III, Bohrson WA, Oscarson RL (1988) The composition of massive sulfide deposits from the sediment-covered floor of Escanaba Trough, Gorda Ridge: Implications for depositional processes. Can Mineral 26:655–673.

Scott SD (1985) Seafloor polymetallic sulfide deposits: Modern and ancient. Marine Mining 5: 191–212.

Strickler MD (1986) Geologic setting of the Turner-Albright massive sulfide deposit, Josephine County, Oregon. Ore Geol 48:115–122.

U.S. Bureau of Mines (1987a) Cost Estimating System Handbook, Part 1. Surface and Underground Mining. Information Circular 9142, 631 pp.

U.S. Bureau of Mines (1987b) Cost Estimating System Handbook, Part 2. Mineral Processing. Information Circular 9143, 566 pp.

Part III Technology

10—Hydrothermal Plume Prospecting: Hydrographic and Geochemical Techniques

EDWARD T. BAKER

Abstract

Hydrothermal plumes, formed by the mixing of hot vent fluids and ambient seawater, can be used to locate, characterize, and quantify sources of seafloor hydrothermal emissions. Vent fluids typically undergo a 10^4-fold dilution as they rise several hundred meters above the sea floor and form neutrally buoyant plumes with heat and chemical anomalies that stretch tens to thousands of kilometers downcurrent of their source. Real-time mapping of these plumes by towing sensitive hydrographic and optical sensors from a surface ship can efficiently locate the plume source, guide discrete chemical sampling of the diluted hydrothermal fluids, and estimate the heat and mass flux of individual vent fields. This paper uses case histories of investigations along the Juan de Fuca and Gorda Ridges to describe strategies for mapping and characterizing hydrothermal plumes at spatial scales ranging from hundreds of meters to hundreds of kilometes.

Introduction

"Black smokers"—sulfide chimneys issuing 300 to 400°C chemical-laden fluids—are the most familiar and spectacular features of sea floor hydrothermal systems. Active chimneys cover only a tiny percentage of the midocean ridge-crest sea floor, however, so finding hydrothermal sites can be a formidable problem. Most commonly, vent sites have been located by the tedious process of slowly towing a deep-tow package such as ANGUS (Spiess and Tyce, 1973) a few meters above the sea floor and monitoring the bottom imagery for direct (vents themselves) or indirect (biological, sedimentological, or hydrographic evidence) signs of hydrothermal activity. A more efficient technique, however, takes advantage of the fact that upon mixing with seawater the hydrothermal fluids are rapidly diluted by a factor of about 10^4 (Lupton et al., 1985), eventually rising to a level of neutral buoyancy and spreading laterally as a distinct layer with anomalous hydrographic and chemical properties and with a spatial scale of tens to thousands of square kilometers (Lupton and Craig, 1981; Reid, 1982; Baker and Massoth, 1986a, 1987).

Plumes were initially used simply as an indication of nearby hydrothermal venting. More recent work has demonstrated that plume studies can address fundamental questions regarding the source strength and chemical composition of seafloor hydrothermal systems (Lupton et al., 1985; Baker and Massoth, 1986a, 1987), the dispersal of vent-specific biological populations (Cowen et al., 1986), and the dispersal of hydrothermal constituents throughout the global ocean (Lupton and Craig, 1981; Reid, 1982). Hydrothermal plumes can serve as Lagrangian tracers of the flow field at ridge-crest depths, and may even dynamically influence currents in the deep sea (Stommel, 1982).

The present chapter uses case histories of hydrothermal investigations along the Juan de Fuca and Gorda Ridges in the northeast Pacific Ocean (Fig. 10.1) to illustrate the hydrographic and geochemical techniques that can be used

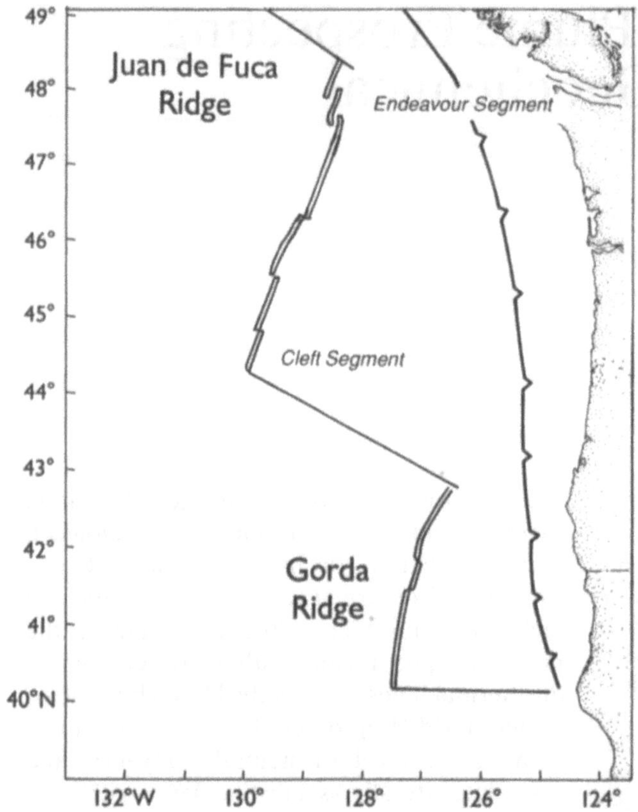

FIGURE 10.1. Location of the Juan de Fuca and Gorda spreading centers in the northeast Pacific Ocean. Location of the Cleft and Endeavour Segments is shown on the Juan de Fuca Ridge.

to locate and interpret hydrothermal plumes. The paper begins with a short discussion of sampling techniques and the merits of various kinds of tracers. Prospecting strategies are then examined on the scale of a ridge system, a single tectonic segment, and a single vent field.

Methods and Data Gathering

Buoyant hydrothermal fluids rise through and mix with ambient seawater until they reach an equilibrium density level and spread laterally as a discrete plume layer. In oceanic environments such as the deep Pacific, where the salinity field is stably stratified, hydrothermal plumes can be identified by positive deviations from the normally linear relationship between potential density and potential temperature or salinity (Weiss et al., 1977; Lupton et al., 1985) (Fig. 10.2). The method is very sensitive and can detect hydrothermal temperature anomalies <0.01°C if a high-resolution CTD (conductivity, temperature, depth probe) is used. Temperature anomalies calculated in this way can be used to estimate the thermal power of the underlying vent source (Baker and Massoth, 1986a, 1987), although the results will generally be a minimum estimate because they do not account for "negative heat" added to the plume by colder, near-bottom seawater entrained and carried upward in the buoyant plume (Speer, 1988). More accurate thermal power calculations depended crucially on the entrainment rate as a function of plume height, a rate that is poorly known. Deducing hydrothermal heat flux from hydrographic measurements is even more difficult where the salinity gradient is unstable, as in the deep Atlantic, because in that case the plume is expressed as a temperature and salinity minimum (Speer and Rona, 1987).

The optical signature of a hydrothermal plume, caused by mineral preciptation within the vented fluids, can also be a sensitive hydrothermal indicator (Fig. 10.2). Because op-

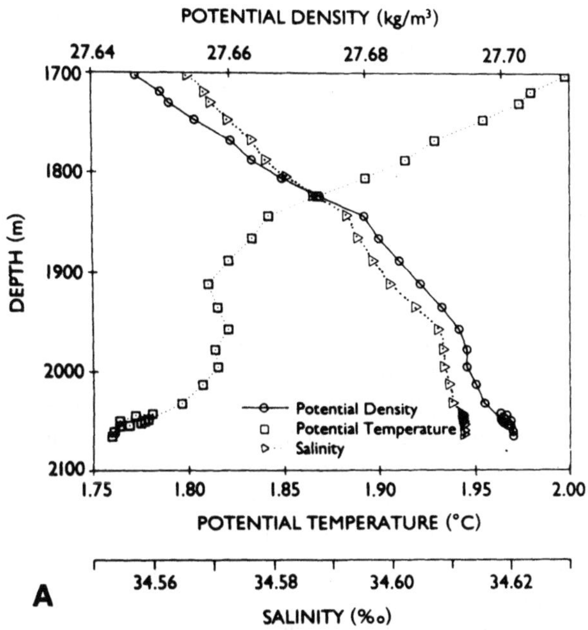

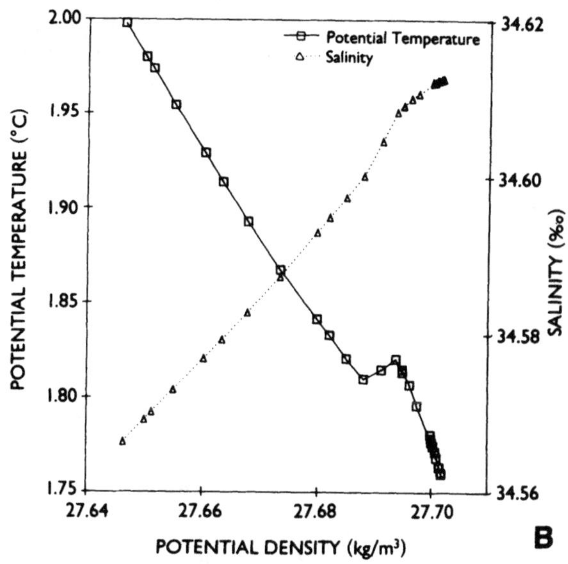

FIGURE 10.2. (A) Vertical profiles of potential temperature, salinity, and potential density through a hydrothermal plume centered between 1,900 and 2,100 m. (B) Potential temperature and salinity plotted against potential density. These plots are linear in the deep Pacific except where modified by the hydrothermal addition of salt (from the entrainment of higher salinity bottom water up into the plume horizon) and heat. (C) Replotting the temperature data as the temperature anomaly, the deviation along the potential temperature axis of the normally linear potential density-potential temperature line in (B), most clearly reveals the vertical distribution and concentration of the hydrothermal plume. Note that the plume is described equally well by the distribution of suspended hydrothermal precipitates, here expressed as a profile of light attenuation.

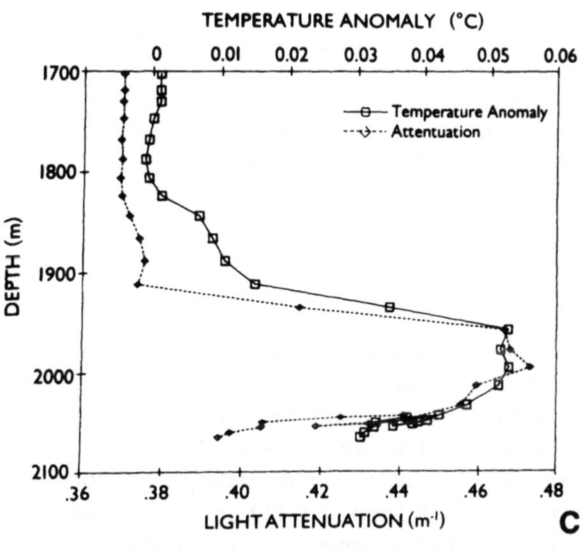

tical anomalies measure the concentration of nonconservative tracers (i.e., particles in the plume can be formed by precipitation and lost by deposition), the simultaneous collection of hydrographic and optical measurements is a powerful tool for studying the physical and chemical evolution of plumes in time and space. For example, the light attenuation/temperature ratio within different plumes on the Juan de Fuca Ridge is a strong function of the Fe/S ratio of the plume particulates and the original vent fluids (Baker and Massoth, 1987). Fe-dominated fluids form a plume rich in very fine-grained Fe oxides that disperse quasiconservatively, whereas S-rich fluids precipitate large, dense, Fe sulfide particles that settle rapidly from the plume. Thus the light attenuation/temperature ratio is areally uniform in plumes with high Fe/S ratios but steadily decreases with distance from the source in plumes with low Fe/S ratios (Fig. 10.3).

Data collection falls into two general groups:

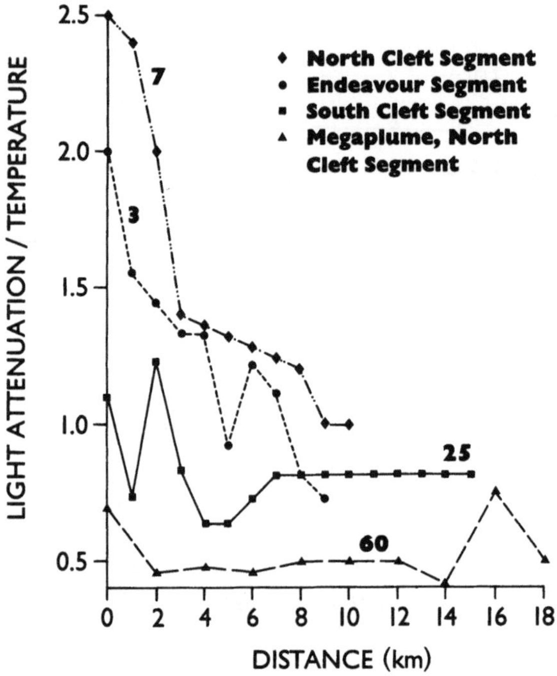

FIGURE 10.3. Plot of the light-attenuation anomaly/temperature anomaly ratio against distance from the plume origin for four Juan de Fuca plumes with different Fe/S molar ratios. Each plume was sampled along a constant density surface corresponding to the center of the neutrally buoyant plume layer.

real-time continuous data processed as it is collected, and discrete data that must be analyzed on board or ashore after collection. For hydrothermal studies, the first category is currently restricted to hydrographic and optical information derived from a CTD interfaced with a beam transmissiometer or other optical device, although recent advances in instrumentation demonstrate that certain chemial analyses can be successfully performed in situ (Johnson et al., 1986, 1988). Because hydrothermal plumes are dynamic features with steep horizontal and vertical gradients, it is crucial to adopt a sampling scheme that will reduce the problem of temporal and spatial aliasing as much as possible. Data return should therefore be maximized from the plume horizon and minimized from waters above or below the plume where little useful informations is present. An efficient strategy in mapping and sampling neutrally buoyant plumes is thus to conduct lengthy "tow-yos," cyclic raisings and lowerings of a CTD/bottle package through the near-bottom waters while slowly steaming (Fig. 10.4). The towed package can be navigated either directly, by using a relay transponder in an acoustic navigation net, or indirectly, by periodically calculating its range and bearing from the known ship's position. The result is a series of two-dimensional transects that can be combined to form a systematic three-dimensional plume survey, as first described by Baker and Massoth (1986a). Note that this strategy is fundamentally different from that employed for deep-tow camera-sled surveys used to image chimneys and sample buoyant plumes (e.g., Detrick et al., 1974; Weiss et al., 1977).

The second category, discrete sample collection, comprises a large number of chemical and mineralogical analyses that require a sample returned to the ship. The usefulness of the various chemical tracers for tracking and interpreting plumes depends on the hydrothermal enrichment relative to seawater, on chemical reactions within the plume, and on the ease of measurement. Figure 10.5 summarizes the hydrothermal enrichments of several chemical species found in vent fluids. The most widely used hydrothermal indicators have been Mn and ^{3}He. Dissolved Mn is useful because of its exceptional concentration in vent fluids (e.g., Edmond, 1980; von Damm, 1988), its quasi-conservative behavior in plumes, and its amenability to shipboard analysis. Helium-3 (specifically the enrichment of the $^3He/^4He$ ratio relative to air; see Craig and Lupton, 1981) is an unequivocal hydrothermal indicator whose conservative behavior allows it to be a plume tracer of global scale (Lupton and Craig, 1981). The complexity of He isotope analyses, however, means that ^{3}He cannot be used as a real-time or shipboard tracer. The plume distribution of some other highly enriched constituents must be carefully interpreted because of their markedly nonconservative behavior, caused either by rapid biological consumption (e.g., H and CH_4) or precipitation and deposition (e.g., Fe).

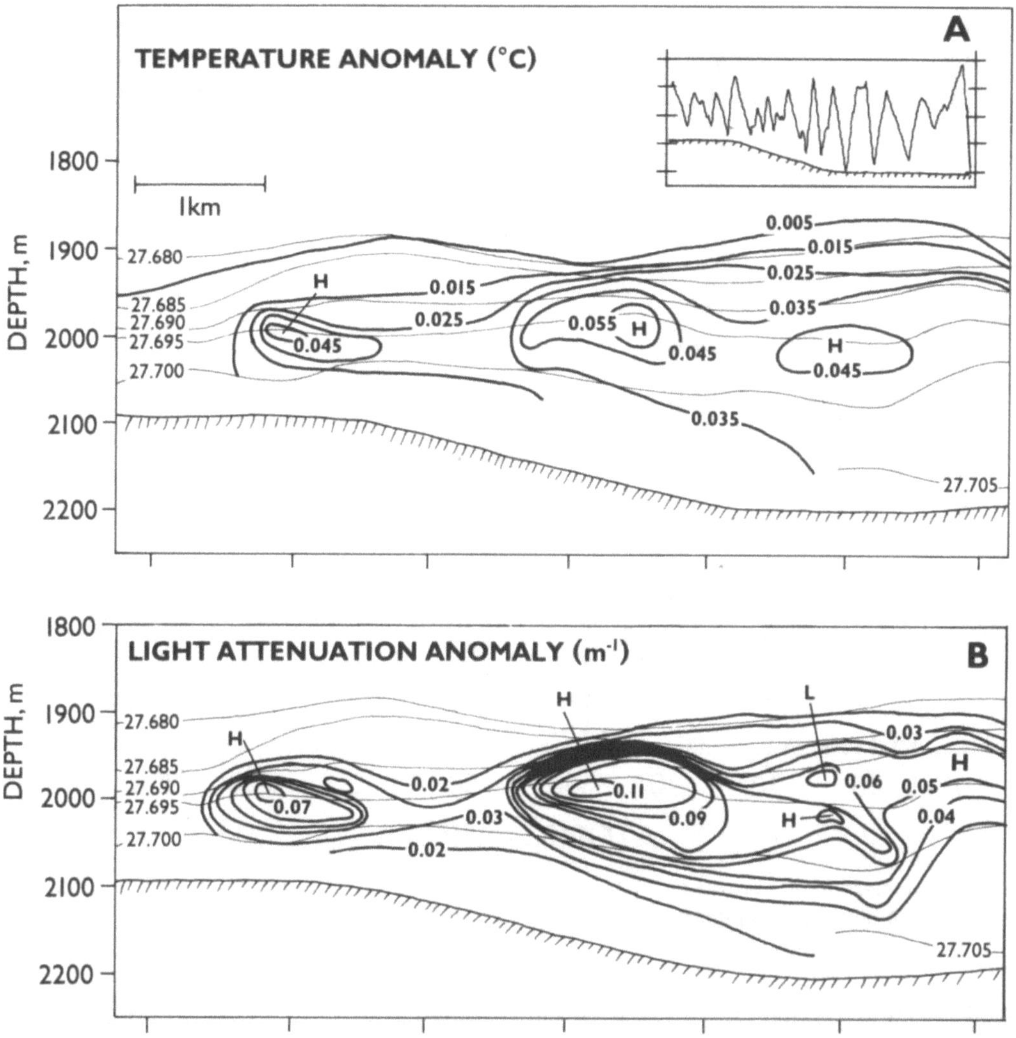

FIGURE 10.4. A CTD tow-yo (see inset) produced this two-dimensional distribution of (A) temperature and (B) light-attenuation anomalies along the crest of Endeavour Segment at about 48° N. The anomalies define a neutrally buoyant plume centered on the 27.695 potential-density surface and advected south (to the right) by the local currents. The two principal maxima overlie two different vent fields. After Baker and Massoth (1987).

Plume Prospecting

Ridge Scale

Large-scale hydrographic and geochemical surveying of the waters overlying the global ridge system lags far behind the extensive bathymetric, structural, and petrographic coverage already compiled by geophysical and geological studies. Early attempts at ridge-scale hydrothermal plume surveys were piggyback operations using a 50-m-long low-resolution (~0.01°C) thermistor chain hung underneath a side-scan sonar during lengthy tows of the entire Juan de Fuca Ridge (Crane et al., 1985) and the East Pacific Rise from 13° N′ N to 8° 20′ N (Crane et al., 1988). These pioneering efforts were often successful in locating major geothermal fields along the ridge crest, but suffered from the limited vertical coverage and lack of salinity (and therefore density) data imposed by use of a thermistor chain. These

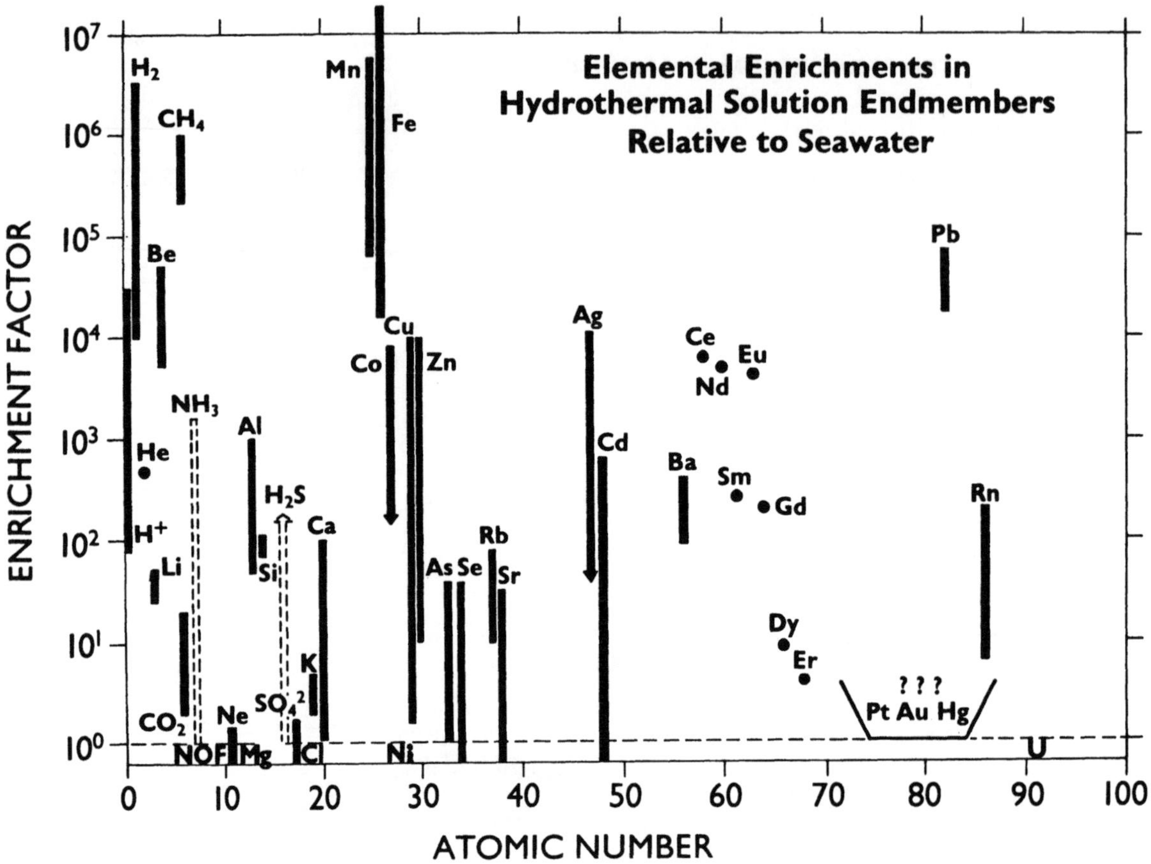

FIGURE 10.5. Diagrammatic illustration of the range of the enrichment factor for a variety of elements and compounds in hydrothermal fluids.

unrefined techniques led to a substantial overestimation of the hydrothermal temperature anomalies. For example, Crane et al. (1985) calculated temperature elevations of 0.04°C to 0.3°C above the five geothermal fields they identified along the Juan de Fuca Ridge. Subsequent detailed CTD tow-yo transects through the same areas (Baker and Massoth, 1986b, 1987; Baker, 1988, unpublished data), however, have shown that temperature anomalies in the neutrally buoyant plumes are more typically in the 0.02 to 0.06°C range and that a major vent field at the north end of Cleft Segment was missed (or was not active) during the survey of Crane et al. (1985). These studies demonstrate that large-scale but spatially detailed hydrographic surveys are an efficient technique for locating active hydrothermal vent fields.

The axial valley of the 320-km-long Gorda Ridge was systematically surveyed in 1985 (Baker et al., 1987) by occupying eight hydrographic and chemical sampling stations. Vertical profiles of light attenuation (Fig. 10.6A), dissolved Mn (Fig. 10.6B), particulate Fe (Fig. 10.6B), and other parameters demonstrated unequivocally the presence of active and apparently high-temperature hydrothermal venting in the northern third of the ridge. On the basis of this and subsequent chemical and hydrographic surveys (Collier and Baker, this volume), sea floor investigations were focused at site GR14 (Fig. 10.6) where a high-temperature vent field was discovered by the Navy submersible *Sea Cliff* in September 1988 (P. Rona, personal communication). Since the survey by Baker et al. (1987), extensive deposits of massive sulfides (Zierenberg et al.,

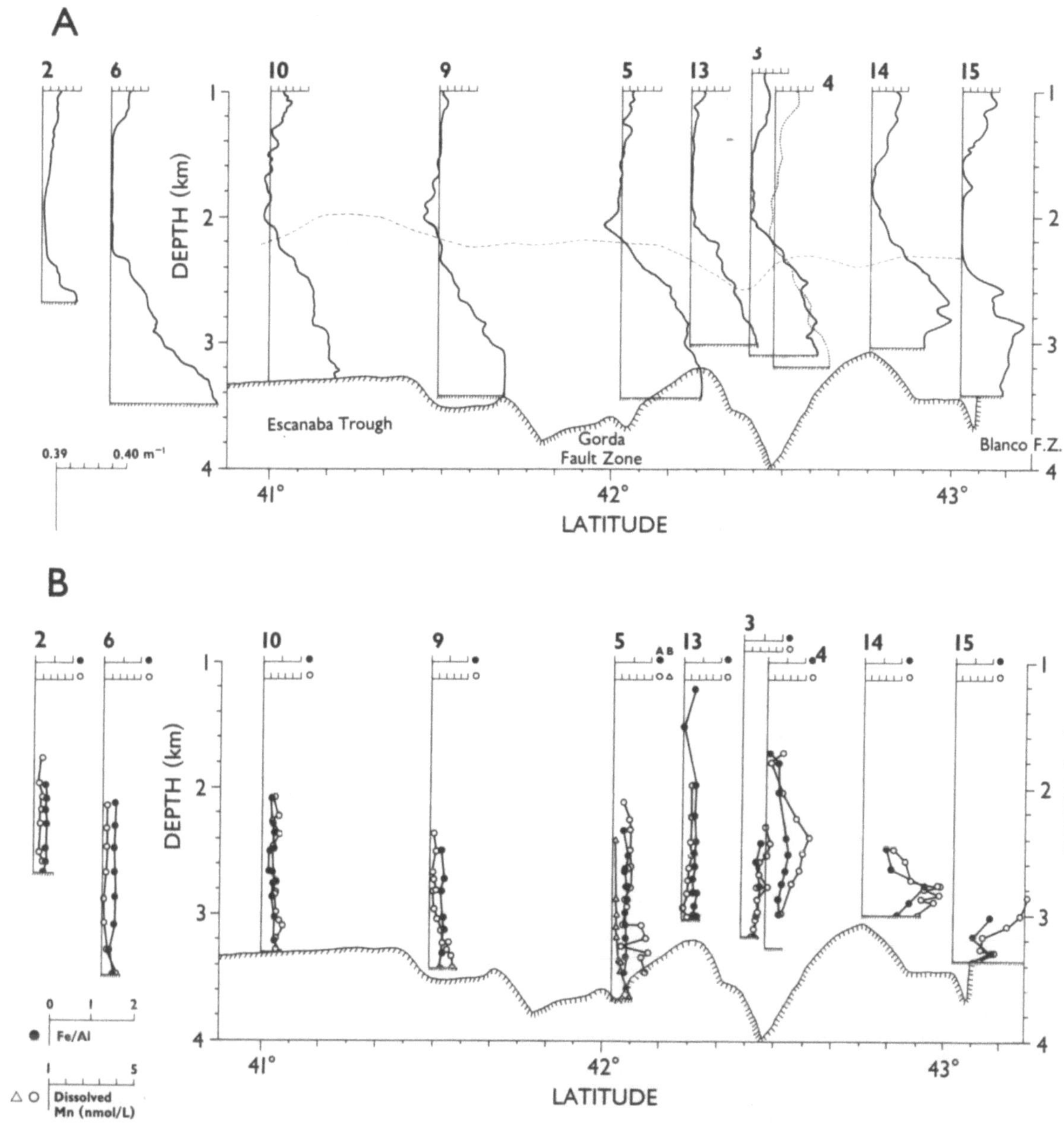

FIGURE 10.6. Profiles of light attenuation (A) and particulate Fe/Al ratios and dissolved Mn (B) along the axial valley of the Gorda Ridge and at off-axis stations about 35 km to the east (Sta. 6) and west (Sta. 2). The dashed line at about 2,000 m in (A) follows the approximate height of the east and west ridge crests. Above-bottom light-attenuation peaks and large Fe/Al and Mn anomalies at stations 14 and 15 indicate the presence of hydrothermal plumes. (After Baker et al., 1987.)

1986, 1988) and numerous active venting sites (Campbell et al., 1988) have been discovered in the sediment-filled southern Gorda Ridge at 41° N. In contrast, station GR10 at 41° N (Fig. 10.6) showed no significant increases in dissolved Mn or particulate Fe, although a very intense methane plume was observed in the bottom 400 m of the water column (Baker et al., 1987). These results are consistent with observations that the vented fluids at that site are clear and of moderate temperature (~220°C) (Campbell et al., 1988), the dis-

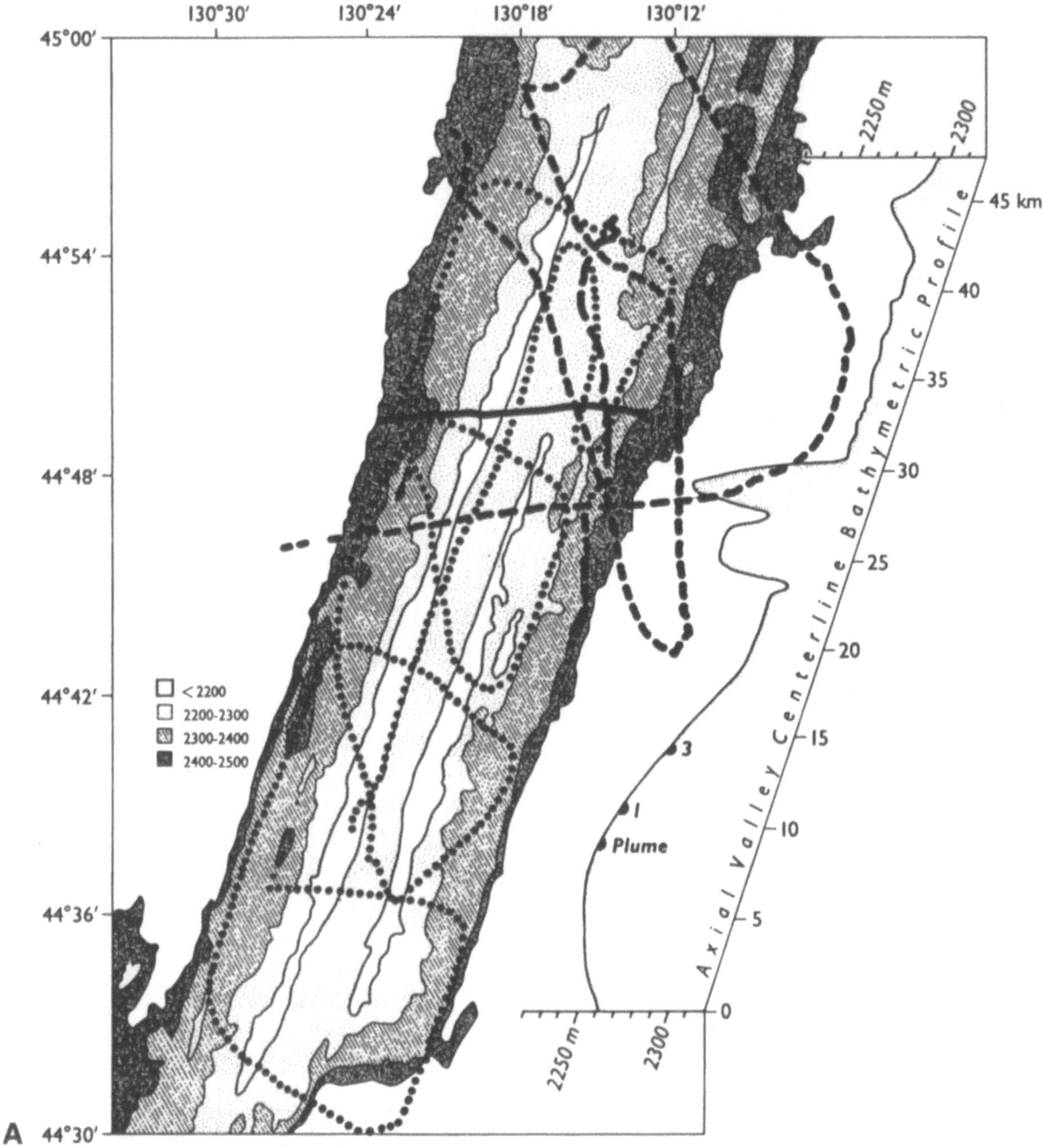

FIGURE 10.7. (A) Bathymetry of the Cleft Segment of the Juan de Fuca Ridge overlain by CTD/transmissometer tow-yo track lines from a 1986 survey. The along-axis profile shows the location of the known (as of 1986) high-temperature vent fields, clustered around 44° 40′ N at the bathymetric high of the segment (the peak at 44° 48′ N is an isolated volcanic cone). (B) Plan view of light-attenuation anomalies greater than 0.04 m^{-1} above background on the 27.70 potential density surface (~2,050 m) over the Cleft Segment. The vent field at 44° 40′ N was producing a plume extending to the north-northwest, similar to previous years (Baker et al., 1985; Baker and Massoth 1986a, 1987). A much larger field, extending from about 44° 50′ N to 45° N, was discovered during this survey and confirmed by camera and submersible observations in 1987 and 1988 (Embley et al., 1988).

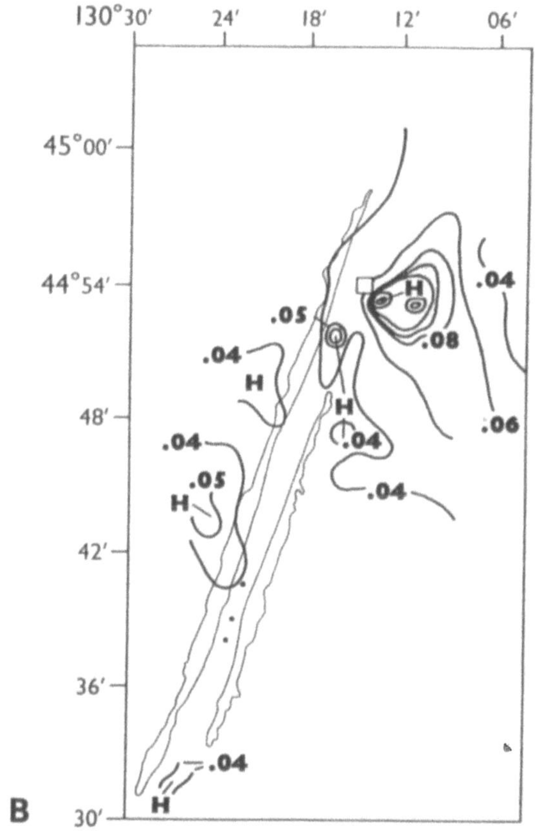

FIGURE 10.7.

solved metals apparently having been lost during passage through the thick sediment cover. It is also possible, however, that the single cast in the southern Gorda Ridge during that survey missed the plume—a good example of the interpretative difficulties caused by widely spaced vertical casts.

Segment Scale

The global midocean ridge system is regularly segmented on a 30 to 80-km length scale (Schouten et al., 1985). Large offsets between segments are expressed as transform fault zones, but the transition zone between most segments is manifested more subtly as an overlapping spreading center, a change in axial linearity, a saddle point, or other feature (Macdonald and Fox, 1988). The along-axis bathymetric profile of an individual segment is typically that of a central high deepenig toward either end. Francheteau and Ballard (1983) hypothesized that such a profile reflected the presence of a centrally located magma chamber and that hydrothermal fields should preferentially be located on the bathymetric high. Guided by this hypothesis, exploration for hydrothermal fields has concentrated on the shallowest portions of segments.

The Cleft Segment of the Juan de Fuca Ridge (Fig. 10.1) exhibits many typical segment characteristics. This 55-km-long segment is bounded on the south by the Blanco Fracture Zone and on the north by an offsect zone where the adjacent segment is displaced ~5 km to the west. An extensive hydrothermal field on the bathymetric high at ~44° 40′ N was discovered by camera tows (Normark et al., 1983) and subsequently studied by submersible (USGS Juan de Fuca Study Group, 1986) and water column surveys (Baker and Massoth, 1986a, 1987). In 1986, however, the entire segment was carefully surveyed with a series of CTD tow-yo tracks (Baker and Massoth, 1986b), and a very different picture of hydrothermal activity on the ridge emerged. In addition to the plume emanating from the known field at 44° 40′ N, the discovery of a more extensive plume with higher temperature and light-attenuation anomalies indicated the existence of a much larger field near 44° 54′ N (Fig. 10.7). This new field is of particular interest because it occurs not on the shallowest but on the deepest part of the segment, almost at its northern termination. For this segment, at least, the vent field distribution does not strictly follow the implications of the Francheteau and Ballard (1983) central injection hypothesis. Vent field distributions along a given segment may change as the segment develops, but since we cannot monitor the development of a single segment through geological time, we must instead survey in space, mapping the distribution of vent fields along many segments to construct a reliable statistical base. A systematic approach to segment surveying will be necessary to determine the relationship between hydrothermal activity, segment structure, and the underlying distribution of magma.

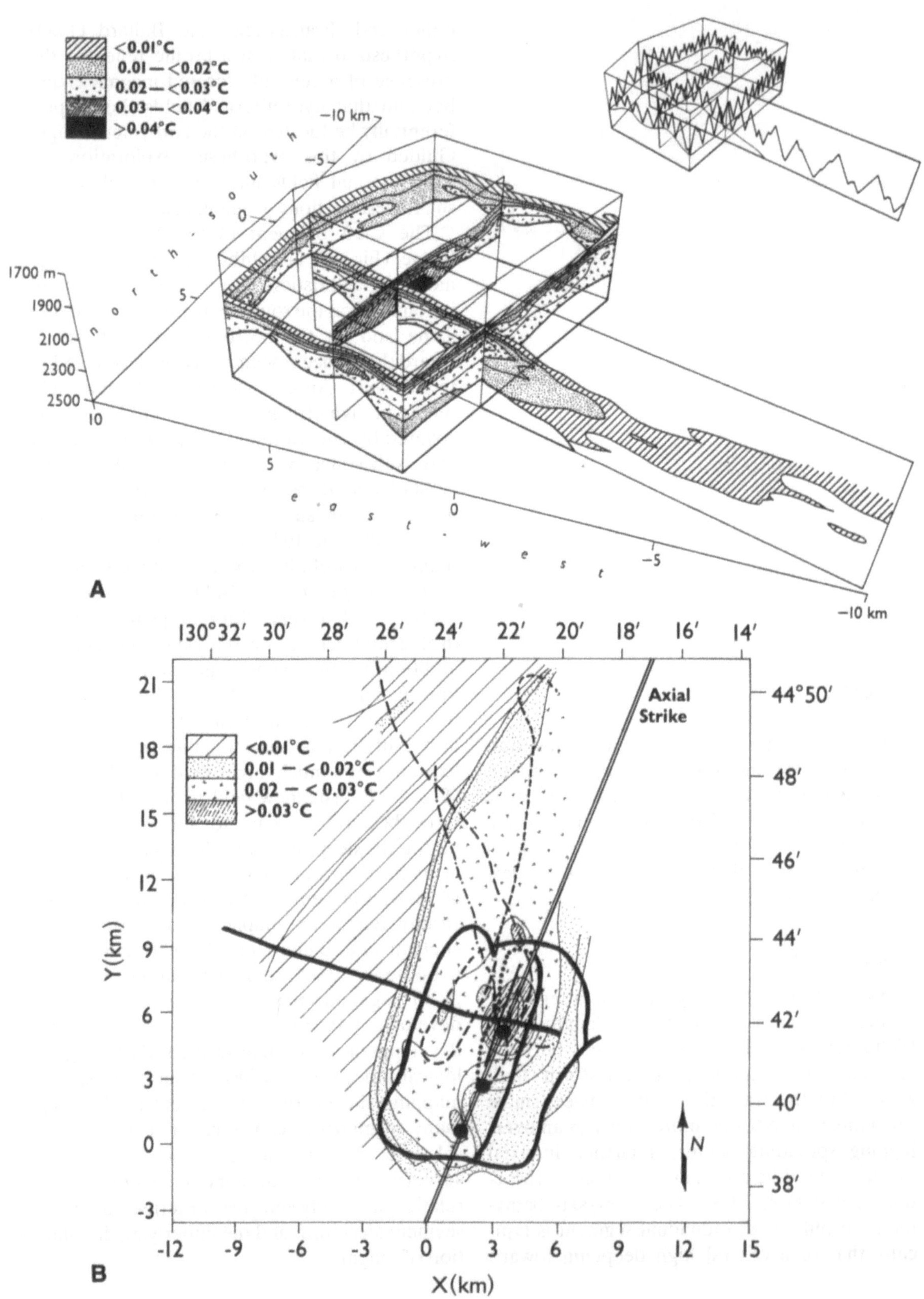

<0.01°C
0.01 — <0.02°C
0.02 — <0.03°C
0.03 — <0.04°C
>0.04°C
1700 m
1900
2100
2300
2500
north - south
east - west
A
130°32′ 30′ 28′ 26′ 24′ 22′ 20′ 18′ 16′ 14′
44°50′
48′
46′
44′
42′
40′
38′
Axial Strike
<0.01°C
0.01 — < 0.02°C
0.02 — < 0.03°C
>0.03°C
Y(km)
X(km)
N
B

Vent-Field Scale

Once the distribution of hydrothermal plumes over a segment has been determined, investigations can be efficiently focused on individual vent fields. Plume prospecting can be used to close range on the vent locations and thus reduce the search area for near-bottom camera imagery, to guide the collection of discrete water samples for chemical and mineralogical analyses, and to quantify the advective transport of heat and mass in the plume and thus estimate the flux of heat and mass from the underlying vent field. The vent field at 44° 40′ N on the Cleft Segment of the Juan de Fuca Ridge has been the site of these kinds of investigations since 1984.

A set of systematic tows around and through a vent field can be combined to yield a three-dimensional view of the plume and to close range on the loation of the high-temperature vents that form the plume. At the 44° 40′ N site during a survey in June 1985, a series of CTD/transmissometer tow-yos were used to image the neutrally buoyant plume three-dimensionally. A box diagram constructed from three separate tows taken over a 30-hour period is shown in Figure 10.8A. Inspection of the vertical sections revealed that the core of the neutrally buoyant plume was centered around the 27.70 potential-density surface. Its regional distribution could thus be mapped by an areal plot of temperature or light-attenuation anomalies on that density surface. Such a map (Fig. 10.8B) clearly shows that the plume originates from the area of known vent sites around 44° 40′ N. The sharp lateral boundaries of the plume and the agreement of the plume axis with the long-term net current direction at the site also emphasize the advective character of the plume. If the plume is advectively controlled, then the transport of a conservative plume constituent, such as heat, in the plume equals the total vent-field flux of that constituent into the plume. Baker and Massoth (1986a, 1987) estimated the heat transport, Q, by the mean flow at this vent field as the difference between plume transport through a vertical cross section downcurrent of the vent field and normal to the plume axis, and the transport through an identical upcurrent cross section:

$$Q = \rho C_p A \bar{U}(\Delta\bar{T}_D - \Delta\bar{T}_u)$$

where ρC_p = specific heat, A = area of the cross section, $\bar{U}$ = mean velocity along the plume axis, and $\Delta\bar{T}_D$ and $\Delta\bar{T}_u$ = mean temperature anomaly of the down- and upcurrent cross sections, respectively. The use of this equation implies that the heat transport is dominated by the mean flow and that time-dependent fluctuations are minor. This assumption is supported by the stability of the temperature anomaly in the core of the plume (Baker and Massoth, 1987).

The net heat transport in this plume during the survey was 580 ± 351 MW, about equivalent to 50 to 500 black smokers of the type studied by Converse et al. (1984) and Little et al. (1987). The actual number of black smokers may be much smaller, however, if widespread diffuse heat sources contribute importantly to the total flux. The measured hydrothermal heat flux from this one field is of the same order as that calculated for the ridge axis of the entire segment (~550 MW) by Morton and Sleep (1985) on the basis of the heat losses

FIGURE 10.8. (A) Three-dimensional representation of the hydrothermal temperature-anomaly plume emanating from the vent field at 44° 40′ N on the Cleft Segment during June 1985. The temperature anomaly is contoured at 0.005°C intervals and shaded at 0.01°C intervals. This view, constructed from CTD tow-yos along the axial valley, around the perimeter of the vent field, and across the valley to the west, looks toward the vent field from the north-northwest. Inset shows the tow path along each transect. (B) Areal map of the plume in (A) constructed by plotting the temperature anomaly on the 27.70 potential density surface from seven tows conducted during a 14-day period. Tows in (A) are shown as heavy solid lines; additional tows are shown as dashed or dotted lines. Contour interval and kilometer scales as in (A). Vent field locations (solid dots) correspond to locations shown in Figure 10.7. (After Baker and Massoth, 1986a.)

necessary to balance a magma chamber depth of 2.4 km.

Other known hydrothermal heat sources on this segment include the newly discovered vent field extending from 44° 54′ N to 45° N (Fig. 10.7) and the megaplumes—cataclysmic releases of $\sim 10^8$ m^3 of hydrothermal fluid (Baker et al., 1988a,b)—found above the ridge axis in 1986 and 1987. Although heat flux from the northerly vent field has not yet been quantified, the intensity and dimensions of the plume are similar to those of the plume from the vent field on Endeavour Segment, whose heat flux was calculated to be $1{,}700 \pm 1{,}110$ MW based on plume heat transport (Baker and Massoth, 1987) and 1,000 to 5,000 MW based on $^{222}Rn/^{3}He$ measurements in the plume (Rosenberg et al., 1988). A single megaplume provides an annually equivalent heat flux of ~2,000 MW.

The total present-day heat flux from steady and episodic sources on the Cleft Segment is thus about 4,500 MW, or nearly a factor of 10 higher than that calculated by the thermal model of Morton and Sleep (1985). This imbalance suggests that the segment may currently be releasing heat faster than the long-term average derived from the geophysical data, or that the thermal model underestimates the true heat flux. Because of the broad error bounds on the plume heat flux calculations, however, it is also possible that no discrepancy exists. Resolution of uncertainties between modeled and measured hydrothermal heat fluxes awaits data from a larger number of ridge segments.

Acknowledgments. This paper was written with the support of the NOAA VENTS Program. G.J. Massoth provided ongoing discussions, chemical data, and a cogent review of the paper. Contribution No. 975 from Pacific Marine Environmental Laboratory/NOAA.

References

Baker ET (1988) A multi-segment view of the distribution of hydrothermal venting along the Juan de Fuca Ridge. EOS 69:1499.

Baker ET, Massoth GJ (1986a) Hydrothermal plume measurements: A regional perspective. Science 234:980–982.

Baker ET, Massoth GJ (1986b) The along-strike distribution of hydrothermal activity on a spreading segment of the Juan de Fuca Ridge. EOS 67:1027.

Baker ET, Massoth GJ (1987) Characteristics of hydrothermal plumes from two vent fields on the Juan de Fuca Ridge, northeast Pacific Ocean. Earth Planetary Sci Lett 85:59–73.

Baker ET, Lavelle JW, Massoth GJ (1985) Hydrothermal particle plumes over the southern Juan de Fuca Ridge. Nature 316:342–344.

Baker ET, Massoth GJ, Collier RW, et al. (1987) Evidence for high-temperature hydrothermal venting on the Gorda Ridge, northeast Pacific Ocean. Deep-Sea Res 34:1461–1476.

Baker ET, Massoth GJ, Feely RA (1988a) Cataclysmic hydrothermal venting on the Juan de Fuca Ridge. Nature 329:149–151.

Baker ET, Massoth GJ, Feely RA, Lupton JE (1988b) Cataclysmic venting, steady venting, and the heat flux from the southern Juan de Fuca Ridge. EOS 69:304.

Campbell AC, German C, Palmer MR, Edmond JM (1988) Preliminary report on the chemistry of hydrothermal fluids from the Escanaba Trough. EOS 69:1271.

Converse DR, Holland HD, Edmond JM (1984) Flow rates in the axial hot springs of the East Pacific Rise (21°N): Implications for the heat budget and the formation of massive sulfide deposits. Earth Planetary Sci Lett 69:159–175.

Cowen JP, Massoth GJ, Baker ET (1986) Bacterial scavenging of Mn and Fe in a mid- to far-field hydrothermal particle plume. Nature 322:169–171.

Craig H, Lupton JE (1981) Helium-3 and mantle volatiles in the ocean and the oceanic crust. In: Emiliani C (ed) The Sea, Vol 7. New York: Wiley, pp 391–428.

Crane K, Aikman III FA, Embley R, Hammond S, Malahoff A, Lupton J (1985) The distribution of geothermal fields on the Juan de Fuca Ridge. J Geophys Res 90:727–744.

Crane K, Aikman III FA, Foucher J-P (1988) The distribution of geothermal fields along the East Pacific Rise from 13°10′N to 8°20′N: Implications for deep-seated origins. Marine Geophys Res 9:211–236.

Detrick RS, Williams DL, Mudie JD, Sclater JG (1974) The Galápagos spreading center: Bottom-water temperatures and the significance of geothermal heating. J Geophys R Astron Soc 38:627–637.

Edmond JM (1980) The chemistry of the 350° hot springs at 21°N on the East Pacific Rise. EOS 61:992.

Embley RW, Hammond S, Murphy K, et al. (1988) Submersible observations of the "megaplume" area: Southern Juan de Fuca Ridge. EOS 69:1497.

Francheteau J, Ballard R (1983) The East Pacific Rise near 21°N, 13°N and 20°S: Inferences for along-strike variability of axial processes of the Mid-Ocean Ridge. Earth Planetary Sci Lett 64:93–116.

Johnson KS, Beehler CL, Sakamoto-Arnold CM, Childress JJ (1986) In-situ measurements of chemical distributions in a deep-sea hydrothermal vent field. Science 231:1139–1141.

Johnson KS, Childress JJ, Sakamoto-Arnold CM, Beehler CL (1988) The biogeochemistry of hydrothermal vent communities in the Galápagos Rift. EOS 69:1270.

Little SA, Stolzenbach KD, Von Herzen RP (1987) Measurements of plume flow from a hydrothermal vent. J Geophys Res 92:2587–2596.

Lupton JE, Craig H (1981) A major helium-3 source at 15°S on the East Pacific Rise. Science 214:13–18.

Lupton JE, Delaney JR, Johnson HP, Tivey MK (1985) Entrainment and vertical transport of deep-ocean water by buoyant hydrothermal plumes. Nature 316:621–623.

Macdonald KC, Fox PJ (1988) The axial summit graben and cross-sectional shape of the East Pacific Rise as indicators of axial magma chambers and recent volcanic eruptions. Earth Planetary Sci Lett 88:119–131.

Morton JL, Sleep NH (1985) A mid-ocean ridge thermal model: Constraints on the volume of axial hydrothermal heat. J Geophys Res 90:11,345–11,353.

Normark WR, Morton JL, Koski RA, Clague DA, Delaney JR (1983) Active hydrothermal vents and sulfide deposits on the southern Juan de Fuca Ridge. Geology 11:158–163.

Reid JL (1982) Evidence of an effect of heat flux from the East Pacific Rise upon the characteristics of the mid-depth waters. Geophys Res Lett 9:381–384.

Rosenberg ND, Lupton JE, Kadko D, Collier R, Lilley MD, Pak K (1988) Estimating heat and chemical fluxes from a seafloor hydrothermal vent field using radon measurements. Nature 334:604–607.

Schouten H, Klitgord KD, Whitehead JA (1985) Segmentation of mid-ocean ridges. Nature 317:225–229.

Speer K (1988) The influence of geothermal sources on deep ocean temperature, salinity and flow fields. Ph.D. Thesis, MIT-WHOI Joint Program, Woods Hole, MA, p 146.

Speer K, Rona P (1987) A model of an Atlantic hydrothermal plume. EOS 68:1325–1326.

Spiess FN, Tyce RC (1973) The Marine Physical Laboratory DEEP-TOW instrumentation system. Scripps Institution of Oceanography reference 73–74, 37 pp (unpublished document).

Stommel H (1982) Is the South Pacific helium-3 plume dynamically active? Earth Planetary Sci Lett 61:63–67.

USGS Juan de Fuca Study Group (1986) Submarine fissure eruptions and hydrothermal vents on the southern Juan de Fuca Ridge: Preliminary observations from the submersible *Alvin*. Geology 14:823–827.

Von Damm KL (1988) Systematics of and postulated controls on submarine hydrothermal solution chemistry. J Geophys Res 93:4551–4561.

Weiss RF, Lonsdale P, Lupton JE, Bainbridge AE, Craig H (1977) Hydrothermal plumes in the Galápagos Rift. Nature 267:600–602.

Zierenberg RA, Koski RA, Shanks III WC, Rosenbauer RJ (1986) Form and composition of sediment-based sulfide-sulfate deposits, Escanaba Trough, Southern Gorda Ridge. EOS 67:1782.

Zierenberg RA, Koski RA, Ross SL, Shanks III WC, Slack JF (1988) Preliminary results of *Alvin* dives on active sediment-hosted sulfide deposits in the Escanaba Trough, southern Gorda Ridge. EOS 69:1488.

11—Acoustic Techniques for Imaging the Seafloor

CHRISTOPHER G. FOX

Abstract

The recent development of sophisticated sonar systems for seafloor imaging allows analyses of terrain beyond simple bathymetric mapping. The SEABEAM multibeam sonar system produces full-coverage 12-kHz soundings, and much of the Gorda Ridge has been surveyed by the National Oceanic and Atmospheric Administration and the U.S. Navy. Complete-coverage digital bathymetry provides the basic data for statistical modeling of seafloor roughness through spectral and fractal mathematical techniques. In addition to soundings, the SEABEAM system can provide fundamental information about bottom properties through evaluation of the backscattered acoustical signal. Side-scan sonar systems allow more detailed observations of the variability of acoustic properties. Available side-scan sonar data range from the broad (60-km swath) complete coverage of the GLORIA system collected by the U.S. Geological Survey to 10-km swath SeaMARC II and 5- to 0.5-km-wide SeaMARC I imaging, which produces pixel spot sizes of as little as 25 cm. The full-coverage, digital nature of these data allows the application of terrain classification techniques developed for satellite imagery. The validity of remote-imaging techniques can only be evaluated when compared to observational ground truth. A recently developed technique for geological mapping from bottom photography facilitates such comparisons.

Introduction

With the initial availability of the Narrow Beam Echo Sounder (NBES) in 1964, deep-sea marine geology entered the digital computing era. Until that time, only wide-beam, analog sonars had been available to the scientific community. Succeeding years saw the introduction of numerous advanced instruments such as SEABEAM (1975) swath-mapping sonar, GLORIA II (1977), SeaMARC I (1980) SeaMARC II (1982), and other side-scan sonars, all of which obtain acoustic data in digital form (see Tyce, 1986, for a complete discussion). Simultaneous with the availability of quality digital data was an explosion in computer technology, allowing the rapid compilation and analysis of collected data sets. Additionally, improved navigation aids have allowed the colocation of multiple data sets as well as the use of sophisticated, spatially dependent analysis techniques. Even new numerical techniques such as the Fast Fourier Transform (Cooley and Tukey, 1965) and fractal geometry (Mandelbrot, 1982) have been developed in the same period. These and other methods allow interpretation of seafloor relief to be extended beyond simple bathymetric mapping. Recently, the Office of Naval Research has recognized the importance of such lines of investigation and introduced a research initiative in quantitative geomorphology. In some cases, the development of these techniques is taking place at Gorda Ridge, and techniques developed at other locations can be directly applied to the study of Gorda Ridge. This chapter reviews some of the existing and developing acoustic techniques for seafloor remote sensing, with application to Gorda Ridge.

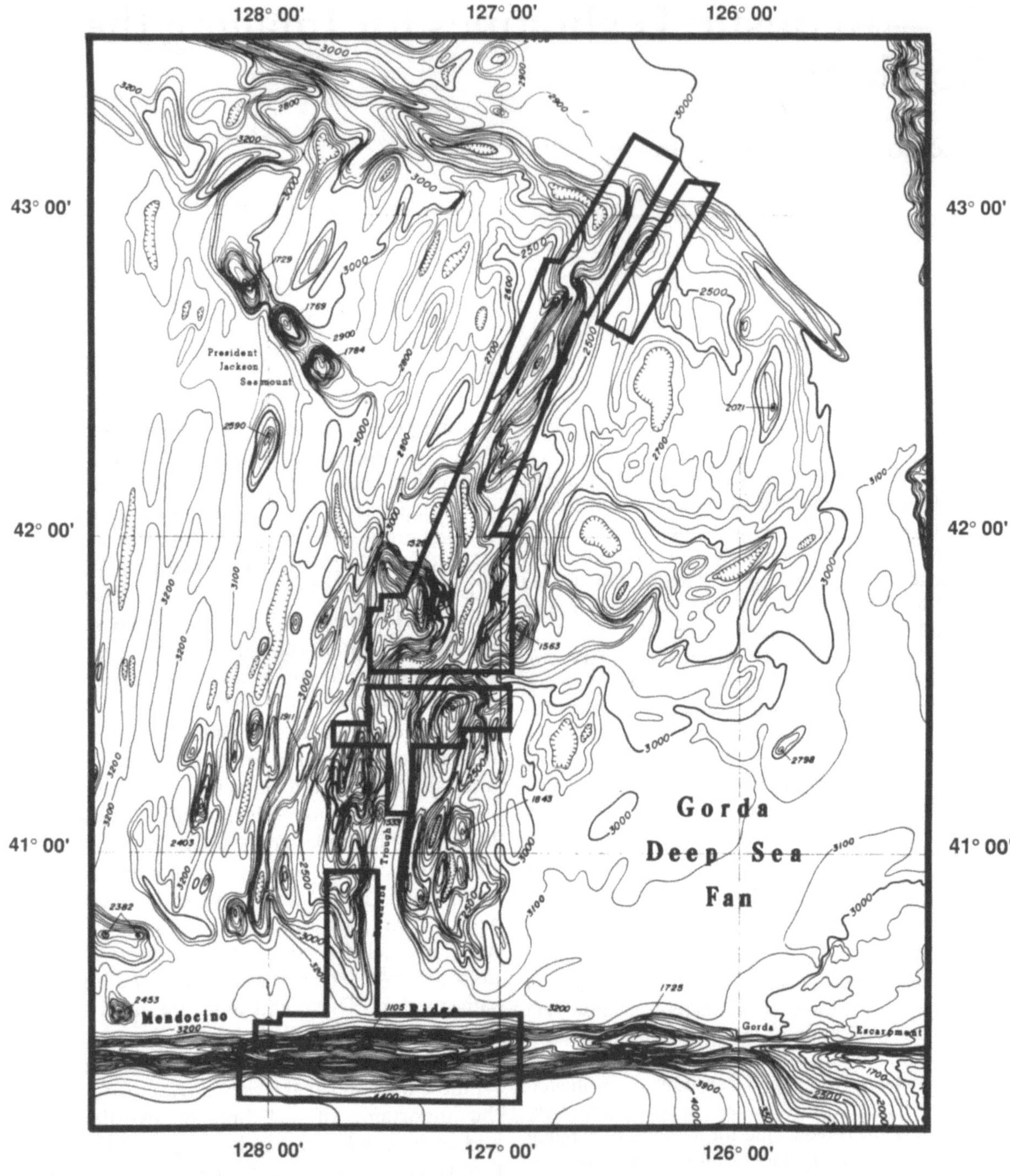

FIGURE 11.1. Geographic coverage of NOAA SEABEAM bathymetric data on Gorda Ridge. Boxed areas have complete-coverage digital soundings. The base map is reproduced from Chase et al. (1981).

Surface Sonars

Hull-mounted sounding systems allow regional surveys of seafloor terrain to be conducted at full ship's speed and relatively low cost. As numerous sonar transects are accumulated, these randomly oriented profiles can be compiled into bathymetric charts, such as that produced in the Gorda Ridge area by Chase et al. (1981) (which forms the base map in Fig. 11.1). The statistical interpretation of randomly oriented profiles is difficult, since the extent of contour control varies widely with data spacing. The use of regularly spaced survey tracks, such as those used by the National Oceanic and Atmospheric Administration's (NOAA) 1971 SEAMAP surveys (Elvers and Potter, 1971), simplifies the variability problem somewhat; however, the uncertainty between tracklines effectively precludes quantitative two-dimensional spatial analyses. Digital, gridded data bases such as SYNBAPS (van Wyckhouse, 1973) and the new Digital Bathymetric Data Base (DBDB) are derived from profile-based bathymetric charts and are useful only in regional, deterministic studies of ridge morphology.

Full-coverage bathymetric surveys are available from multibeam sonar systems such as SEABEAM. With proper trackline spacing, it is possible to obtain a complete bathymetric chart of the seafloor in a relatively short period. The adjacent beam spacing for SEABEAM equals 4.65% of water depth, allowing horizontal resolution of 75 to 150 m for water depths found at Gorda Ridge. Since 1980, SEABEAM-equipped NOAA survey vessels have collected a large (although incomplete) data set from Gorda Ridge for the areas shown in Figure 11.1, and these data are currently undergoing processing to produce detailed bathymetric charts of the selected areas. A small section of this data set is illustrated in Figure 11.2 and compared to the earlier bathymetry from Chase et al. (1981). Malahoff (1985) presents a portion of the data from the northern Inner Rift. When the data set is complete, the resulting charts should aid tremendously in geological and tectonic studies of the ridge as well as in supporting other operations, such as ocean-bottom drilling or submersible dive programs.

Digital soundings collected from SEABEAM can also be used for statistical characterization of seafloor physiography. Fox and Hayes (1985) applied Fourier spectral techniques to similar data and produced quantitative measures of seafloor roughness for northern Gorda Ridge. The patterns observed using these techniques provide new insights into the relief-forming processes acting on the ridge, which are not obvious from examination of the bathymetry alone. In addition, a method was derived to allow the anisotropy of the seafloor fabric to be quantitively assessed. Fractal techniques may also be applied to measure the variability (or roughness) of the terrain. There is much current research to develop other tools for the statistical description of seafloor relief, such as that by Shaw and Smith (1987).

Both bathymetric charting and statistical modeling of terrain are based on the measured travel time of acoustic energy in the water column. In practice, the returning echo from the seafloor is digitized and analyzed for the time delay of maximum energy received, and the digitized signal is discarded. There is, however, a wealth of information about bottom properties represented in the time-dependent distribution of the returned acoustic energy. DeMoustier (1986) pioneered techniques for acquiring this information and applying it to understanding the physical properties of the seafloor. The interaction of sound with the seafloor is a complex function of incident angle, reflectivity of the material, and microroughness of the surface within the footprint. Interpretation of bottom properties is further complicated by the dependence of the signal on transducer characteristics, the beam-forming algorithm, acoustic frequency, and side-lobe interference, to name but a few. Figure 11.3 illustrates a typical gather from a SEABEAM ping being analyzed for backscatter character. The analysis of the return pulses in terms of peak amplitude, variance, total energy, and other statistical measures reflects the acoustic physiographic properties of the seafloor within the insonified footprint of each beam. The

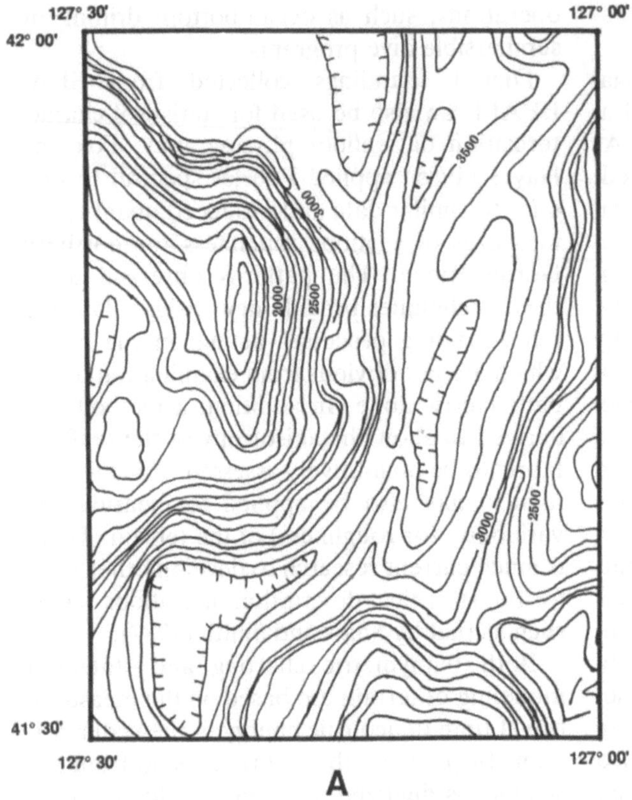

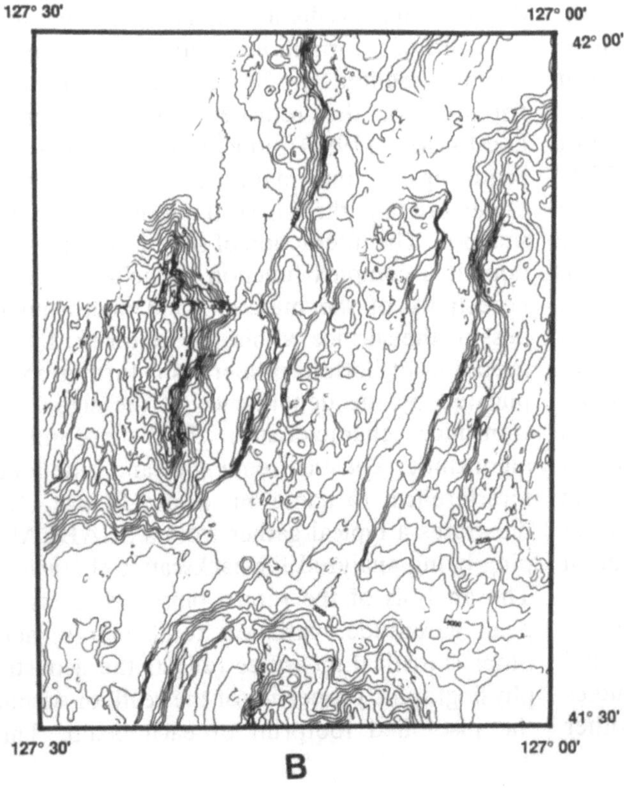

FIGURE 11.2. Comparison of Central Gorda Ridge bathymetric charts generated with different techniques. Map A, from Chase et al. (1981), was created by manually contouring randomly oriented sonar profiles from several sources. The identical area is represented in map B; however, the data are from full-coverage SEABEAM data and are automatically contoured and drawn using computers. The increase in detail is apparent.

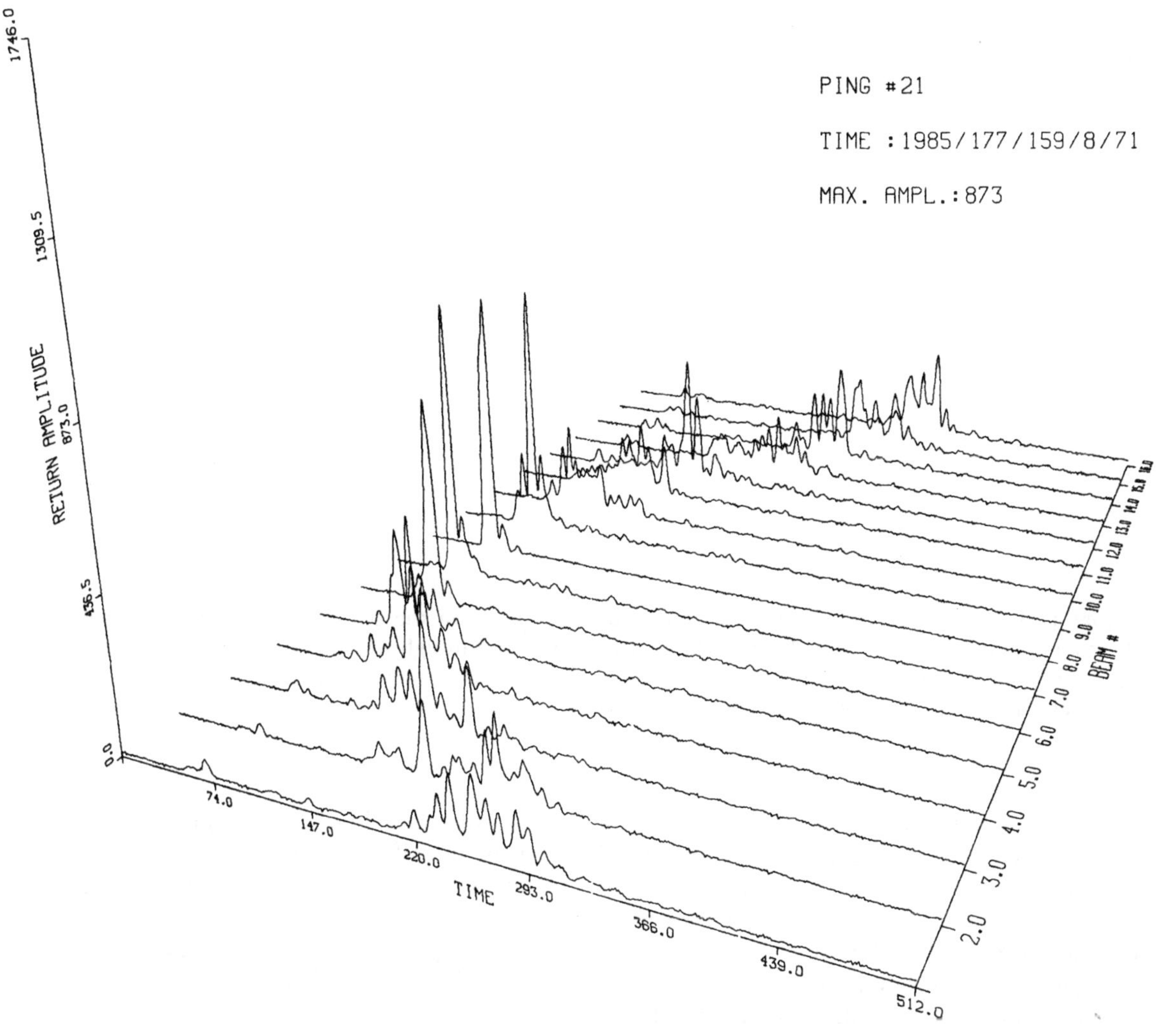

FIGURE 11.3. The acoustic response of the SEABEAM system through the time of a single sounding. The Z-axis represents amplitude, and each of the 16 formed beams is shown versus time in milliseconds. The vertically incident beam is number 8 with each adjacent beam spaced at a 2.67° angle.

additional information provided by examining the full signal returned from SEABEAM will greatly increase the understanding of the processes at work on the seafloor through an inexpensive remote-sensing approach.

Side-scan Sonar

Seafloor acoustic backscatter can be measured directly using towed side-scan sonar instruments. The systems generally used in deep ocean research are GLORIA II, SeaMARC I, and SeaMARC II. GLORIA II is a low-frequency (6.2/6.8-kHz) sonar capable of ensonifying swaths of up to 60 km width, with a spatial resolution of 30 m. As part of its program to map the U.S. Exclusive Economic Zone (EEZ), the U.S. Geological Survey (USGS) has imaged the entire Gorda Ridge (EEZ SCAN Scientific Staff, 1986). SeaMARC II surveys have also been completed for the Gorda Ridge axis north of the Escanaba Trough (Clague and Holmes, 1987). This 11/12-kHz system insonifies a 10-km-wide swath with 5-m spatial resolution. A SeaMARC I sur-

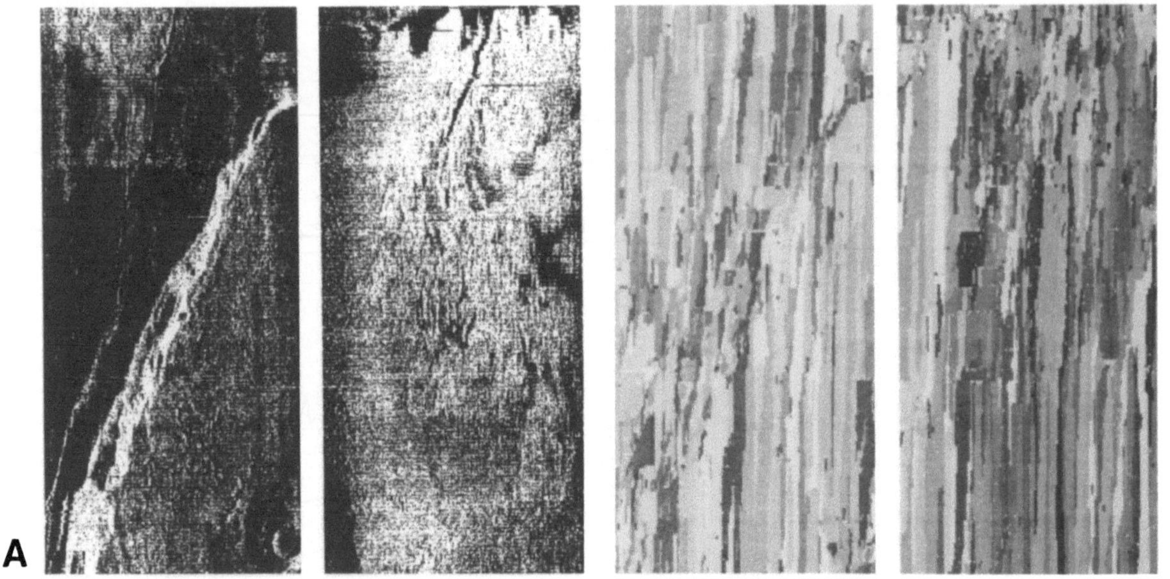

FIGURE 11.4. SeaMARC I side-scan sonar image of the floor of Axial Caldera, Juan de Fuca Ridge. Image A represents strong acoustic returns as light; weak returns as dark. The pixel size corresponds to 1 square on the seafloor. Image B represents one scheme through which the sonar image can be classified into regions of similar statistical properties.

vey was conducted in the northern Gorda Ridge rift valley by NOAA in 1987. SeaMARC I is a deep-towed 27/30-kHz system capable of resolutions of 0.25 to 2.5 m with swath widths ranging from 512 to 5,180 m.

High-resolution side-scan sonar imaging is an invaluable tool for interpreting the distribution of small-scale features and fabrics of the seafloor. The sonar scan lines (which are digitized to 2,048 pixels and eight-bit precision for the SeaMARC systems) are similar to monochromatic microwave scan lines from earth-orbiting satellites. The availability of the backscattered energy values in digital form allows both the enhancement of the image and the application of objective classification algorithms. Enhancement techniques include false coloring, digital filtering, and histogram equalization of the image intensity distribution. Classification techniques attempt to aggregate areas of an image into regions of similar attributes (Fig. 11.4). For LANDSAT satellite images, these regions may represent land use, such as forest, water, urban, agricultural, and even subdivisions of agricultural usage into individual crop types. The objectives in classifying sonar images are similar; however, it is not yet clear which statistical parameterizations are appropriate or what geological processes control the areal distributions of the statistically derived regions.

Calibration of Remote-Sensing Results

Before remote-sensing results can be applied to physical problems, it is necessary to determine the environmental conditions that are controlling the spatial distribution and numerical parameters of the statistical classification scheme. The LANDSAT researcher must determine if there are true physical boundaries corresponding to those inferred through the mathematical manipulation of the multispectral imaging, or the technique has no application. The standard classification techniques do not provide a means to determine at what level of subdivision the results are adequate. The investigator must select a level of subdivision appropriate to the particular problem being addressed. In addition, an observer on the ground must determine the nature of the classified regions—coniferous forest versus a corn field, for example—before the technique can be applied to areas that are not directly observed. It is important to stress that this calibration of the remote-sensing results does not need to be performed over a large region—only in sufficient areas to determine the various end members of the physical system under study.

Satellite researchers have a considerable advantage over marine geologists in obtaining ground-truth information. The difficulties of observing the seafloor beneath thousands of meters of water become a major concern to calibration of acoustic remote-sensing techniques. Only through bottom photography and manned or unmanned submersible observations can the seafloor be visually inspected, and then only for limited areas. Also critical to these observations is a navigation system adequate to construct accurate geological maps of a region.

As part of NOAA's VENTS research program, an extensive geological and geophysical investigation has been undertaken at Axial Seamount (46° N, 130° W) on Juan de Fuca Ridge. Within the caldera of Axial Seamount, data have been collected from SeaMARC I at 2.5-m, 1.0-m, and 0.5-m resolutions. Additional data collection in 1987 included 0.25-m resolution SeaMARC I as well as an extensive SEABEAM backscatter survey. SEABEAM bathymetric data have been collected over an 8-year time period, and GLORIA II and SeaMARC II data are also available. Finally, over 20,000 bottom photographs have been collected, and three, full-submersible dive programs have been conducted over the past several years. All of the data have been collected within a bottom-mounted transponder network.

The Axial Seamount data sets provide a unique opportunity to develop and calibrate acoustic remote-sensing techniques. Two approaches are available to confirm the location of calculated regional boundaries. By collecting several types of data of differing resolutions at the same location, the various sensors can be intercalibrated by studying the repeatability of patterns generated by different methods. Also, the accurate positioning of bottom photographs within a transponder network allows a geological map to be constructed and compared to remote-sensing results. To support this mapping effort, a computerized system was developed to project the large quantity of photographically derived information on to a base map for determining geological boundaries (Fox et al., 1988). In addition, this computerized system categorizes the photographs in terms of lava type, morphology, and sediment cover to allow interpretation of the physical factors controlling the remote-sensing results. The results obtained from these calibration efforts should assist the interpretation of remote-sensing data at other ridge-crest areas where less-complete data sets are available, such as at Gorda Ridge.

Applications to Hydrothermal Systems

A primary area of interest to researchers on Gorda Ridge and other ridge crests is the fundamental processes at work in hydrothermal systems. In addition to general scientific interest, these systems are responsible for concentrating sulfide and other minerals as well as influencing the midwater geochemistry of the open ocean. One application of acoustic remote sensing is in attempting to identify and map the areal distribution of these systems. Most of the vent fields discovered to date do not exceed 100 m in diameter. When one attempts to locate such sites on Gorda Ridge, it is roughly equivalent to searching for a football field in a large mountain range. Obviously, remote-sensing techniques could aid considerably in this effort.

At present, the preferred method for locating new vent fields on the seafloor is through the detection of hydrographic and geochemical anomalies in the water column (Baker, this volume). Temperature anomalies, on the order of hundredths of a degree, indicate an active high-temperature venting site in proximity on the seafloor. Vent fields that are temporarily inactive for any reason, or those with relatively low discharge rates or low temperatures, cannot be detected. To close the location of the hydrothermal fluid source to a position that can be studied by submersible depends at present on observing evidence of hydrothermal activity on bottom photographs. To continue the terrestrial analogy, one can imagine searching for a football field with a camera flown 5 m above the ground that is being dangled from a hot air balloon flying 3 km above. Although there have been many sites discovered and localized using these techniques, there are undoubtedly many more that have eluded observation.

Acoustic remote-sensing techniques may help close the gap between the broad view of the water column methods and discrete bottom photographs. For hydrothermal fields to be directly detectable acoustically, they must be within the resolution of the sensors and provide a signature that is unique among surrounding geological environments. The physiography and mineralogy of known hydrothermal fields indicate that remote identification may be possible. Hydrothermal chimneys can rise 10 or more meters above the surrounding terrain and are often several meters in diameter, which is well within the resolution of SeaMARC I. The highly variable relief of hydrothermal terrains may be detectable within a SEABEAM footprint. Also, the contrasting mineralogies of hydrothermal chimneys and surrounding basalts or lavas could yield a clear acoustic signal. If the vent sites are not directly identifiable from sonar, it is possible that their occurrences may coincide with particular geological conditions that are detectable. All methods described will be developed over the coming years for the ASHES vent field at Axial Seamount and then applied to new or existing data from other areas, such as Gorda Ridge. If the techniques are adequate, the tools can be used not only to aid in the discovery of vent fields, but also for mapping their distributions to allow quantitative estimates of the extent of mineral formation and the impact of these systems on ocean chemistry.

In addition to studying the spatial distributions of hydrothermal fields, passive acoustic remote-sensing techniques may also be applied to monitoring the temporal variability of hydrothermal activity. The discovery of a large, ephemeral thermal anomaly in the water column over Juan de Fuca Ridge in 1986 (Baker et al., 1987a) indicates that hydrothermal processes may occur in large, catastrophic episodes as well as in quasi-steady-state venting processes. To monitor such activity requires the emplacement of long-term instrumentation to record over the period of years. Acoustic noise, if correlated with vent discharge volume, may provide an inexpensive means of quantifying the temporal variability of hydrothermal activity at a vent field. Preliminary work by Little et al. (1986) indicates that such a correlation should exist, and experiments are currently planned for the ASHES field to obtain high-frequency samples of the acoustic output of individual chimneys.

Episodic venting may be an important effect at northern Gorda Ridge. Baker et al. (1987b) found clear evidence of hydrothermal activity at the GR14 site on Gorda Ridge in summer of 1985. When the area was revisited in summer of 1986, a much weakened signal was detected (Collier and Baker, this volume). Although this apparent fluctuation in the level of hydrothermal activity may be an artifact of sampling, it may also reflect a true change in fluid output at the site. With the development of a long-term acoustic monitoring instrumentation and identification of the source field, it may be possible to confirm or deny this important effect.

Discussion

Because of the low transmissivity of light in seawater, acoustic energy represents the best physical medium for imaging the seafloor. The development in recent years of advanced, digital sonar systems, high-speed computers, accurate navigation systems, and appropriate mathematical methods allows satellite remote-sensing technology to be applied to the study of the seafloor. To calibrate the various systems, it is necessary to compare results from several instruments collected in a common location. Such a study is ongoing at Axial Seamount on Juan de Fuca Ridge. In addition to providing general geological/geophysical information, acoustic remote-sensing tools can be applied to the study of ridge-crest hydrothermal systems. It is anticipated that the acoustic remote-sensing techniques being developed at the Juan de Fuca Ridge will be applicable to the study of the geological and hydrothermal processes at work on Gorda Ridge.

Acknowledgments. Thanks to Mary Gosla, Susan Hanneman, Andy Lau, Maria Restrepo, and Marijke van Heeswijk. The author is supported through NOAA's VENTS research program. PMEL contribution number 979.

References

Baker ET, Massoth GJ, Feely RA (1987a) Cataclysmic hydrothermal venting on the Juan de Fuca Ridge. Nature 329:149–151.

Baker ET, Massoth GJ, Collier RW, et al. (1987b) Evidence for high temperature hydrothermal venting on the Gorda Ridge, northeast Pacific Ocean. Deep-Sea Res 34:1461–1476.

Chase TE, Wilde P, Normark WR, Miller CP, Seckins DA, Young JD (1981) Offshore Topography of the Western United States between 32° and 49° North Latitudes, Plate 2. U.S. Geological Survey Open-File Map 81:433.

Clague DA, Holmes ML (1987) Geology, petrology, and mineral potential of the Gorda Ridge. In: Scholl DW, Grantz A, Vedder JG (eds) Geology and Resource Potential of the Continental Margin of Western North America and Adjacent Ocean Basins—Beaufort Sea to Baja California. Houston: Circum-Pacific Council for Energy and Mineral Resources Earth Science Series 6:563–580.

Cooley JW, Tukey JW (1965) An algorithm for the machine calculation of complex Fourier series. Math Comput 19/90:297–301.

DeMoustier C (1986) Beyond bathymetry: Mapping acoustic backscatter from the deep seafloor with SEABEAM. J Acoust Soc Am 79:2.

EEZ SCAN Scientific Staff (1986) Atlas of the Exclusive Economic Zone, Western Conterminous United States. U.S. Geological Survey Miscellaneous Investigations Series I-1792, 152 pp, scale 1:500,000.

Elvers DJ, Potter KW (1971) NOS SEAMAP Planning Report. International Hydrographic Bulletin.

Fox CG, Hayes DE (1985) Quantitative methods for analyzing the roughness of the seafloor. Rev Geophys Space Phys 23/1:1–48.

Fox CG, Murphy KM, Embley RW (1988) Automated display and statistical analysis of interpreted deep-sea bottom photographs. Marine Geol 78:199–216.

Little SA, Purdy GM, Stolzenbach KD (1986) Acoustic monitoring of hydrothermal vent fields. EOS 67/44:1232.

Malahoff A (1985) Hydrothermal vents and polymetallic sulfides of the Galapagos and Gorda/Juan de Fuca Ridge systems and of submarine volcanoes. Bull Biol Soc Wash 6:3–18.

Mandelbrot B (1982) The Fractal Geometry of Nature. San Francisco: W.H. Freeman, 460 pp.

Shaw PR, Smith DK (1987) Statistical methods for describing seafloor topography. Geophys Res Lett 14:1061–1064.

Tyce RC (1986) Deep seafloor mapping systems—a review. Marine Technol Soc J 20/4:4–16.

van Wyckhouse RJ (1973) Synthetic Bathymetric Profiling System (SYNBAPS). Bay St. Louis, MS: Naval Oceanographic Office Technical Report TR-233.

12—Technologies for Geophysical Exploration on the Ocean Bottom

JOHN A. HILDEBRAND

Abstract

Several approaches have been used for ocean bottom geophysical exploration including magnetic, gravity, and seismic methods. Magnetic anomalies near the ocean bottom are measured with a nuclear precession magnetometer deployed from a deeply towed vehicle. They reveal fine-scale crustal magnetization indicating geomagnetic field reversals and zones of varying lithology. Gravity anomalies of the ocean bottom are measured with a gravimeter placed on the seafloor. The seafloor gravimeter is more precise than sea surface gravimeters owing to greater platform stability, and it has improved capability for revealing crustal density structures because of its closer proximity to crustal sources. Ocean bottom seismic methods use on-bottom seismographs or hydrophones and on-bottom shots. Use of on-bottom seismic instrumentation allows characterization of shallow inhomogeneous crustral structure. Accurate positioning is needed for fine-scale ocean bottom geophysical exploration and may be accomplished with acoustic transponders and satellite navigation.

Introduction

About 62% of the surface of the earth is covered by water of more than 1 km depth, yet the abyssal seafloor has been explored or sampled in only a limited number of areas. Geophysical exploration tools adapted to the deep ocean bottom have great potential for providing fine-scale surveys of the oceanic sediments and crust. These techniques give information on the nature and spatial distribution of the materials that comprise the ocean bottom.

These are two advantages to geophysical exploration conducted at or near the ocean bottom instead of at the ocean surface. The first is proximity to the materials under study. By observing the magnetic field, gravity field, or seismic waves at or near the ocean bottom, a great advantage is gained in the size of the observed signals. The second advantage is enhanced spatial resolution for ocean bottom techniques. The spatial resolution is roughly proportional to the spacing between the source and the detector, resulting in kilometer resolution for ocean surface techniques and tens of meters' resolution for ocean bottom techniques. Commensurate with their enhanced resolution, ocean bottom exploration techniques require a precise means of sea floor positioning. Three techniques of ocean bottom geophysical exploration will be described: near-bottom magnetometry, ocean bottom gravimetry, and ocean bottom seismology.

Near-Bottom Magnetometry

Near-bottom magnetic field measurements may be collected from a deeply towed instrumentation package (e.g., Spiess and Mudie, 1970; Macdonald et al., 1980) or from a submersible (e.g., Macdonald et al., 1983). Proton precession magnetometers have proved to be very reliable for these applications because of their high sensitivity, ease of operation, and ease of calibration. When operated with a deeply towed vehicle, the sensing coils of the magnetometer are towed 30 m behind the main instrument pressure case, and the associated magnetometer electronics (power supply, amplifiers) are contained inside the pressure case to avoid contamination of the magnetic field signal (Mudie, 1964). The proton magnetometer sensor consists of a coil immersed in a hydrogeneous liquid such as oil or water. During the polarization stage of operation a direct current of several amperes is passed through the coils for several seconds to align a small fraction (1 part in 10^6) of the protons. Use of a toroidally shaped coil ensures that some portion of the aligned protons are not parallel to the local ambient magnetic field (Acker, 1971). When the direct current is turned off, the polarized protons will precess about the earth's ambient field with a frequency that is proportional to the field amplitude. The precessing protons produce a voltage in the coil whose frequency can be measured and converted to field strength by use of the gyromagnetic ratio. One hertz corresponds to 23.5 nT (nano-Tesla) of field strength, so that 2 kHz resonance is experienced for typical 50,000 nT fields. To obtain 1 nT sensitivity it is necessary to measure the frequency to a fraction of a hertz. The limitation to accurate determination of the precession frequency is the exponential decay of precession with time; typically, the precession signal can be measured for about 6 sec. A total magnetic field measurement with an accuracy of about 1 nT at a repetition rate of 10 sec or more is achieved by the proton precession magnetometer.

New nuclear precession magnetometer designs may provide improved measurements of magnetic fields in the deep ocean environment (Slocum and Marton, 1974). The new magnetometer uses precession of helium-3 gas to give the total magnetic field. Its increased precision stems from very long (>100 hours) relaxation times for the helium-3 gas and results in more than 100 times the sensitivity of proton precession systems. Because of the long relaxation times, the precession frequency may be accurately determined by counting over long periods. Long relaxation times necessitate operation with low-frequency (<1 Hz) magnetic signals, but this is not a limitation for deeply towed magnetometers which are limited to slow towing speeds (signals of 0.01 Hz or less). The sensitivity of the helium-3 magnetometer is beyond the level of the geomagnetic background noise, even when operated on the seafloor. To benefit from the enhanced sensitivity, it may be necessary to operate it with a pair of sensors configured as a gradiometer.

Measurements of magnetic field anomalies close to the ocean bottom produce high amplitude signals (1,000 nT anomalies) with short wavelengths (<1 km) owing to the close proximity of the magnetic sources. These large-amplitude anomalies can result both from geomagnetic field reversals recorded by the oceanic crust and from source-detector path length variations due to relief of the oceanic crust. Figure 12.1 shows a near-bottom magnetic profile made across the Gorda Ridge at 41.5° N (Atwater and Mudie, 1973) which was inverted to obtain the magnetization amplitude as a function of distance from the spreading center (Klitgord et al., 1975). A large peak in the deep-tow magnetic anomaly is located over a 1 km region at the center of the rise crest. The magnetization inversion reveals a highly magnetized zone with amplitude 25 A/m associated with this central anomaly.

Ocean Bottom Gravimetry

Gravimeters adapted to work on the deep ocean bottom allow more accurate and more detailed gravity measurements to be made than is possible from surface ships or submarines. Recent work at Scripps Institution of Oceanography has adapted a commercially available La Coste and Romberg underwater gravimeter for use at water depths of up to several kilometers. An overview of the system is shown in Figure 12.2. The two major components are the gravimeter package and an intermediate sonar package. The sonar package provides acoustic navigation, a conductivity-temperature-depth instrument, precision measurement of adjacent topography, and termination of the main electromechanical wire.

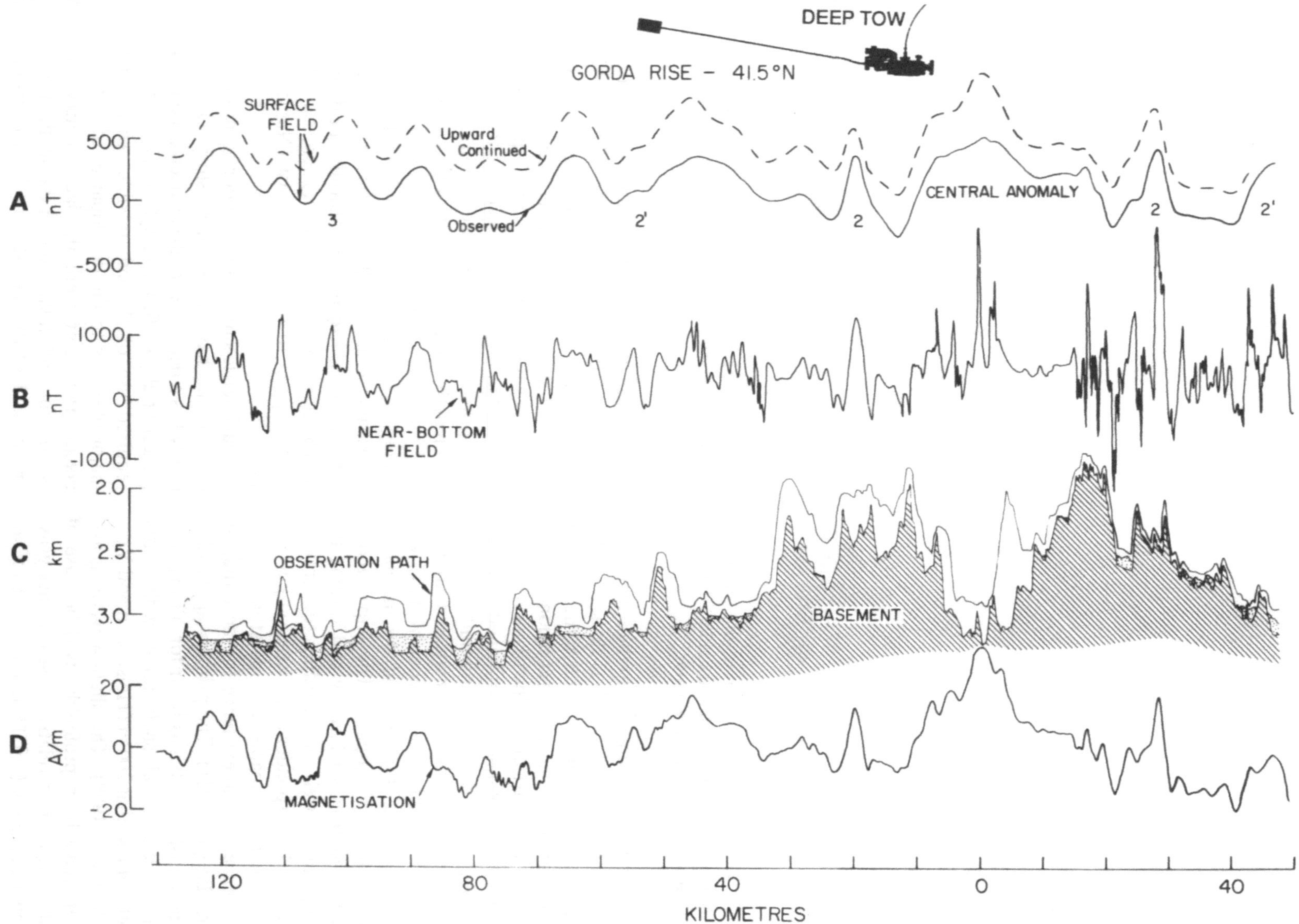

FIGURE 12.1. Geophysical profile across the Gorda Ridge at 41.5° N projected along azimuth 100. The profiles show (A) the observed and upward continued magnetic fields at the sea surface, (B) the near-bottom magnetic field, (C) the bathymetry and observation path for the deep-tow vehicle, and (D) the calculated magnetization of the oceanic crust. (After Klitgord et al., 1975.)

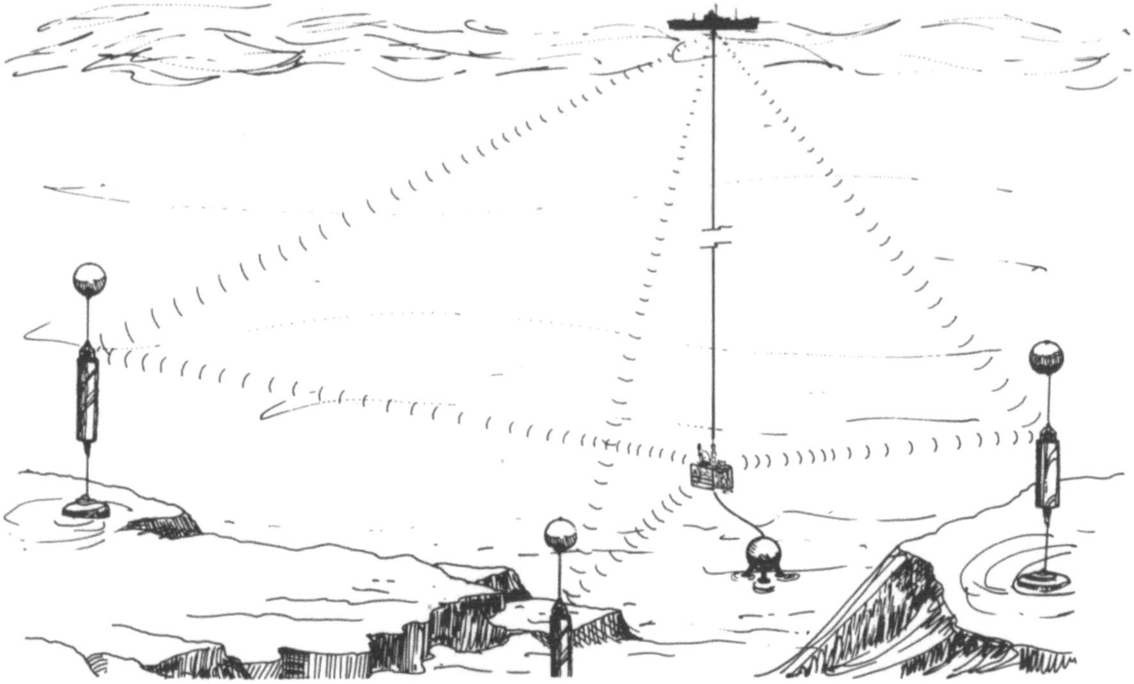

FIGURE 12.2. Artist's conception of the ocean-bottom gravimeter package in operation on the seafloor. Acoustic transponders are shown deployed encircling the gravimeter, and a sonar package is shown on the electrical wire above the gravimeter.

The gravimeter is connected to the sonar package by a flexible tether which is slack when the meter is on the bottom. The tether is constructed from a 1-inch-diameter nylon mooring line with an electrical conductor inserted into its core. The gravimeter is contained in a pressure case along with a leveling unit and associated electronics for telemetry and control. On board the ship are the control unit, data recorders, sonar displays, and the navigation and data logging computer. The ship's laboratory has controls for leveling and zeroing the gravimeter as well as recording the gravity, depth, and transponder navigation outputs.

Gravimeters capable of operating underwater have been in existence for more than 40 years (Pepper, 1941). The primary application of these instruments has been for oil exploration in continental shelf areas with water depths of less that 200 m (Heiskanen and Vening Meinesz, 1958). The shallow-water gravimeters have been profitably used because of their high accuracy relative to sea surface gravity measurements (~0.02 mGal vs. ~1 mGal) and because of their decreased source to sensor separation (Beyer et al., 1966; Wing, 1967; Neuman and Talwani, 1972). The proximity between the gravimeter and the source region determines the wavelength limit of information contained in the gravity measurements. A seafloor gravimeter allows observation of short-wavelength components of the oceanic gravity field and aids in determining the structure of geological features in the upper few kilometers of oceanic crust. Shallow water gravimeters have been used to find oil-bearing salt domes which are a few thousand meters across and lie in the upper 10 km of the crust (Nettleton, 1976).

Gravimetric surveys on the floor of the deep-ocean basins have no parallel history. Only recently has a deep-ocean survey been conducted (Spiess et al., 1980; Luyendyk, 1984), and this required use of the submersible *Alvin*. The major obstacles to the development of a deep-ocean gravimeter were the lack of accurate positioning and navigation on the seafloor and imprecise bathymetry. Recent improvements

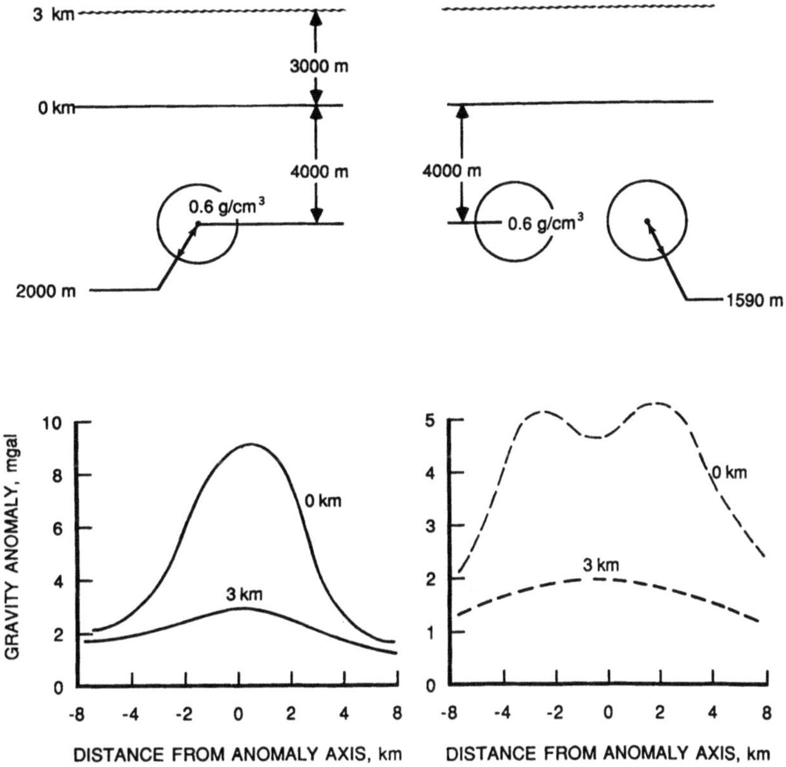

FIGURE 12.3. Comparison between the gravity anomalies produced by (a) a single spherical body and (b) two spherical bodies of equivalent total mass. Profiles are computed for a sea surface (solid line) and a seafloor observation plane (dashed line).

in seafloor positioning (e.g., Spiess et al., 1983) and in topographic mapping (SEABEAM) make the development of a seafloor gravimeter a practical goal.

The importance of close source-sensor proximity in discerning short wavelength crustal features is illustrated by the following model calculation. A comparison between calculated gravity profiles over a spherical body and over a pair of spherical bodies, whose centers are located at a depth of 4 km within the ocean crust, is shown in Figure 12.3. Profiles were calculated at the sea surface and on the seafloor assuming a water depth of 3 km. The anomalies measured from the sea surface differ by less than 1 mGal for the two models, and the sea surface measurements do not resolve the two-lobed structure of the pair of spherical bodies. By contrast, the on-bottom gravity profiles reveal as much as a 4 mGal difference between the two models, and the measurements clearly resolve the pair of spherical bodies. These models show that significant features visible with a gravimeter on the deep-ocean bottom may be undetectable by a shipboard instrument.

A seafloor gravimeter will be useful for several problems concerning sea floor shallow crustal structure on the scale of a kilometer or less. For example, the rise crests are thought to contain significant fine-scale crustal variations due to the presence of magma chambers and hydrothermal circulation. An *Alvin*-based East Pacific Rise gravity study delimited a body of low-density, hydrothermally fractured rock at the rise axis (Luyendyk, 1984). The Bouguer anomaly associated with this feature was on the order of 2 mGal and would not have been observed on a shipboard gravimeter. Further rise crest seafloor gravity studies might constrain the presence and extent of magma chambers and hydrothermal circulation. Seafloor gravity surveys at high points along the ridge (locations of excess magma supply) may be compared to surveys at low points (locations of scarce magma supply) to understand along strike variation in ridge crest magma chambers. Another area of investigation is the variation of gravity at active areas on the ridge or at active submarine volcanoes. Experiments on active subaerial volcanoes have revealed temporal variations in their gravity fields that provided information on magma movement and surface distortions (Dzurisin et al., 1980; Jachens and Eaton, 1980; Sanderson, 1982). A gravity survey was recently conducted at Axial Volcano, an active seamount on the Juan de Fuca Ridge, revealing details of the subsurface structure (Stevenson et al., 1988).

Ocean Bottom Seismometry

Ocean bottom seismic receivers and ocean bottom seismic sources may be used to characterize the structure of the deep seafloor. There are several advantages to using seismic receivers and sources located close to the seafloor. First, seafloor placement allows measurement of shallow crustal structure within the top kilometer of sediment or crust (Purdy, 1986). Second, propagation paths within the ocean water column are eliminated, removing ambiguities in water velocity from the observed seismic arrival. Third, seismic shear waves and other modes may be more easily generated and detected. Ocean bottom seismic receivers are typically operated as autonomous vehicles (see Fig. 12.4) and use either geophones to sense particle velocity (e.g., Moore et al., 1981) or hydrophones to sense pressure (e.g., Koelsch and Purdy, 1979). Ocean bottom seismic sources may be either explosive shots or mechanical resonators. Seafloor explosive shots require special techniques to initiate the detonation at the high pressures found on the ocean bottom. Two systems have recently been developed for deployment of seafloor shots of moderate size (Dorman et al., 1986).

The principal approach to studying ocean bottom seismic structure has involved collection of seismic refraction profiles using single instruments (e.g., Orcutt et al., 1976). The data are interpreted in terms of laterally homogeneous, horizontally layered structures or simple heterogeneous structures using travel time inversion and synthetic seismogram modeling techniques. These techniques are inadequate, however, when large-scale lateral inhomogeneity is present, such as in a rise crest environment. New approaches are being developed that may allow mapping of the velocity and attenuation structure of laterally inhomogeneous crust. For example, arrays of instruments may be used and the seismic signal from individual receivers combined to yield maps of the directional properties of the arrival. Likewise, the travel times to an array of instruments can be used as a collection of path integrated measurements of acoustic velocity from which an image may be constructed by tomographic techniques (e.g., Hilderbrand et al., 1987; Toomey et al., 1988).

Seismic array experiments require greater precision in instrument placement and navigation than for traditional single-instrument experiments because of the need to combine the signals from a number of sensors. Precision navigational aids include acoustic transponders and GPS (global positioning system) satellite navigation. The GPS navigation system gives highly accurate horizontal position information (~50 m accuracy) but is currently available for only about 5 hours per day. By dropping a seismic instrument from the ship during a period of GPS navigation it may be positioned to within a few hundred meters. A more accurate approach is to lower the instrument to the seafloor on a wire and release it at a location navigated within a set of acoustic transponders (Schreiner et al., 1987). With this method seafloor instruments have been deployed with about 1 m lateral position accuracy. A similar technique has been used for accurate ocean bottom shot deployment (Dorman et al., 1986). A sonar package served as a seafloor shooting box for navigation, detonation timing, and position control of the shots that were suspended beneath it.

Ocean Bottom Positioning

Accurate ocean bottom positioning is a key aspect of near-bottom geophysical exploration techniques. The variability of fine-scale features on the seafloor make precise positioning an absolute necessity. Ocean bottom naviga-

FIGURE 12.4. Ocean bottom seismograph package of the Scripps Institution of Oceanography.

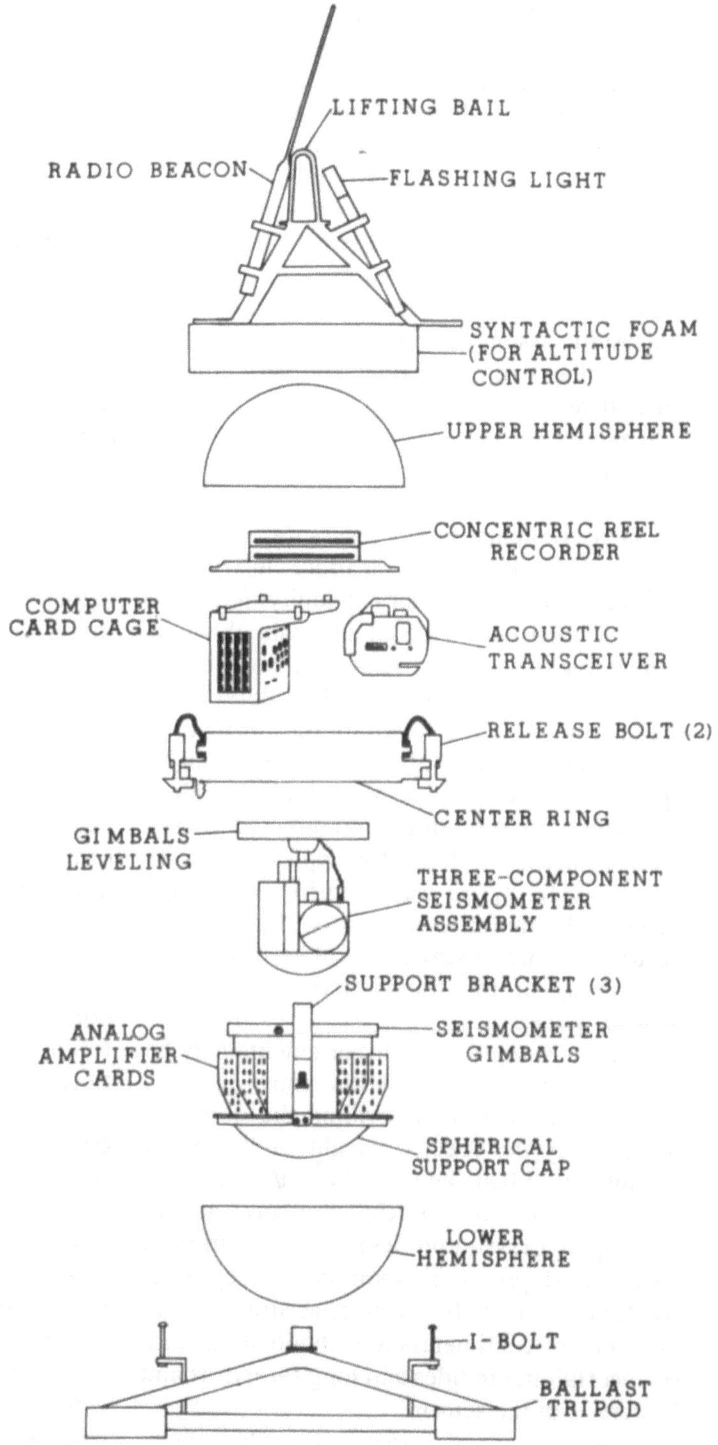

tion requires specialized techniques; radar and optical-based systems used for land positioning are replaced by acoustic and pressure-based systems in the deep ocean environment. Acoustic positioning systems typically measure pulse travel times to determine distances. These systems are sensitive to variations in acoustic propagation velocity owing to changes in ocean temperature, pressure, and salinity. Pressure systems measure depth under the hydrostatic assumption and must be corrected for the effects of ocean tides and variations in water density.

Seafloor instrument positioning requires both vertical and horizontal information. Instrument depth is determined either by acoustic measurements or by pressure measurements. An autonomous seafloor instrument with acoustic transponder capability may have its depth measured by locating a ship above it, minimizing the acoustic range between the ship and the instrument. This method determines the instrument depth to within tens of meters. Acoustical depth may also be obtained from an up-looking echo sounder, an acoustic transducer mounted on the vehicle and directed toward the sea surface. Such systems typically operate near 20 kHz (1.3 m wavelength) and resolve depth to within a few meters. Inaccuracies occuring owing to scattering off surface waves and timing of the transmitted and received acoustic pulse.

More accurate depth measurements may be obtained from sensitive pressure gauges. Depth is calculated by dividing the observed pressure by the specific weight of the water (nominally 1,030 $kg/m^3 \times 9.8$ m/sec^2). The most accurate commercial pressure gauges (e.g., quartz crystal probes) have resolution of 69 Pa (0.01 psi), corresponding to depth resolution of 7 mm. In practice, accuracies of at best tens of centimeters are obtained because of uncertainties in tides and long-term drift and hysteresis of the sensor.

Measurements of horizontal seafloor position are made with several different scales of accuracy and complexity. The simplest method is to locate a surface ship directly above the sea floor instrument by minimizing the acoustic range. Positioning the ship using satellite (GPS, transit) or radar (Loran) navigation gives an estimate of the seafloor instrument position that is accurate to 100 m at best. A variation of this method is to use multibeam bathymetry to position the ship in a "map-matching" mode; again, accuracies of about 100 m are achievable.

A more precise method for horizontal positioning uses acoustic transponder beacons anchored near the seafloor. By combining the ranges to three or more transponders, an estimate of horizontal position is calculated by triangulation. Two transponders suffice if an independent depth estimate is available. This system requires an acoustic path and adequate transmission band width for accurate timing. Available acoustic paths are a severe limitation for near-bottom systems, since increasing velocity with depth results in upward-refracted rays. When both the transmitter and receiver are mounted within 2 m of the seafloor, the usable range is limited to about 1 km horizontal range, whereas for transmitter and receiver located 100 m above the seafloor the usable path is 8 km. The band width available for acoustic transmission is another limitation. For propagation paths of a few kilometers, acoustic attenuation in seawater limits maximum usable frequencies to approximately 30 kHz. These frequencies result in 30-μsec rise times and centimeter errors in position. For propagation paths of 100 km, usable frequencies are limited to approximately 1 kHz, providing 1 msec rise times and meter errors in position.

Another requirement for seafloor transponder systems is knowledge of the environmental sound speed. The three seawater parameters most relevant to sound speed variation are temperature, pressure, and salinity. The effects of small changes in these parameters are: for temperature, 5 m/sec/°C; for pressure, 0.016 m/sec/m depth; and for salinity, 1.33 m/sec/part per thousand.

Knowledge of seawater temperature, pressure, and salinity is a limiting factor for precise positioning. For measurements accurate to 1 m over 10 km range, sound speed must be determined to 0.15 m/sec. This accuracy requires knowledge of temperature to 0.03°C, of depth (pressure) to 10 m, and of salinity to 0.1 part per thousand. For 10 cm positioning precision these seawater parameters must be known with 1 order of magnitude greater precision, or equivalently the sound speed must be directly measured to 0.015 m/sec. Lateral temperature variations of 0.005°C over kilometer ranges are typical for the seafloor environment and for extreme cases (e.g., near hydrothermal vents) may be considerably larger. Tidal excursions of a meter or more are typical, leading to variations in pressure at depth. Salinity variations may be important, particularly in coastal areas. Surveys of environmental parameters and/or sound speed are needed for precision better than 1 m in position.

Precise seafloor positioning also depends on acoustic transponder technology for travel time determination (Spiess, 1980). Two aspects of transponder technology are important: consistent recognition response time, and signal wave forms that minimize reverberation interference. Current transponders have recognition response circuits that vary by about 1 msec (1.5 m range), depending on received signal power. Prototype transponders are being developed with 2 μsec response time uncertainty, independent of signal power, allowing centimeter range resolution (Spiess, 1985). These transponders must be mounted on the seafloor for position stability and therefore require short (1 msec or less) pulse lengths to discriminate against reverberation.

Current practice achieves seafloor position measurements accurate to about 1 m over a range of several kilometers. Transponder beacons are typically deployed with the aid of GPS satellite navigation to give initial location estimates. To further constrain their locations, a set of simultaneous ranges to three or more transponders are collected. The transponder locations are subsequently adjusted to minimize the error of the range data. A sound speed profile, assumed to be horizontally layered, is included in the calculations. Vehicle motion and ray curvature are typically ignored. Under these conditions, range crossing errors are often as small as 1 m. Positioning precision of less than 1 m requires more precise environmental sound speed data than are currently collected.

Other systems that are potentially useful for seafloor positioning include optical and inertial types. Optical positioning techniques use short pulses of light in a ranging mode. These techniques are severely restricted in range owing to the high attenuation and scattering of light by seawater; under optimal conditions the absorption length (distance for signal strength to fall by 63%) in seawater is 25 m. A system could be constructed with at best a few centimeters accuracy over a few hundred meters range. Inertial system use gyrostabilized accelerometers to provide navigational extrapolation capability. They are initialized by reference to known positions but are subject to gradual drift, which may accumulate to unacceptable levels. These systems are currently used for submarine navigation and for borehole well logging. Closure error of about 5 m has been reported for usage in 3-km-deep wells. There are continuing programs to improve the accuracy of inertial systems. They may find application for seafloor positioning if their precision is improved and if their cost is not excessive.

Summary

This chapter has described several techniques for ocean bottom geophysical exploration: (1) magnetic anomalies near the ocean bottom may be measured with a nuclear precession magnetometer deployed from a deeply towed vehicle; (2) gravity anomalies of the ocean bottom may be measured with a gravimeter placed on the seafloor; (3) ocean bottom seismic methods use on-bottom seismographs or hydrophones and on-bottom shots. These ocean bottom geophysical exploration techniques produce large-amplitude signals and give fine-scale lateral resolution. The technical challenge involved in on-bottom exploration is development of remotely operated instrumentation and methods of ocean bottom positioning. The state of the art for ocean bottom positioning uses acoustic transponders and pressure gauges to obtain 1 m precision over a few kilometers range, a precision adequate for many applications of ocean bottom geophysical exploration.

Acknowledgments. I thank Alan Chave and Peter Lonsdale for helpful reviews of this manuscript. Research support was provided by the Office of Naval Research contracts N00014-87-K-0010 and N00014-80-C-0220 (ocean bottom gravity) and National Science Foundation grant OCE84-15704 (ocean bottom seismology).

References

Acker FE (1971) Calculation of the signal voltage induced in a toroidal proton-precession magnetometer sensor. IEEE Trans Geosci Electronics GE-9:98–103.

Atwater TM, Mudie JM (1973) Detailed near-bottom geophysical study of the Gorda Rise. J Geophys Res 78:8665–8686.

Beyer LA, Von Huene RE, McCulloh TH, Lovett JR (1966) Measuring gravity on the seafloor in deep water. J Geophys Res 71:2091–2100.

Dorman LM, Hildebrand JA, Hammer PTC, Sauter AW, Schreiner AE (1986) Seismic study of a seamount interior. EOS 67:1083.

Dzurisin D, Anderson LA, Eaton GP, et al. (1980) Geophysical observations of Kilaeua volcano, Hawaii. 2. Constraints on the magma supply during November 1975–September 1977. J Volcanol Geotherm Res 7:241–269.

Heiskanen WA, Vening Meinesz FA (1958) *The Earth and its Gravity Field.* New York: McGraw-Hill, p 470.

Hildebrand JA, Dorman LM, Hammer PTC, Schriener AE, Sauter AW, Cornuelle B (1987) Seismic tomography of Jasper Seamount. Geophys Res Lett (submitted).

Jachens RC, Eaton GP (1980) Geophysical observation of Kilauea volcano, Hawaii. 1. Temporal gravity variations related to the 29 November 1975, M-7.2 earthquake and associated summit collapse. J Volcanol Geotherm Res 7:225–240.

Klitgord KD, Mudie JD, Huestis SP, Parker RL (1975) An analyis of near-bottom magnetic anomalies: Sea-floor spreading and the magnetized layer. Geophys J R Astron Soc 43:387–424.

Koelsch DE, Purdy GM (1979) An ocean bottom hydrophone instrument for seismic refraction experiment in the deep ocean. Marine Geophys Res 4:115–125.

Koelsch DE, Witzell WE, Broda JE, Wooding JW, Purdy GM (1986) A deep towed explosive source for seismic experiments on the ocean floor. Marine Geophys Res 8:345–361.

Luyendyk BP (1984) On-bottom gravity profile across the East Pacific Rise crest at 21° north. Geophysics 49:2166–2177.

Macdonald KC, Miller SP, Huestis SP, Spiess FN (1980) Three-dimensional modeling of a magnetic reversal boundary from inversion of deep-tow measurements. J Geophys Res 85:3670–3680.

Macdonald KC, Miller SP, Luyendyk BP, Atwater TM, Shure L (1983) Investigation of a Vine-Matthews magnetic lineation from a submersible: The source and character of marine magnetic anomalies. J Geophys Res 88:3403–3418.

Moore RD, Dorman LM, Huang C, Berliner DL (1981) An ocean bottom, microprocessor based seismometer. Marine Geophys Res 4:451–477.

Mudie JD (1964) Deep-towed proton magnetometer. MPL Technical Memorandum 150: Scripps Institution of Oceanography, La Jolla, CA.

Nettleton LL (1976) *Gravity and Magnetics in Oil Prospecting.* New York: McGraw-Hill.

Neuman LD, Talwani M (1972) Accelerations and errors in gravity measurements on surface ships. J Geophys Res 77:4330–4338.

Orcutt JA, Kennett BLN, Dorman LM (1976) Structure of the East Pacific Rise from an ocean bottom seismometer survey. Geophys J R Astron Soc 45:305–320.

Pepper TB (1941) The Gulf underwater gravimeter. Geophysics 6:34–44.

Purdy GM (1986) A determination of the seismic velocity structure of sediments using both sources and receivers near the ocean floor. Marine Geophys Res 8:75–91.

Sanderson TJO (1982) Gravimetric defection of magma movements at Mount Etna. Nature 297:487–490.

Schreiner AE, Dorman LM, Hildebrand JA, Lahav D, Spiess FN (1987) Wavelength and correlation length of deep ocean ambient seismic noise. EOS 68:1373.

Slocum RE, Marton BI (1974) A nuclear free precession magnetometer using optically polarized He^3 gas. IEEE Trans Magnetics 10:528–531.

Spiess FN (1980) Acoustic techniques for marine geodesy. Marine Geodesy 4:13–27.

Spiess FN (1985) Suboceanic geodetic measurements. IEEE Trans Geosci Remote Sens 23:502–510.

Spiess FN, Mudie JD (1970) Small scale topographic and magnetic feature. In: AE Maxwell (eds) *The Sea, 4, Part I.* New York: John Wiley and Sons, pp 205–250.

Spiess FN et al. (1980) East Pacific Rise: Hot springs and geophysical experiments. Science 207:1421–1432.

Spiess FN, Cox CS, Hays EE, Porter RP, Roberts

FA (1983) Seafloor referenced positioning: Needs and opportunities. Washington: Panel on Ocean Bottom Positioning of the National Research Council's Committee on Geodesy National Academy Press, p 54.

Stevenson JM, Hammer PTC, Hildebrand JA, Fox CG (1988) The sea-surface and seafloor gravity field of Axial Volcano, Juan de Fuca Ridge. EOS 69:1467.

Toomey DR, Purdy GM, Solomon SC (1988) First results from an East Pacific Rise seismic tomography experiment. EOS 69:1474.

Wing CG (1967) An experimental deep-sea-bottom gravimeter. J Geophys Res 72:1249–1257.

13—Seafloor Electromagnetic Exploration Methods

ALAN D. CHAVE

Abstract

Electrical exploration of the marine environment is just beginning, but interest in the application of electromagnetic approaches is rising rapidly. A brief review of seafloor electromagnetic exploration principles and techniques is given, with an emphasis on the midocean ridge environment. The physical processes governing the electrical conductivity of rocks are discussed. Two types of seafloor electromagnetic exploration systems are then described. The first uses a frequency-domain controlled source to inject energy into the seafloor and has been used for both shallow and deep sounding of the oceanic lithosphere. The second uses a time-domain controlled source and is under development as a tool for rapid, near-bottom surveying.

Introduction

Recent developments in submarine geology and scientific instrumentation have resulted in increased interest in the use of electromagnetic methods for the exploration of the seafloor. Little attention had previously been given to their application in the marine environment because of a pervasive misunderstanding of the physical meaning of rock electrical conductivity information, a widespread belief that the high electrical conductivity of seawater precluded the application of electromagnetic approaches, and a lack of economic interest in the deep seafloor. However, marine electromagnetic probing of the structure of the deep earth has progressed substantially in academic circles over the past decade. The focus of marine electromagnetic work will increasingly be on the detailed structure of midocean ridges in coming years. In addition, the presence of deposits of polymetallic sulfides with economic potential in the near-surface regions and around hydrothermal vents on midocean ridges has brought increasing attention to the principal geophysical tools of the land mineral explorationist—the electrical methods.

In this chapter, a brief review of seafloor electromagnetic exploration methodology is given. This begins with a discussion of the physical interpretation of rock electrical conductivity information, emphasizing the near-bottom, midocean ridge environment. The remainder of the chapter focuses on active source electromagnetics using two complementary approaches. The first, frequency-domain controlled source electromagnetics (CSEM), is especially suitable for detailed, regional and local studies of electrical structure. The second, time-domain electromagnetics (TDEM), shows tremendous potential as a rapid mapping tool when deployed from a near-bottom, deep-towed vehicle.

Electrical Conductivity in Rocks

Electrical conductivity covers a greater range than any other physical variable in the earth, spanning at least seven decades. This should be contrasted to seismic velocity, which varies by less than a factor of 10 throughout the earth.

The electrical conductivity of seawater is linearly proportional to its temperature under oceanic conditions, with a very small salinity and pressure effect; it varies from as much as 5 S/m near the surface to about 3.2 S/m belows the main thermocline. The conductivity of marine sediments is about an order of magnitude lower and is controlled by water content and porosity. The conductivity of rocks at subsolidus temperatures is usually much smaller than that of seawater and is dominated by the effects of interstitial fluids in cracks and pores at low temperatures (<600°C) and by mineral conduction at temperatures above 1,000°C. At temperatures exceeding the solidus, the electrical conductivity of rocks is determined by the fraction and connectedness of partial melt and can easily approach or exceed the seawater value. Because all of the relevant physical processes controlling electrical conductivity are thermally activated with a distinct activation energy and $e^{-1/T}$ temperature dependence, a single mechanism must be dominant for a given set of environmental parameters (e.g., temperature, pressure, volatile content). The interpretation of electromagnetic data, as for other geophysical measurements, depends on both support from laboratory experiments made under controlled conditions and corroborating evidence from other geophysical studies.

It is known from large-scale resistivity experiments in Deep Sea Drilling Project Hole 504B that the bulk conductivity of the pillow lava region at the top of seismic layer 2 is about 0.1 S/m (Becker et al., 1982; Becker, 1985). Based on measurements in other boreholes, the conductivity appears to decrease with the age of the underlying lithosphere, probably because of crack sealing by chemical precipitates. A sharp decrease in conductivity by a factor of 100 is observed in hole 504B near the base of layer 2 at a depth of about 1 km (Becker et al., 1982; Becker, 1985). A decrease in conductivity with depth is also required by the results of three controlled source electromagnetic experiments at other locations (Young and Cox, 1981; Cox et al., 1986).

This leads to the following simplified model for ocean crustal electrical conductivity. In the upper oceanic crust, the conductivity is largely controlled by fractures, cracking, and the presence of seawater, varying with the size and connectedness of the fluid passages and with the fluid temperature. As a result, electrical conductivity is an excellent indicator of active hydrothermal circulation and its depth of penetration. The conductivity decreases at depth in the crust as the porosity falls owing to crack closure and sealing. The intrinsic conductivity of the dry silicates making up the crust is ordinarily quite low. In the deeper parts of the crust, the electrical conductivity is expected to be of order 0.001 S/m, but over hot, young lithosphere near ridge axes, high temperatures may lead to an increase of as much as a factor of 10. However, conducting minerals, principally sulfides and magnetite, do occur in the oceanic crust, usually in association with hydrothermal regimes; the presence of grain boundary magnetite has been proposed as an explanation for the high conductivity of serpentinites (Stesky and Brace, 1973). These minerals will usually appear as small-scale, near-surface inclusions in a much lower conductivity crustal matrix, although serpentinized rocks are sometimes found in association with deep-seated rock assemblages. In addition, a layer of relatively conductive sediment may blanket the oceanic crust, although sediment cover is usually thin on midocean ridges.

Below the seismic Moho, the electrical conductivity is generally much lower than in the crust because of reduced hydrothermal activity, lower water content, and the absence of conductive mineral phases. Based on a variety of laboratory and field data, it is expected that the conductivity of dry lithospheric rocks is generally below 0.001 S/m in the upper 20 km below the sea floor and then rises sharply as the temperature increases. It is clear that the thermal structure of the lithosphere has a controlling influence on its conductivity if the volatile content is low or, conversely, the rise in conductivity with depth will outline the upper part of the thermal boundary layer which delimits the oceanic lithosphere. Two controlled source electromagnetic studies over 25-MY-old lithosphere support these statements, giving a conductivity of about 10^{-5} S/m in the upper 20 to

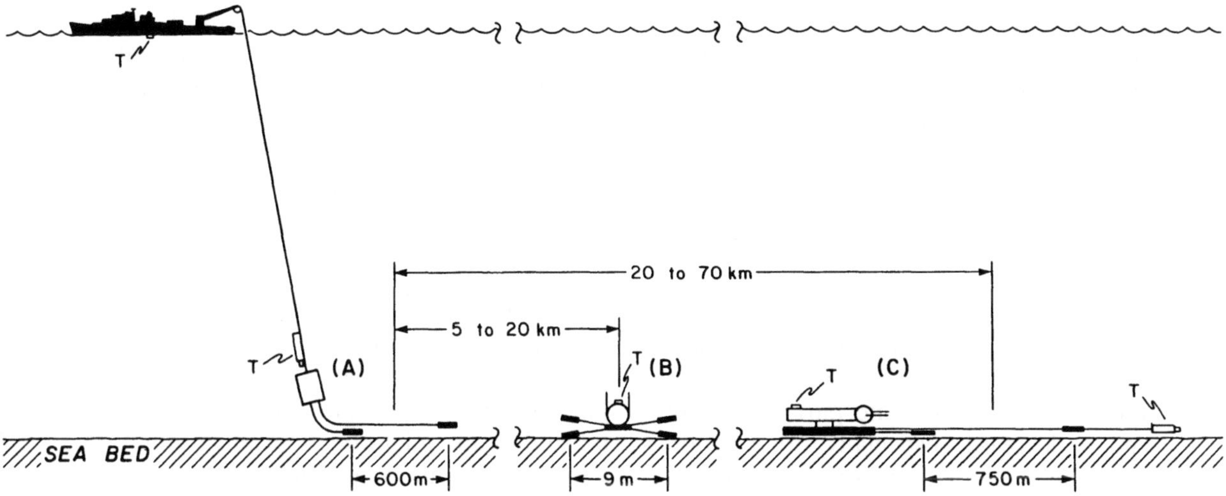

FIGURE 13.1. Layout for a horizontal electric dipole deep-sounding experiment.

30 km of the lithosphere (Cox et al., 1986). This requires a water content of under 0.1% by volume.

The conductivities of partially molten silicate rocks are much higher, approaching 0.1 S/m or more for small partial melt fractions. The conductivities of the melt fraction are 100 to 1,000 times higher than that of the host rock (Presnall et al., 1971). This suggests that the use of electrical methods to detect and delineate midocean ridge magma chambers is worthy of serious consideration; the conductivity contrast between magma chamber and host rock is a factor of 10 to 100 or more.

This brief discussion should make it clear that electrical conductivity measurements can provide unique and valuable information about the structure of the seafloor on all spatial scales and for a variety of geophysical purposes. Further background on laboratory measurements of rock conductivity and the effect of melting may be found in Shankland (1981) and Shankland and Waff (1977), respectively.

Frequency-Domain Controlled Source Electromagnetics

Frequency-domain controlled source electromagnetics (CSEM) utilizes a seafloor source and a series of seafloor receivers placed at various ranges from the source. Figure 13.1 shows a schematic picture of the layout of a CSEM experiment. A surface ship provides fixed-frequency AC power to a seafloor transmitter through a single conductor cable with a seawater return, typically at high voltage (1 kV) and low current (10 A). The transmitter (A) steps this down by roughly a factor of 10, increasing the available current proportionally, and switches the half cycles of the input wave form to synthesize signals at frequencies from 0.05 to 64 Hz. This is used to drive an insulated antenna with bared ends of about 600 m in length. Either short-arm (B) or long-wire (C) receivers are placed at ranges of a few kilometers to 100 km or more from the source and use synchronous stacking techniques to narrow the receiver band width and raise the signal-to-noise ratio. The transmitter and receivers function with a prearranged program of frequencies and operating times. Acoustic transponders (T) are used to locate all of the seafloor instruments from a surface ship.

Oceanic CSEM is based on three main principles:

1. Because the conductivity of seawater is usually much larger than that of rock, the portion of the signal propagating through the ocean attenuates rapidly with range. The relevant scale is the electromagnetic skin depth and is about 270 m at 1 Hz in

seawater, decreasing with the square root of frequency. This means that any EM signal that is observed at ranges exceeding a few skin depths in seawater must be traveling through the underlying rocks, and the field behavior with range and frequency is diagnostic of the conductivity structure beneath the seafloor.

2. CSEM depends on diffusive propagation through low-conductivity material. This means that the maximum range that can be achieved at a given frequency increases as the conductivity of some part of the lithosphere decreases, especially if that region is near the seafloor. Conversely, if material with a high enough conductivity is present, no propagation will be observed at range, although an intervening low relative conductivity zone can channel the fields like a leaky wave guide at intermediate distances. If high conductivity zones are buried at increasing depths, their influence will be oberved at larger source-receiver separations. The penetration depth of a CSEM system in a uniform medium is of order half the source-receiver spacing.
3. The natural noise level at the sea floor at frequencies of 0.05 to 100 Hz is extremely low and is typically millions of times smaller than would be observed on land. This means that very weak signals from a source can be detected at considerable ranges. For measurements of seafloor electric noise spectra, see Cox et al. (1978) and Webb and Cox (1986).

Two types of oceanic CSEM systems have been constructed and used to date. In Canada, a method has been developed using a vertical electric dipole (VED) source stretching from the sea surface to the seafloor and a series of horizontal magnetic field receivers placed on the seafloor (Edwards et al., 1981, 1985). The technique is based on the magnetometric resistivity principle and has been deployed successfully on several occasions and in a variety of locales. For example, soundings have been obtained in the coastal waters of British Columbia and in Middle Valley on the Juan de Fuca Ridge (Edwards et al., 1985; Nobes et al., 1986). A miniaturized version of the equipment has also been constructed and deployed from the submersible *Alvin* for small-scale studies in a polymetallic sulfide region (Wolfgram et al., 1986). While the VED method is (in a relative sense) logistically simpler than the method described below and does yield information on subsurface structure at very small source-receiver spacing because it can operate in the DC limit, its depth of penetration into the crust and lithosphere is not as great as for other techniques. It does have important applications in near-surface exploration and crustal studies and might be especially useful in the neovolcanic zone, where the topography is rugged and sediment cover is sparse, making the deployment of horizontal seafloor antennas fraught with danger. However, it is difficult to envision using a VED system in a towed mode.

An alternative approach using a horizontal electric dipole (HED) source energized at frequencies above 0.05 Hz and a series of horizontal electric field receivers, all deployed on the seafloor, has been developed at the Scripps Institution of Oceanography. A similar system is being constructed at the University of Cambridge in England. An HED experiment is depicted in Figure 13.1. The object of this type of experiment is detection of the deep (usually subcrustal) electrical conductivity beneath the seafloor. However, a deep sounding experiment must yield some information on intermediate (i.e., crustal) regions, since the electromagnetic signals must cross such zones to reach the deeper ones. A description of HED equipment and operations may be found in Chave et al. (1989).

Since the HED system is based on the propagation of AC electromagnetic disturbances from the source, it can require considerable horizontal offsets of source and receiver, depending critically on the underlying conductivity structure. In a recent experiment, a 24-Hz signal was easily detected at a range of 70 km over moderate-age lithosphere (Cox et al., 1986). Figure 13.2 illustrates the behavior of the HED radial electric field as a function of range at a frequency of 1 Hz for a simple buried layer in a half space of conductivity 0.05 S/m overlain by a seawater half space of con-

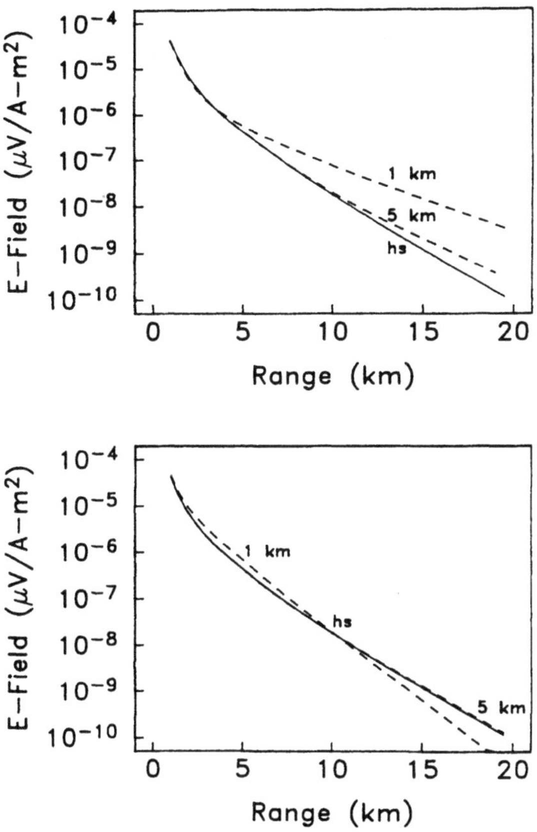

FIGURE 13.2. Radial electric field produced by an HED source as a function of range at a frequency of 1 Hz for an ocean half space overlying a rock half-space containing 1-km-thick layers centered at 1 and 5 km depth.

ductivity 3.2 S/m. The layers are 1 km thick and centered at either 1 km or 5 km depth. In the upper panel, the layer has a low relative conductivity of 0.005 S/m, while in the lower panel the conductivity of the layer is high (0.05 S/m). If the layer has a low relative conductivity, then propagation along the zone is observed, and the field amplitude is larger than would be expected for a uniform half space at the higher conductivity at all ranges beyond a critical value. As the layer is buried at greater depths, the critical distance increases, and the size of the effect decreases owing to attenuation in the overlying rock. Conversely, if the buried layer has a high relative conductivity, the observed field is initially larger than the half space value owing to propagation in the low (relative) conductivity channel lying between the ocean and the buried layer and is attenuated at longer ranges owing to losses in the high-conductivity zone. Similar variations are seen as a function of frequency for a fixed source-receiver offset. This example is intended to be illustrative and does not reflect a realistic earth structure. In many instances of practical interest, the earth will be markedly two- or three-dimensional, and additional current channeling effects will be observed. These can be sorted out by modeling, but knowledge of the field variations on both small and large scales is essential to achieve this. For example, discrimination of a highly conductive magma chamber from surrounding host rock requires measurements both at ranges comparable to the depth to the anomalous body and at ranges of many skin depths in the resistive matrix.

Two experiments using the apparatus depicted in Figure 13.1 have been conducted near 29° N, 122° W over 25-MY-old lithosphere in water 4,300 m deep. Data were collected at source-receiver ranges of 10 to 70 km and frequencies of 0.06 to 24 Hz. Figure 13.3 shows typical data at a range of 60 km using a LEM-type receiver. The source sent nearly equal amplitude components at 8 and 24 Hz. The upper panel shows a stacked recording of the signal obtained by averaging 449 data sequences of 4 sec duration each. The bottom panel shows the Fourier amplitude spectrum. An interpretation of these experiments was presented by Cox et al. (1986) and yields the following characteristics:

1. The effective crustal conductivity is of order 0.001 S/m and extends to a depth of 3 to 7 km. The integrated conductivity of the crustal section is much lower than has been observed in situ in the upper crust and allows penetration of the source signals into the lithosphere at rather high frequencies.
2. The conductivity drops to an extremely low value in the upper mantle. If the low conductivity persists to very great depths in the mantle, the results suggest an effective conductivity of 3×10^{-5} S/m. If the conductivity rises appreciably below 30 km (this is not resolved by the data), then the minimum conductivity must be smaller than 10^{-5} S/m.

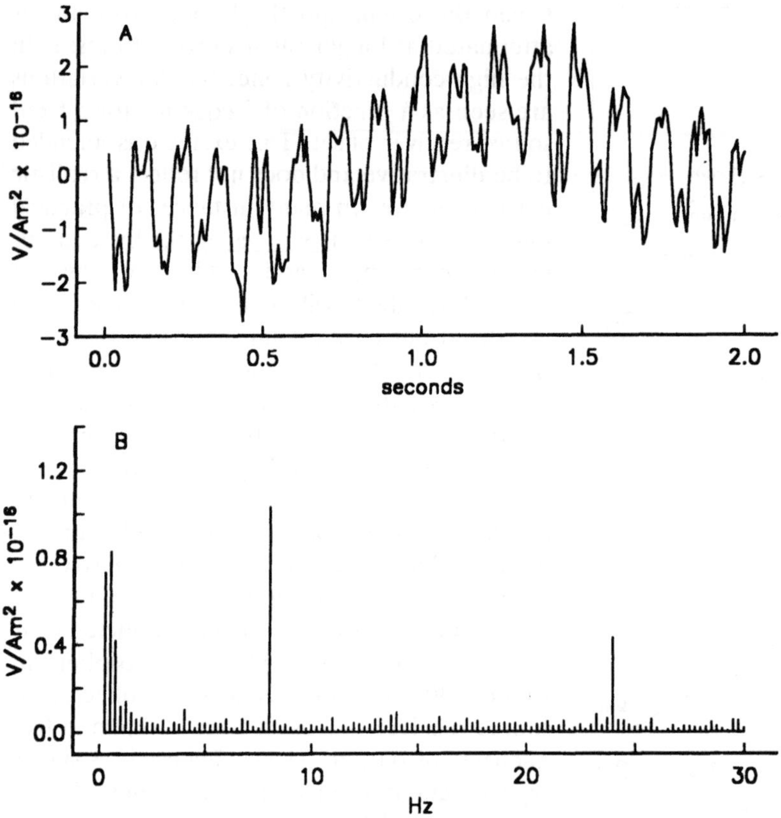

FIGURE 13.3. Data collected by a long-wire electric field instrument in the North Pacific near 29° N, 122° W in 4,300 m of water at a source-receiver separation of about 60 km.

3. The received signals scatter more with increasing range. This scatter is not due to background or instrument noise and may be due to anisotropy in some part of the lithosphere, irregularity in the shape of the resistive zone, or inclusions within the resistive zone.

A variety of geophysical problems can be attacked using CSEM, including (1) studies of the vertical and lateral extent of hydrothermal alteration in the oceanic crust and its evolution with crustal age, (2) studies of the changes in thermal structure of the lithosphere as a function of plate age, and (3) examination of the structure under fracture zones. Most of these are relatively deep targets that require the penetration ability of an HED system. However, problem (1) might best be treated with the Canadian VED system on the ridge axis because of the technical problems associated with working in rough topography. Preliminary development of a towed HED system has recently been accomplished; this may find application in rapid surveying of the near-bottom electrical structure beneath the sea.

Time-Domain Controlled Source EM

Transient or time-domain electromagnetic methods (TDEM) are used extensively for terrestrial exploration; they are the principal surface geophysical tool of the minerals exploration industry. TDEM systems typically utilize an electric or magnetic source energized with an impulsive or steplike current and receivers that measure the response of the surrounding medium to it. The time of propagation of an electromagnetic transient in a conductive medium is just the electromagnetic diffusion time and is proportional to both the electrical conductivity and the square of the source-receiver separation. The time delay between turn-on (or turn-off) and the arrival of a signal propagating through the earth and the shape and amplitude of the transient are functions of the conductivity structure. Since the conductivity of seawater is generally larger than that of rock, that portion of a transient traveling through the ocean usually arrives later than that in the seafloor and is well separated in time for reasonable values of the conductivity contrast. In the less common instance of a relatively conductive seafloor, the sense of the arrivals is reversed.

TDEM systems offer advantages over CSEM when the source-receiver spacing is small in terms of skin depths in seawater. In this instance, the CSEM receiver will see a strong waterborne wave, and the much weaker part of the signal propagating through the underlying rock will be overwhelmed and difficult to detect. This type of complication can be avoided by raising the frequency of a CSEM system; this limits the penetration depth. Since the responses to the ocean and rock are distinct in time, TDEM does not suffer from this limitation at arbitrarily small source-receiver spacings; in any case, most TDEM systems involve measurements made when the source is off. In addition, topographic variations and near-surface heterogeneity of electrical structure can have strong effects on CSEM fields, especially at small source-receiver offsets. These complications will alter the transient amplitude but affect its shape to a lesser extent, since transient shape is an integral function of the properties of a volume of material. Seafloor TDEM systems are also self-calibrating in the sense that the response at late times, which is solely a function of the known conductivity of seawater, is obtained every time a measurement is made. The principal disadvantages of TDEM systems are a need for a relatively high signal-to-noise ratio and some theoretical complications. Since the energy from a TDEM source is spread over a wide range in frequency, the stacking time to achieve a good signal-to-noise ratio can be prohibitive at large ranges where the signal is attenuated, and a CSEM system using only a few frequencies is preferred. However, when the range is small and the signal is large, TDEM can provide more information about the conductivity structure in a fixed amount of time, since it is roughly equivalent to operating a CSEM system at many frequencies simultaneously. The theory for TDEM is somewhat more complicated than that for CSEM, but this problem is diminishing in importance as numerical algorithms improve.

Cheesman et al. (1987) have performed an extensive investigation of TDEM systems for seafloor exploration. After studying a variety of possible source and receiver types, they showed that a system composed of a horizontal magnetic dipole (HMD) source and a coaxial magnetic receiver offered the best sensitivity to seafloor conductivity under realistic conditions. Figure 13.4 shows the step-on response for such a configuration with two half spaces in contact (inset) as a function of the conductivity contrast between seawater and rock for the case where seawater is more conductive than rock. The dashed line is the analytical solution; the solid line shows the numerical result. The step-on response is normalized by the late-time or DC magnetic field, while the time scale is normalized by the electromagnetic diffsion time in seawater. A second, dimensional time scale for a source-receiver separation of 100 m is shown below this. The initial rise of the curves is due to propagation through rock, followed by a final rise to the DC magnetic field value when the portion of the signal traveling in the ocean arrives; this always occurs at the same dimensionless time. The HMD system is equally sensitive to rock conductivity when the rock is more conductive than seawater; the sense of the arrivals of the seafloor and seawater parts of the signal is just reversed. This is a property that is not shared by other system types. In particular, an HED system is not useful when the conductivity of the seafloor is high since the signal decreases in amplitude as the seafloor conductivity rises.

Cheesman et al. (1987) also investigated the use of a TDEM system to study two- and three-dimensional structures, demonstrating the utility of the method to detect buried geological bodies. Since the portion of the transient traveling in the ocean is distinct from that in rock, seawater can be assumed to be perfectly

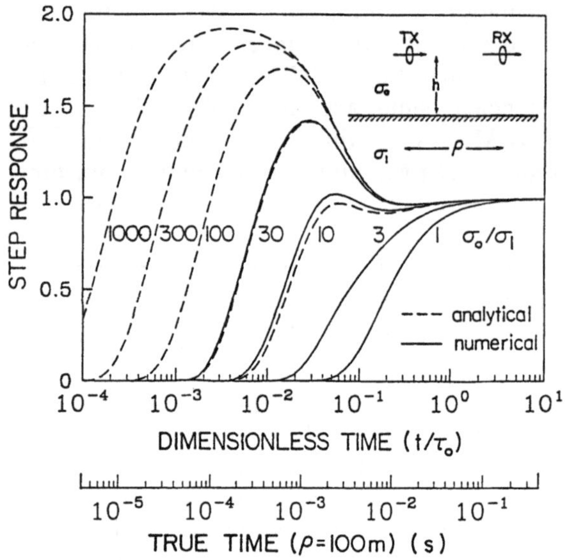

FIGURE 13.4. The step-on response for the coaxial magnetic dipole-dipole system for two half spaces in contact.

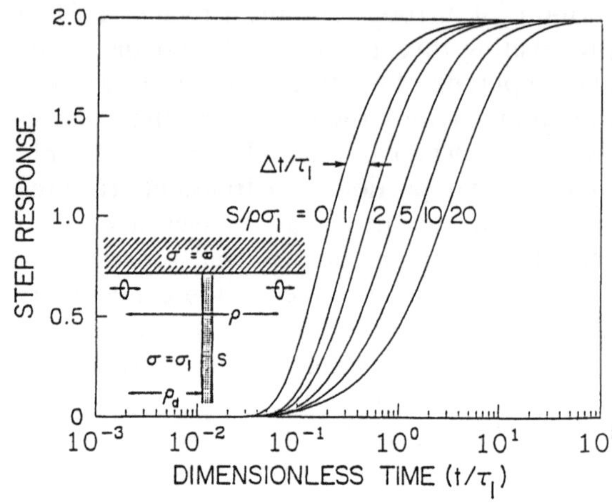

FIGURE 13.5. The transient step-on response of the double half space model shown in the inset modified by the inclusion of a semiinfinite conductive vertical dike.

conducting for modeling purposes, simplifying the salient physics. This simply moves the oceanic part of the transient out to infinite time without greatly affecting the geophysically interesting seafloor part. Figure 13.5 shows one such idealized model, consisting of a semi-infinite conductive vertical dike having a conductance (conductivity-thickness product) S embedded in a double half space (inset) and located between the source and receiver. The curves show the part of the transient due to propagation through the seafloor as a function of dike conductance normalized by the product of source-receiver separation ρ and halfspace conductivity σ_1. As the conductance of the dike rises, the transient is increasingly impeded in time; the incremental delay is linearly proportional to the dike conductance.

An HMD transient system is currently under joint development between Scripps and the University of Toronto. Figure 13.6 shows a sketch of the planned system attached to the Marine Physical Laboratory Deep Tow vehicle (Spiess and Tyce, 1973). The unit consists of a controller and batteries located on the sled, which is towed about 100 m off the seafloor by a surface ship, and two components suspended beneath the vehicle and near the seafloor. The source is a large, solenoidal coil encased in a protective housing and energized with a step current at periodic intervals by the controller. The housing is designed to withstand contact with the seafloor. The receiver is a three-component set of induction coils which are sensitive to the time rate of change of the magnetic field. The transmitter-receiver distance is determined by detecting an acoustic signal sent from the source with a hydrophone at the receiver. In practice, the receiver must be stationary when making a measurement. This can be accomplished by lowering the Deep Tow system momentarily; a single measurement is easily made in a few seconds at most. Over rough topography, it will be essential to employ this approach since neither source nor receiver is likely to survive being dragged along the bottom. Data quality can be assessed in real time by sending it up the sled towing cable. Use of the Deep Tow has many advantages, since additional data from up-down, forward, and side-scan sonars, cameras, magnetometers, and temperature probes are available. Precision navigation of the entire system is performed by means of a network of acoustic transponders. It should be noted that this type of approach is analogous to airborne

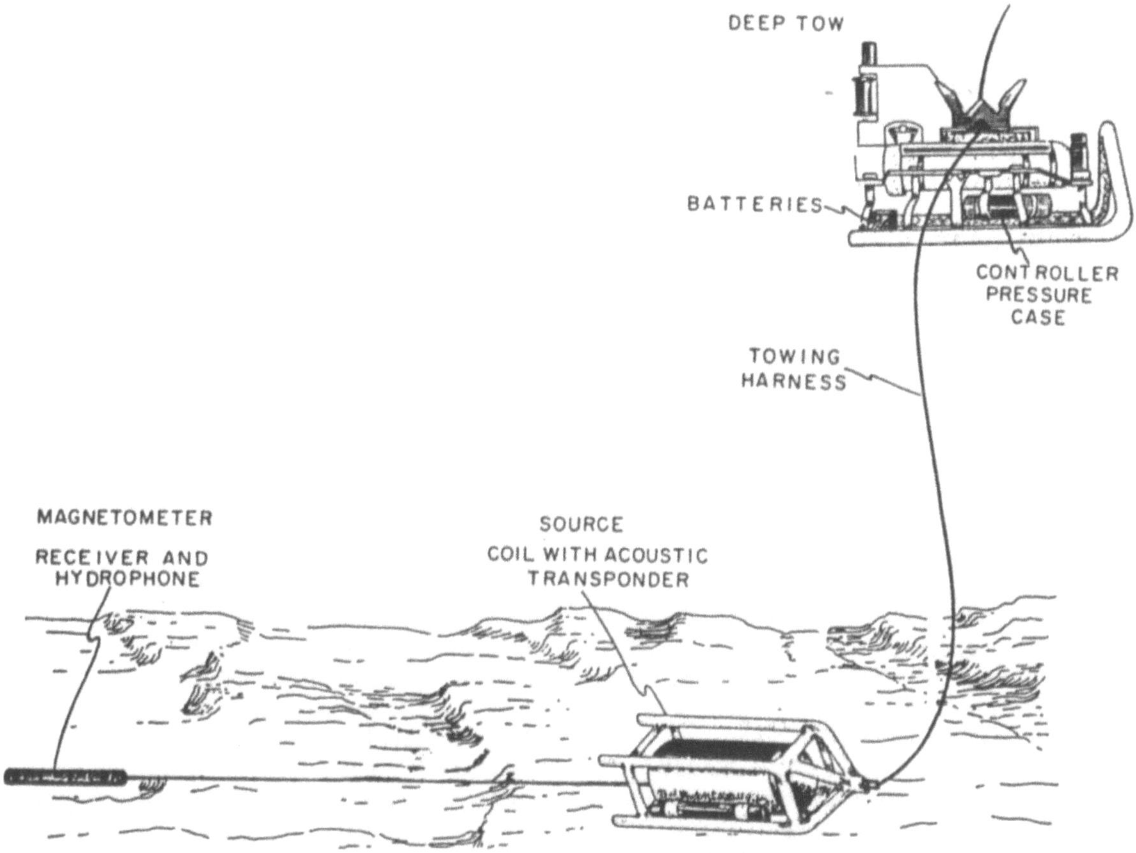

FIGURE 13.6. Sketch of a planned seafloor TDEM system.

electromagnetic surveying on land, in which large areas are covered rapidly by use of a frequency- or time-domain system located on an aircraft.

It is likely that TDEM systems will be most useful in studies of shallow structure and the upper 1 km of the ocean crust. Some prospective targets include sulfide bodies, ridge crest hydrothermal systems, and hydrothermally altered zones in the oceanic crust. While sulfide bodies can be located visually, and their surfaces can be mapping using acoustic methods, estimates of their depth extent are often lacking or imprecise. This is especially true on well-sedimented ridge crests such as the Juan de Fuca Ridge. TDEM is a well-established method for the delineation of buried sulfide mineral bodies on land and can be adapted to the study of submarine deposits. Little is known about the depth extent and plumbing geometry of midocean ridge hydrothermal systems; this is another instance where inadequate surface observations are all that are available. Since electrical conductivity is sensitive to the presence of pore fluids and the connectedness of the matrix containing it, TDEM surveys over ridge crest hydrothermal systems can give valuable subsurface information. The spatial and temporal variability of hydrothermal alteration in the upper ocean crust can also be studied using TDEM. All of these goals require the development of a rugged, towed system that can be operated in a survey mode. This will require good bottom navigation, since repeated and intersecting tracks are essential for gaining sufficient spatial covarage to resolve two- and three-dimensional subsurface structures.

Acknowledgments. Many of the ideas contained in this work result from ongoing collaborative research with C.S. Cox, R.N. Edwards, S.C. Webb, and S.C. Constable. J.A. Hildebrand provided a careful review of the manuscript.

References

Becker K (1985) Large scale electrical resistivity and bulk porosity of the oceanic crust, Hole 504B, Costa Rica Rift. In: Anderson R, Honnorez J, et al. (eds) Initial Reports of the Deep Sea Drilling Project, Vol 83. Washington: U.S. Government Printing Office, pp 419–427.

Becker K, Von Herzen RP, Francis TJG, et al. (1982) In situ electrical resistivity and bulk porosity of the oceanic crust, Costa Rica Rift. Nature 300: 594–598.

Chave AD, Constable SC, Edwards RN (1989) Electrical exploration methods for the seafloor. In: Nabighian MN (ed) Electromagnetic Methods in Applied Geophysics, Vol 2. Tulsa, Society of Exploration Geophysicists (in press).

Cheesman SJ, Edwards RN, Chave AD (1987) On the theory of sea floor conductivity mapping using transient EM systems. Geophysics 52: 204–217.

Cox CS, Kroll N, Pistek P, Watson K. (1978) Electromagnetic fluctuations induced by wind waves on the deep-sea floor. J Geophys Res 83: 431–442.

Cox CS, Constable SC, Chave AD, Webb SC (1986) Controlled-source electromagnetic sounding of the oceanic lithosphere. Nature 320: 53–54.

Edwards RN, Law LK, DeLaurier JM (1981) On measuring the electrical conductivity of the oceanic crust by a modified magnetometric resistivity method. J Geophys Res 86: 11609–11615.

Edwards RN, Law LK, Wolfgram PA, et al. (1985) First results of the MOSES experiment: Sea sediment conductivity and thickness determination, Bute Inlet, British Columbia, by magnetometric off-shore electrical sounding. Geophysics 50: 153–160.

Nobes DC, Law LK, Edwards RN (1986) The determination of resistivity and porosity of the sediment and fractured basalt layers near the Juan de Fuca Ridge. Geophys J R Astron Soc 86: 289–318.

Presnall DC, Simmons CL, Porath H (1971) Change in electrical conductivity of a synthetic basalt during melting. J Geophys Res 76: 5665–5672.

Shankland TJ (1981) Electrical conduction in mantle materials. In: Evolution of the Earth, Geodynamics Series Vol 5. Washington: American Geophysical Union, pp 256–263.

Shankland TJ, Waff HS (1977) Partial melting and electrical conductivity anomalies in the upper mantle. J Geophys Res 82: 5409–5417.

Spiess FN, Tyce RC (1973) Marine Physical Laboratory Deep Tow instrumentation system. Scripps Inst Oceanogr Ref Ser 73–4, 37 pp.

Stesky RM, Brace WF (1973) Electrical conductivity of serpentinized rocks to 6 kilobars. J Geophys Res 78: 7614–7621.

Webb SC, Cox CS (1986) Observations and modeling of seafloor microseisms. J Geophys Res 91: 7343–7358.

Wolfgram, PA, Edwards RN, Law LK, Bone MN (1986) Polymetallic sulfide exploration on the deep sea floor: The feasibility of the mini-MOSES technique. Geophysics 51: 1808–1818.

Young PD, Cox CS (1981) Electromagnetic active source sounding near the East Pacific Rise. Geophys Res Lett 8: 1043–1046.

14—An Electric Drill for Deep-Sea Coring

PATRICK J.C. RYALL

Abstract

A series of drills has been developed at the Bedford Institute of Oceanography for collecting cores of hard rock from the ocean floor. Early deep-sea drills were hydrostatically powered and were principally employed on the Mid-Atlantic Ridge. However, these drills have been replaced by a deep-water version of an electric rock-core drill originally designed for continental shelf work. Part of the design philosophy behind these drills is operability from any research ship. To that end, the cable is mounted on a portable winch, and the workshops and controls are built into standard 20-foot containers.

The drill is driven by a 2,400-V electric motor powered and controlled from the ship through a contrahelically armored triaxial cable. Information from the drill, such as tilt angle, extension rate, down load, and flushing pressure, is multiplexed and transmitted up the cable. A television camera and light mounted on a pan and tilt unit have been added to the drill to aid site selection. Problems with interference and lighting have not been completely solved. The drill has been proved at depths of 3,500 m and on slopes of up to 30°. The drill was used in May 1986 to recover a short core of massive pyrite and pyrrhotite from the Middle Valley of the northern Juan de Fuca Ridge.

Introduction

The need to develop a rotary diamond drill in order to recover hard rock samples from the sea floor was recognized by scientists at the Bedford Institute of Oceanography (BIO) in the mid 1960s. Their early interest was in coring the basalts of the Mid-Atlantic Ridge, which was then being studied at 45° N by BIO (e.g., Loncarevic et al., 1966; Brooke et al., 1970). Because of the lack of a suitable cable to transmit power to a drill, they designed a self-contained unit utilizing the energy available from the hydrostatic pressure of seawater at depth (Brooke and Gilbert, 1968) using general principles first developed by Joly (1914). The motor consisted of an impeller connected to seawater on one side and to an empty chamber on the other. A valve was opened to turn on the unit permitting seawater, driven by hydrostatic pressure, to run through the impeller into the empty chamber. Such a system was able to develop 4 HP for up to 22 min. Power was limited only by the depth of operation and size of the chamber. In practice, the destabilizing effects of the empty chamber on landing and the added weight when full set practical limits. This rig had modest success on the Mid-Atlantic Ridge in 1969 and 1971 (Brooke et al., 1970; Ade-Hall et al., 1973). Because of a redirection of Canadian research effort from the mid-Atlantic toward the shallower waters of the Arctic, the drill was not developed further.

Workers at Woods Hole Oceanographic Institution built a battery-powered drill in the early 1970s for use with the submersible *Alvin* (Davis et al., 1975). This drill was of similar configuration to the BIO hydrostatic drill except that DC electrical power replaced the hy-

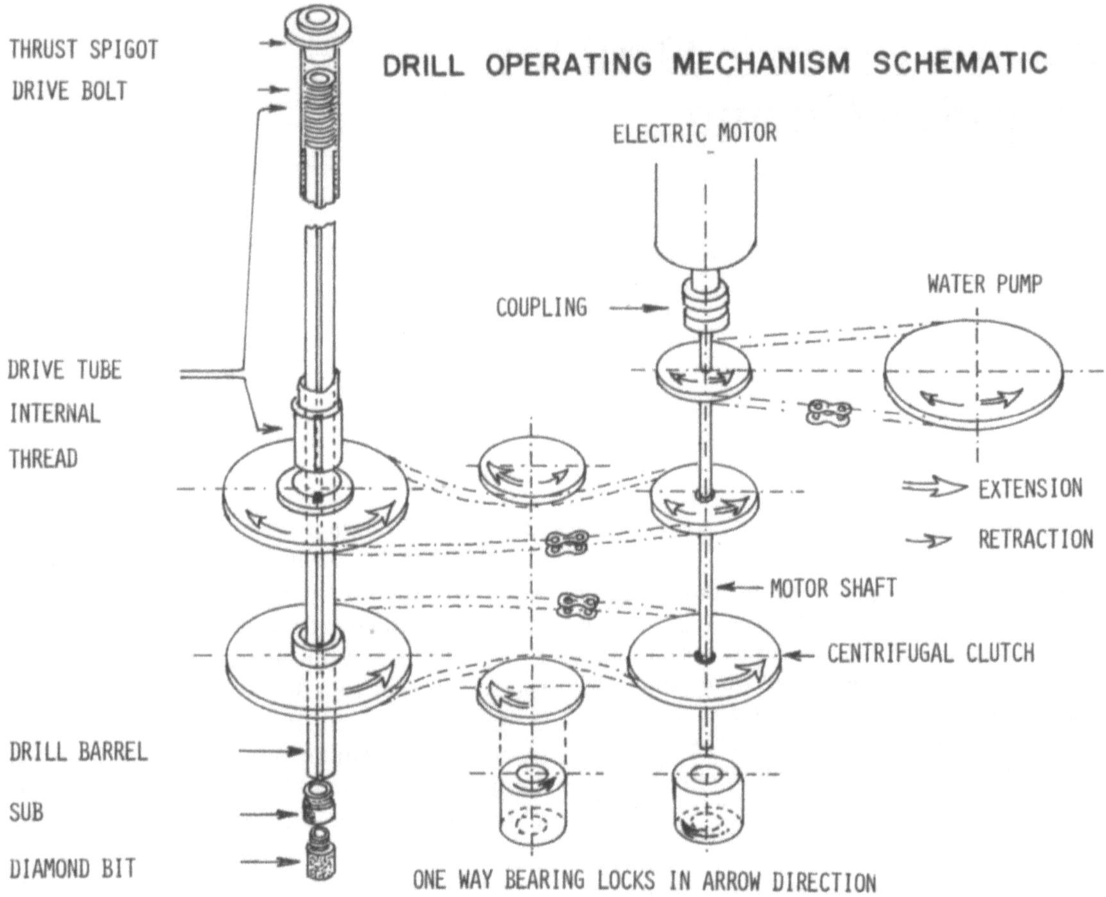

FIGURE 14.1. Schematic of drill mechanism. Large arrows show the direction of rotation on extension; small arrows, the direction on retraction. The drive sprockets on the main shaft have different numbers of teeth to give different rotation rates of the drive tube and drill barrel.

drostatic system. This drill was used in 1974 from a surface ship to obtain four cores from the FAMOUS area and was also employed on the 1977 Galapagos expedition. Other drills were built in the 1970s for use on the continental shelves. The Japanese MD-300PT (Kinoshita et al., 1977) is a self-contained, automatically controlled, battery-powered drill which was followed by the larger MD-500H (Inoue et al., 1978), which is capable of recovering a 6-m-long core. In Britain, the Institute of Ocean Sciences (IOS) built an electric drill powered from the surface (McQuillin and Ardus, 1977), which was the first unit to attempt to use television as an aid in site selection.

BIO/Dalhousie Electric Drill

In addition to the hydrostatic drill, BIO had also developed an electrically powered drill for use on the continental shelf (Fowler and Kingston, 1975). In this drill a 20-foot (6-m) drill rod is attached to a threaded bolt which runs inside an internally threaded plastic drive tube. In the drilling mode, the drill rod and drive tube are rotated at slightly different speeds with the result that the drill rotates at 1,300 rpm and extends at about 18 cm/min (Fig. 14.1). In retraction, the direction of rotation of the motor and drive tube are reversed while a one-way clutch prevents rotation of the drill rod which retracts at about 3.3 m/min. Initially this drill was powered at 575 V with electricity and control signals sent through an umbilical cable and the drill supported on a separate rope. This arrangement made drill deployment and recovery slow and cumbersome and was only suited for shallow depths. Dalhousie University secured funding from the Natural Sciences and Engineering Research Council to modify the drill for use in deeper water. The major change required was to step up the voltage to 2,400 V to reduce ohmic losses and to obtain a triaxial, contrahelically armored cable (Ryall et al., 1981) to supply power and support the weight of the drill. All the signals to and from the drill are multiplexed and coupled to the inner coax of cable. This system allows drill recovery from 3,000-m depth to be performed in less time than from 600 m with the two-part system. Figure 14.2 shows the current configuration of the drill.

Experience has shown that success in drilling operation depends on careful operator control of the drill. The following parameters are displayed for the operator on a six-channel chart recorder:

1. Pitch on two axes—these are required to see if the drill has landed within 30° of vertical.
2. Compass—this was installed with the intention of being able to orient cores for paleomagnetic studies. It has so far only been used to monitor rotation of the drill frame during raising and lowering.
3. Download—the force with which the drill bit is pushing down on the rock. This is virtually zero in sediments and soft rock, is commonly several hundred pounds when drilling in hard rock, and can get to over 1,000 lb just before jamming. Normally the drill is shut off before that happens.
4. Flushing pressure—the pressure developed by the flushing pump, which is pumping water down the drill barrel and past the bit to cool it and remove cuttings.
5. Motor current—currents in all three phases are measured and displayed on meters. One of them is displayed on the chart recorder. At idling this will be just over 1 amp; at stall, over 5 amps. Normally, it is not allowed to go above 3 amps to prevent jamming.

Figure 14.3a shows a typical display of flushing pressure download and motor current while drilling. These variables can change very quickly and hence require immediate reaction from the operator to prevent jamming.

Figure 14.3b shows download and motor current while drilling sulfide under 5 m of sediment. The small increase in download marks the beginning of the sulfide. Compared to the currents and downloads in Figure 14.3a, this shows relatively easy drilling.

All but the motor current are measured on the drill and telemetered up the cable. The motor current is measured at the surface. When drilling, the operator's attention is focused on the motor current and download. As long as these are kept within the limits given above, it is always possible to retract the drill barrel back inside the mast before lifting off the bottom. If the limits are exceeded—which can happen within a second or two—the drill may remain jammed in the rock. In such an eventuality the drill barrel will get bent when the drill is lifted off the seafloor. Although not disastrous, this is to be avoided because of the down time required to fit a new drill barrel.

A major operational concern was the behavior of the armored cable when the drill was put on the seafloor. It is well known that when tension is suddenly removed from a cable it will usually kink and/or tangle up at the bottom end owing to torque imbalance in the cable. We have found that this type of damage can almost be eliminated by using a contrahelically armored cable and by carefully lowering the drill onto the bottom. Typically, the cable is lowered or raised at 40 m/min. Higher speeds have been tried but lead to excessive drill spinning and spooling problems on recovery. A pinger is placed 25 m up the cable from the drill. After the drill is on the bottom—determined by the changes in pitch angles—enough cable is paid out to lower the pinger to within a few meters of bottom. This procedure

FIGURE 14.2. Photograph of drill on deck of CSS *Tully*. Drill base is 8 feet on a side with about a ton of lead enclosed in the base triangle. Drill mast extends out of the picture to a height of 20 feet. Drill mechanism is the black box under the motor (stainless-steel cylinder with black cap on top and compensating cylinder to left side). Pressure cases on the left contain electronics and electrical switch mechanism. Horizontal cylinder on right is pan and tilt mounting television camera and light. In operation drill cable is mechanically attached to the lifting tongue on the back and plugged into top of pressure case.

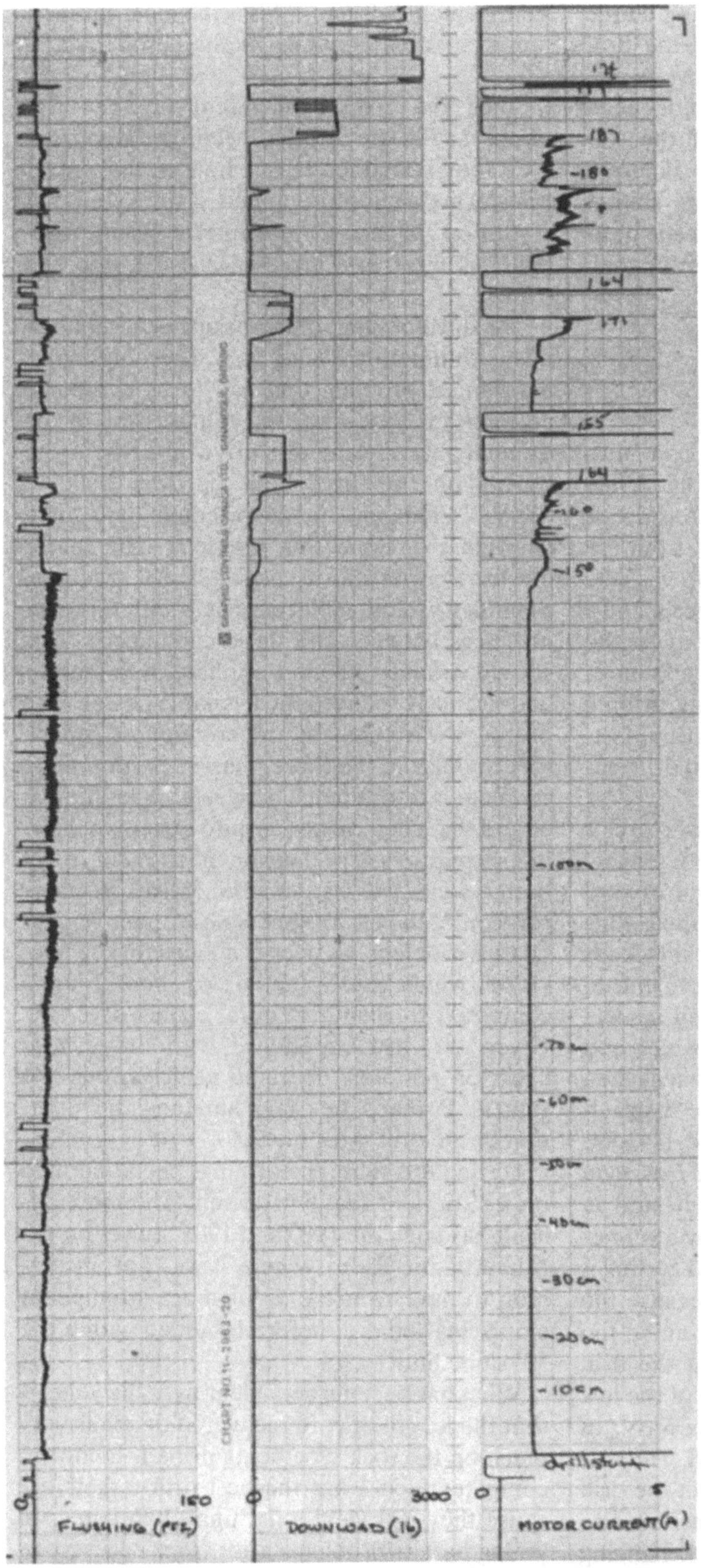

A

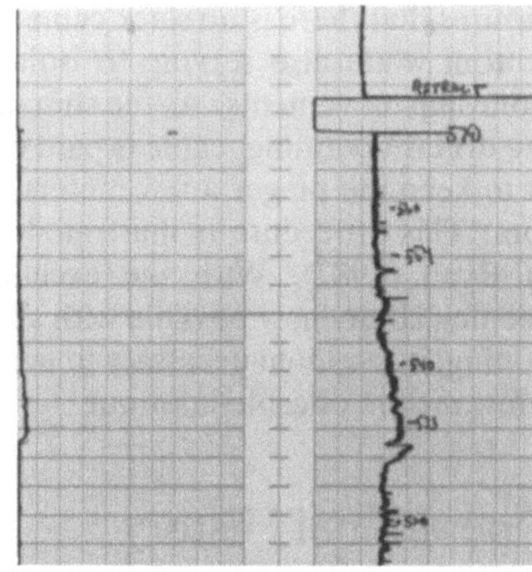

B

FIGURE 14.3. (A) A section of chart recorder display showing the three channels used to monitor drill operation: motor current, down-load, and flushing pressure (from right to left). The other three channels that are continuously monitored show tilt on two axes and compass. These do not normally change once the drill is on the bottom. (B) Record of motor current and down-load while drilling sulfide. From the small down-load increase at about 530 cm it can be seen that the sulfide is easily drillable.

ensures that there is no slack cable lying on the bottom while also serving to isolate the drill from ship movements. As the ship moves while the drill is operating, cable is paid out or taken in to keep the pinger a few meters off the bottom. This procedure is illustrated in Figure 2 in Ryall (1987). With the excellent station keeping ability now possible with acoustic positioning, it is seldom necessary to pay more than a few meters of cable in or out.

Operational History

Since 1979, the drill has been used on mid-ocean ridges at increasing water depths. While only deep-water drilling will be considered here, it should be borne in mind that from the mid-1970s to the mid-1980s, this drill was also being used by the Atlantic Geoscience Centre in their reconnaissance geologic survey of the Canadian continental shelf (e.g., MacLean et al., 1977, 1978). In these operations drilling would be part of a multidisciplinary cruise lasting about a month with as many as 50 drill stations scheduled in a cruise.

Drilling operations were carried out exclusively from the CSS *Hudson* until 1983. Operations on the *Hudson* are very straightforward since the winch is mounted ahead of the bridge with the drum axis fore and aft. The cable leads straight off the winch to a block on the end of a telescopic crane. The drill is picked up and lowered over the side by using the winch and crane in tandem. The boom of the crane is then lowered into a crutch to off-load the hydraulics, and the drill is lowered to the bottom. Operation over the stern of the *Paul Langevin III* in 1983 was slightly more complicated in that the drill weight was supported by a runner from the crane while picking it up and getting it over the side. This was necessary because the large change in lead angle while swinging the drill over the stern prohibited using the drill cable to support the weight. Because of the low free board, a slip hook was not necessary. In 1986, working from the larger CSS *Tully*, a crane with a slip hook was used to get the drill overboard, then the weight was transferred to the drill cable running over a block hanging from a small A-frame. This is the most cumbersome and most damaging method used to date.

Since the 1983 deployment on the *Paul Langevin III*, the drill has been operated from a 20-foot container which houses the power transformers and is equipped as an electronics workshop. Whenever possible, a second container is used as a mechanical workshop and store.

The development of increasing drill capabilities is outlined in Table 14.1. The only major mechanical problem that had to be overcome as depth increased was the compression of the original polyvinylchloride drive tubes that were seizing on the drive bolts at over 2,000 m depth. Changing to chlorinated polyvinylchloride overcame this problem. The major electrical problem was caused by the electrical asymmetry of the cable, especially different capacitances between the three conductors. This caused voltage to be out of phase at the motor and seriously reduced the available power. The problem was cured by the addition of capacitors to balance the three phases. A continuing problem is interference between the multiplexer and the video carrier despite extensive filtering. It requires very delicate adjustment to get both systems working simultaneously. A major revision of the electronics is now under way. A further problem has been the lights for the television, which have a lifetime of about 2 hours compared to manufacturers' claims of 2,000. Given the vast experience with underwater lights on remotely operated vehicles, this is a suprise. Perhaps the rough handling and vibration of the drill are a factor.

During this time, the drill has been operated on a variety of seafloor terrain. The most difficult has been the 1980 and 1982 cruises on the Mid-Atlantic Ridge, where it was not uncommon to have to make 10 or more attempts at emplacing the drill before it landed within the 30° slope limit.

What has become very clear over the years is that the length of core recovered depends very much on the rock type being drilled. Uniform, competent rocks are the best drill targets because they will yield long, unbroken cores. In young basalts problems arise from a variety of

TABLE 14.1. Chronology of BIO/DAL deep-drill development.

Starting Point: BIO Continental Shelf Drill—Up to 20 ft. of 1 inch diameter core
—Stable on slopes <15°
—Two-part buoyed rope/cable—650 m water depth

Date	Location	Modifications	Water depth	Number of stations	Number of cores	Total length	Rock type/ remarks
1979	Azores seamount	More stable frame (30°)	450–650 m	10	10	15 m	Basalt (9 my) & carbonate
1980	Mid-Atlantic ridge	E–M cable (5000 ft)	400–800 m	11	10	6.5 m	Basalt (1–2 my) & carbonate
1981	Mona bank	2300 V motor	1050 m	2	1	1.5 m	Carbonate
	Juan de Fuca ridge	5000 m cable	2250–2725 m	6	0	0	Power Problems
1982	Mid-Atlantic ridge	Solve power problems	540–2720 m	22	7 + (6)[a]	3.6 m	6 Stations to full extensions in sedimen in Median Valley
1983	Bermuda	New CPVC drive tubes	390–1960 m	13	8	5.2 m	Carbonate
	Juan de Fuca ridge	First try of TV	2198–2224 m	10	7 + (2)[b]	0.7 m	Basalt, very young, glassy & highly fractured
1984	Bermuda	None	790–3230 m	16	11	3.6 m	Basalt/carbonate agglomerate
1985	Kane fracture zone	TV working	1850–3170 m	5	5	2.4 m	Basalt—short cores, gabbro—1 m cores
1986	Middle valley Juan de Fuca ridge	None	2460 m	3	2	0.5 m	Polymetallic sulphide, dense mud

[a] Number of extensions to 5.7 m without encountering rock
[b] Number of cores containing pebbles only

factors: (1) the piece of rock has to be large enough and/or well enough embedded to stay still while being drilled; if it is too small the drill will simply spin it around; (2) when the drill bit breaks out of one piece, the core will sometimes break, forming a "mushroom cap," which partly covers the diamonds on the bit, thereby preventing further progress; (3) because of its rapidly quenched state, the basalt is highly stressed and often fractures during drilling, thereby forming effective natural wedges to jam the drill barrel. All of these problems combine to result in very short cores in young basalts. The situation can be further complicated by the presence of drainback structures or collapse pits as described by Ballard et al. (1979) in the Galapagos Rift. On several deployments in the 1983 Juan de Fuca cruise, the drill penetrated a few centimeters of rock and then extended through a void for up to a meter before encountering solid rock again. Under these conditions the drill was not able to spud into the second surface. Older basalts such as those drilled in the 1979 cruise yield longer cores because many of the fractures are cemented.

A photograph (Fig. 14.4) of cores from cruise 85.010 to the Kane Fracture Zone illustrates the results of the variation in rock type. Cores 08 and 09 are basalt and have broken into small pieces, jamming and yielding a short core. Core 11 is a fairly homogeneous gabbro that gave an almost continuous core that had to be broken to remove it from the drill. Core 12 is another gabbro, which was more fractured and veined and was broken up into small pieces while being drilled. However, it did not break into wedge-shaped pieces as readily as the basalt.

Figure 14.5 shows the core drilled from the

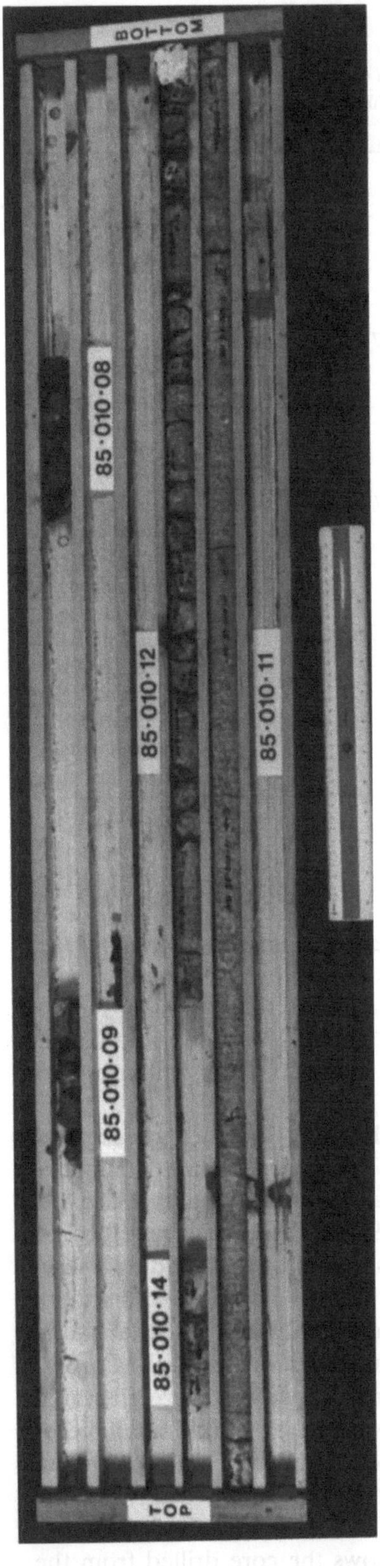

FIGURE 14.4. Box of drill core from Hudson cruise 85-010 to the Kane Fracture Zone. Cores 08 and 09 are basalt. Cores 11, 12, and 14 are gabbro from more than 3,000-m water depth. The fine breaks in core were made when removing it from the drill. Ruler is 12 inches long.

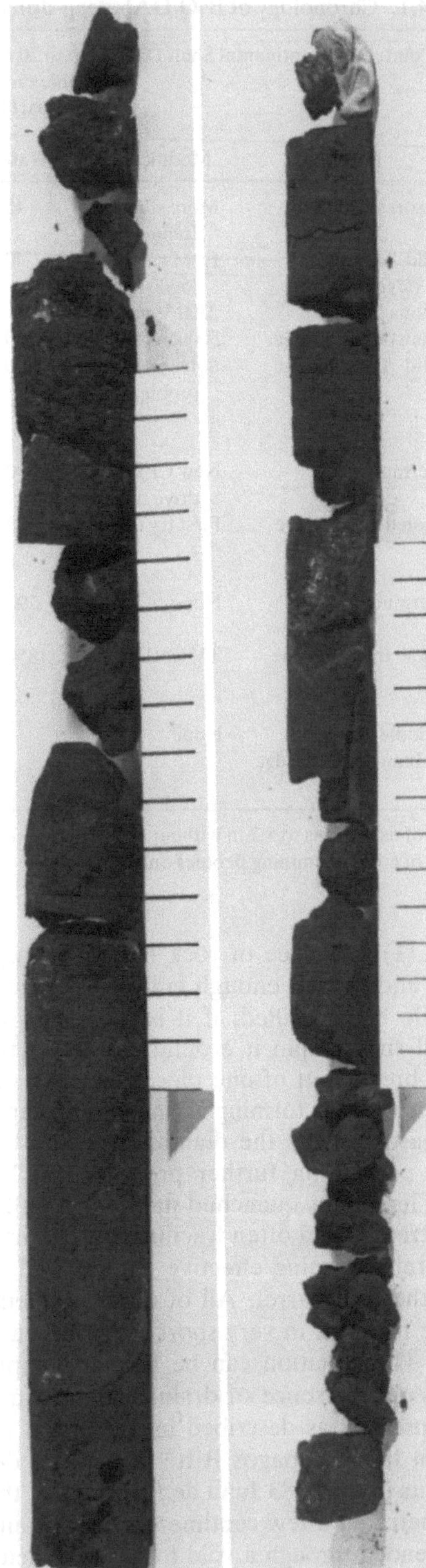

FIGURE 14.5. Sulfide core from North Mound of Middle Valley. Top of the core consisting of stiff mud is at the lower right. Bottom of the core of hard rock is at the upper left. Distance between scale lines is 1 cm. Photo was taken some months after drilling, by which time the mud had started to crumble.

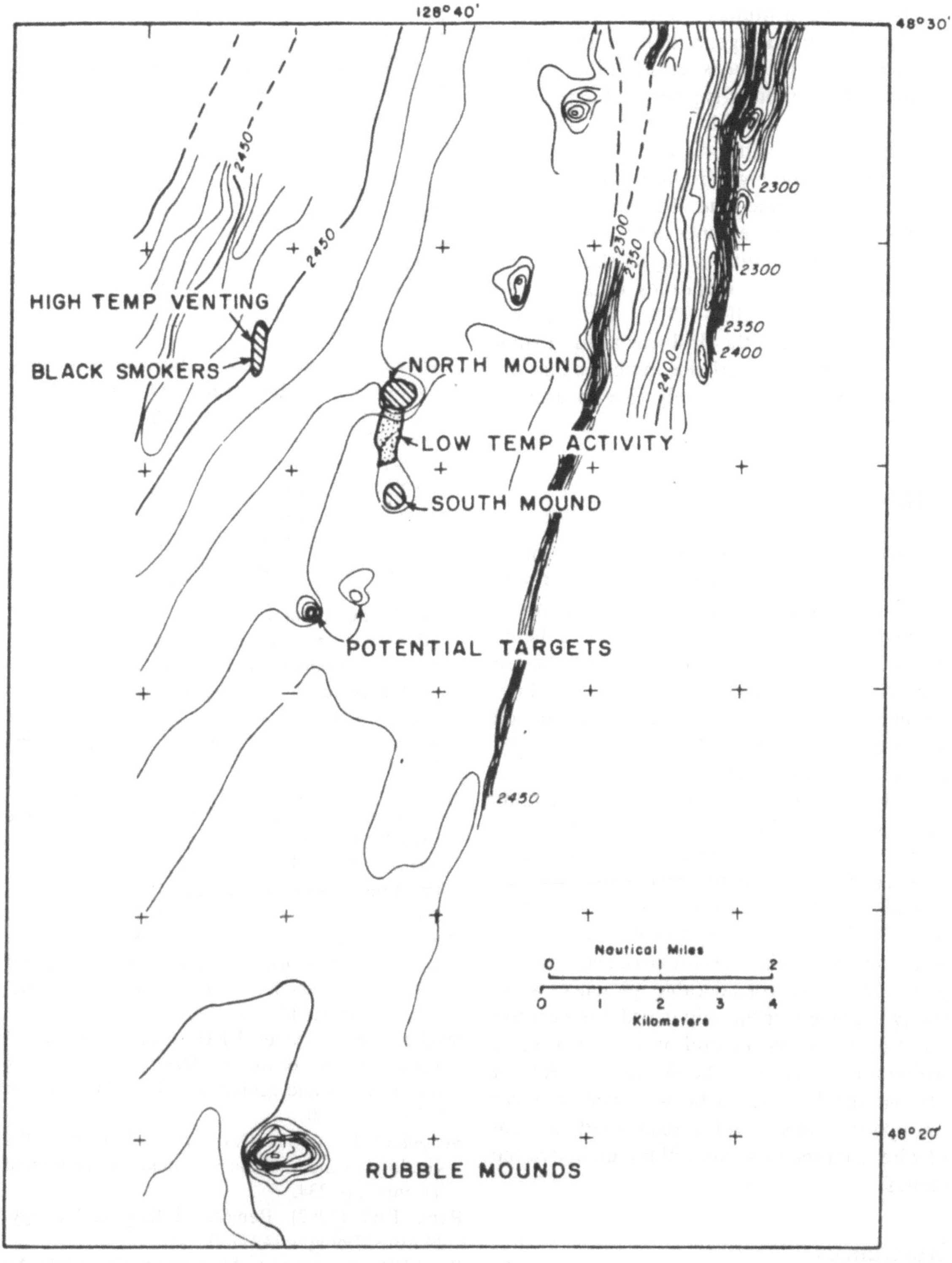

FIGURE 14.6. Location of the sulfide occurrences and active hydrothermal venting in the Middle Valley area of Endeavour Ridge, northern Juan de Fuca Ridge.

North Mound in Middle Valley of the Northern Juan de Fuca Ridge. Sulfides in this region are described by Davis et al. (1987). The location of the mound is given in Figure 14.6. Prior to recovering this core, the drill had penetrated 5 m of soft sediment. The top of the core shown at the lower right of the figure is a stiff, cohesive sulfide mud. The bottom part of the core shown on the upper left is a hard, massive sulfide. Preliminary analysis showed economically significant quantities of zinc (1.4–8.3%) and copper (0.7–1.3%). Drilling was terminated because the drill was at full extension. This limited experience indicates that the massive sulfides are relatively easy to drill, as indeed they are on land.

Summary

The BIO/DAL drill has been developed into a tool capable of coring rock at water depths to 3,500 m. Success in coring depends as much as anything on the nature of the rock being drilled. Some rocks, such as young basalts, are simply not good targets for diamond drilling. Other rocks, which are more competent and less prone to fracture, are more likely to give long cores; gabbros and carbonates are examples. Judging from one attempt, polymetallic sulfides are also likely to be readily cored.

While it is not a factor that has been mentioned so far, it should be obvious that, for successful sampling of sulfides, the overriding necessity is to put the drill in the right place. This means detailed reconnaissance, precise navigation, and considerable patience while trying to maneuver the ship in order to position the drill that is on the end of 3 km of wire to within a few meters of the desired location. A comparable feat would be to lower your car onto your driveway with it suspended on 3 km of wire from the Goodyear blimp up above the clouds!

References

Ade-Hall JM, Aumento F, Ryall PJC, Gerstein RE, Brooke J, McKeown DL (1973) The Mid-Atlantic Ridge near 45°N. XXI. Magnetic results from basalt drill cores from the Median Valley. Can J Earth Sci 10: 679–696.

Ballard RD, Holcomb RT, Andel TH (1979) The Galapagos Rift at 86°W. 3. Sheet flows collapse pits, and lava lakes of the rift valley. J Geophys Res 84: 5407–5422.

Brooke J, Gilbert RLG (1968) Sea drilling technique of Bedford Institute. The development of the Bedford Institute deep sea drill. Deep-Sea Research 15: 483–490.

Brooke J, Irving E, Park JK (1970) The Mid-Atlantic Ridge near 45°N. XIII. Magnetic properties of basalt bore-core. Can J Earth Sci 7: 1515–1527.

Davis EE, Goodfellow WD, Bornhold BD, et al. (1987) Massive sulphides in a sedimented rift valley, northern Juan de Fuca Ridge. Earth Planetary Sci Lett 82: 49–61.

Davis RE, Williams DL, Von Herzen RP (1975) ARPA rock drill report. Unpublished manuscript, Technical Report WHOI 75–28.

Fowler GA, Kingston PF (1975) An underwater drill for continental shelf exploration. J Soc Underwater Technol 1: 4, 18–22.

Inoue E, Maruyama S, Yuasa M, Oushika H (1978) Submerged rock drill MD-500H. Proceedings of 5th International Conference of Ocean Development, September 25–29, 1978, Tokyo, B2: 41–46.

Joly J (1914) On the investigation of the deep-sea deposits. Sci Proc R Dublin Soc N Ser XIV: 256–267.

Kinoshita Y, Oshika H, Inoue E, (1977) A new type of submerged rock drill "MD-300PT." Marine Geol 25: 321–333.

Loncarevic BD, Mason CS, Matthews DH (1966) The Mid-Atlantic Ridge near 45°N. I. The Median Valley. Can J Earth Sci 3: 327–349.

MacLean B, Jansa LF, Falconer, RKH, Srivastava SP (1977) Ordovician strata on the southeastern Baffin Island Shelf revealed by shallow drilling. Can J Earth Sci 14: 1925–1939.

MacLean B, Falconer RKH, Clarke DB (1978) Tertiary basalts of western Davis Strait: Bedrock core samples and geophysical data. Can J Earth Sci 15: 773–780.

McQuillin R, Ardus DA, (1977) Exploring the Geology of the Shelf Seas. London: Graham and Trotman, p. 234.

Ryall PJC (1987) Remote drilling technology. Marine Mining 6: 149–160.

Ryall PJC, Fowler GA, Manchester KS (1981) An electric rock-core drill for deep ocean use. Proceedings of 13th Annual Offshore Technology Conference, Houston, pp 123–126.

15—Mining Technology for Gorda Ridge Sulfides

MICHAEL J. CRUICKSHANK

Introduction

To assess the technical needs for exploitation of any ore deposit it is essential to know the characteristics of the deposit in terms of its location, dimensions, engineering properties, distribution of values within the deposit, and environmental setting. With these data in hand it should be possible to select a method of mining and prepare a mining development plan.

In the case of the Gorda Ridge, the data available are so sparse that deposit characterizations, for the most part, must be assumed in order to prepare the development plan. Such hypothetical scenarios are useful in the early stages of resource evaluation and were used in the preparation of an environmental impact statement addressing a possible lease offering for mineral exploration and development in the Gorda Ridge area (MMS, 1983). Since that time much has been learned about the distribution of minerals of potential economic interest in the Gorda Ridge and a discussion of the types of deposit potentially available and possible approaches to their mining is presented in this paper. In this discussion the critical economic aspects are not addressed and are assumed to be favorable.

Ore Deposit Models

Hydrothermal metalliferous sulfide deposits associated with crustal plate boundaries include at least 13 varieties recognized or postulated at this time (Table 15.1). Eight of the 13 are significantly different in terms of their apparent mining technology requirements and may be predominantly associated with diverging plate boundaries such as the Gorda Ridge.

Classification in Table 15.1 is made in terms of the depth of occurrence of the resource in the seabed. The eight varieties possibly associated with diverging plate boundaries such as the Gorda Ridge are, in descending order of their depth of occurrence, geothermal energy, vent fluids, metalliferous muds, volcanogenic sediments, surficial precipitates, subseabed precipitates, exhalative beds, and deep-seabed massive differentiates. The major characteristics of these varietal forms are discussed briefly below.

Geothermal Energy

The process of plate divergence is associated with the extrusion of fluid magma along the ridge crest at the plate boundaries. In active areas of divergence, temperatures of hundreds of degrees centigrade occur within a few hundred meters of the seabed. Austin (1966) has summarized the many geothermal areas located offshore in continental shelf waters, and more recent work in the active ridge crests suggests a potential worth examining in more detail, perhaps in combination with the recovery of hydrothermal fluids, as recently carried out in the Salton Sea area in California (Maimoni, 1983). The areas suitable for geothermal energy recovery would be very extensive, parallel-

TABLE 15.1. Models of mineral deposits associated with diverging plate boundaries.

Nature of resource model	Depth of occurrence	Examples: Seabed	Land	Grade upper 10%<	Tonnage (median)	No. of deposits examined[a]
Geothermal energy	0–1.5 km	Active zones	Salton Sea			
Hot vent fluids	Surficial	Active zones	Salton Sea			
Metalliferous muds	0–30 m	Red Sea	?	Cu 0.45, Zn 2.07, Au 0.5, Ag 30	664×10^6	1
Volcanogenic sediments (Mn)	Surficial	DSDP hole	Numerous	Mn 49, P 0.03	47×10^3	93
Surficial precipitates						
Individual stacks or mounds	Surficial	Active zones	?			
Coalesced stacks or mounds or cemented tallus	Surficial	Active zones	?			
Subseabed precipitates						
Cyprus massive sulfides	0–0.8 km	Postulated	Cyprus	Cu 3.9, Pb —, Zn 2.1, Au 1.9, Ag 3.3	1.6×10^6 ($3>25 \times 10^6$)	49
Kuroko massive sulfides	0–0.8 km	Postulated	Kuroko[b]	Cu 3.5, Pb 1.9, Zn 8.7, Au 2.3, Ag 100	1.5×10^6 ($3>100 \times 10^6$)	432
Stock works	0.8–1.5 km	Postulated	Cyprus/ Kuroko	Included	Included	Included
Vein systems	1–1.5 km	Postulated	Cyprus/ Kuroko	Included	Included	Included
Exhalative beds	0.4 km	Escanaba Trough (?)	Besshi	Cu 3.8, Pb —, Zn 0.4, Au 0.76, Ag 9.5	2.2×10^5 ($3>25 \times 10^6$)	44
Deep seabed massive differentiates						
Podiform chromatic	8–10 km	Postulated	Numerous	Cr 46, Rh 5, Ir —, Ru —, Pd 2, Pt 4	2.0×10^4	174
Ni-Pt sulfides	>10 km	Postulated	?			

[a] In USGS (1986).
[b] Kuroko-type deposites are formed in felsic volcanics in back-arc environments.

ing the axis or in proximity to volcanic emanations off axis.

Hot Vent Fluids

Highly mineralized vent fluids which are actively forming sulfide precipitates on the seabed are also host to an extraordinary suite of life forms that are unique to these locations. At this time there is a general assumption that, to avoid disruption of the biota, active vents would not be used for the capture of vent fluids. However, the possibility exists that capture of fluids could be achieved without any damaging effects on the ecosystems by judicious drilling into the hydrothermal reservoir and tapping the fluids beneath the surface. The areas available would likely be confined to the vicinity of those locations with active vents.

Metalliferous Muds

The prime example of metalliferous muds associated with plate boundaries is the deposits in the Red Sea (Degens and Ross, 1969). These, being formed in close proximity to continental rocks and thick beds of salt evaporites, are not quite analogous to conditions at mid-ocean ridges. Nevertheless, the formation of muds or fresh sedimentary deposits of similar consistency may be possible in specific environments.

The Atlantis II deposit in the Red Sea covers an area of approximately 40 km^2 with a deposit thickness varying from 7 to 11 m, in a mean water depth of 2,200 m (Guney et al., 1974). The deposit itself consists of banded layers of fine-grained sulfides of a consistency comparable to shoe polish or peat, of which some 16,000 m^3 was removed for test purposes in 1978 (Hahlbrock, 1979).

Volcanogenic Sediments

These are generally small, lenticular or stratiform manganiferous deposits associated with seabed volcanism (Koski, in USGS, 1986). A median deposit size of 47,000 tons and a maximum of about 12 million tons is given by Mosier (in USGS, 1986). Although likely not economically important, such deposits have been recorded in drilling reports of the Ocean Drilling Project.

Surficial Precipitates

These are probably the best known of the seabed metalliferous sulfides, because they are formed in the active vent areas that have been the targets for much of the scientific work on plate boundary minerals. The deposits consist of individual stacks or mounds formed around active hot vents, or they may occur as coalesced stacks covering a considerable area such as at the Galapagos site (Malahoff, 1982), or they may form a cement of sulfide material around a talus matrix at the faulted boundaries of the rift zones. Individual stacks may be up to 5 m in thickness and 35 m high. Coalesced stacks forming a large mound in the Galapagos area were reported to be as much as 200 m wide, 1,000 m long, and 40 m thick. The sulfide minerals sampled had a reported density of 4.1 g/cm^3 (Malahoff, 1982). Surficial samples only of materials from the mounds have been acquired, and the internal structure and features remain speculative.

Subseabed Precipitates

Because of the inability at this time to drill holes into the seabed to sample the sulfide deposits in depth, the character of these deposits, which include massive sulfides overlying stockworks and veins, must remain speculative. Based on land analogs of such deposits in Cyprus (Singer, in USGS, 1986) the deposits are typically ore bodies of about 1.5 million tons, from surface to 800 m in depth, and overlying stockworks and veins to depths of 1.5 km. They are likely to be intermittently located along the strike of the active ridge where they are formed and should occur also in lines parallel to the ridge crest. The maximum size ore body recorded by Singer (in USGS, 1986) is 25 million tons.

Exhalative Beds

Where hydrothermal fluids are discharged into a body of recent sediments, the precipitates may be confined to favorable beds and spread laterally over distances of several kilometers. Surface expressions in the form of mounds with hydrothermal seeps may be present. Such deposits are postulated in the Guaymas Basin, in the Explorer Ridge, and in the Escanaba Trough and have their analogs in the Besshi-type deposits of which Cox and Singer (in USGS, 1986) cite 44 examples from Japan. The ore bodies are thin and sheetlike, have massive to well-laminated sulfides, and tend to cluster in en-echelon pattern. Median size of the deposits cited is 22,000 tons to a maximum of 25 million tons.

Deep-Seabed Massive Differentiates

In the rocks of the upper mantle at depths greater than 5 km and beneath the crustal activity, it is postulated that chromite and nickel/platinum sulfides will be formed by differential crystallization of these minerals (Bonatti, 1978). Land analogs are described by Albers, Singer, and Page (in USGS, 1986) for

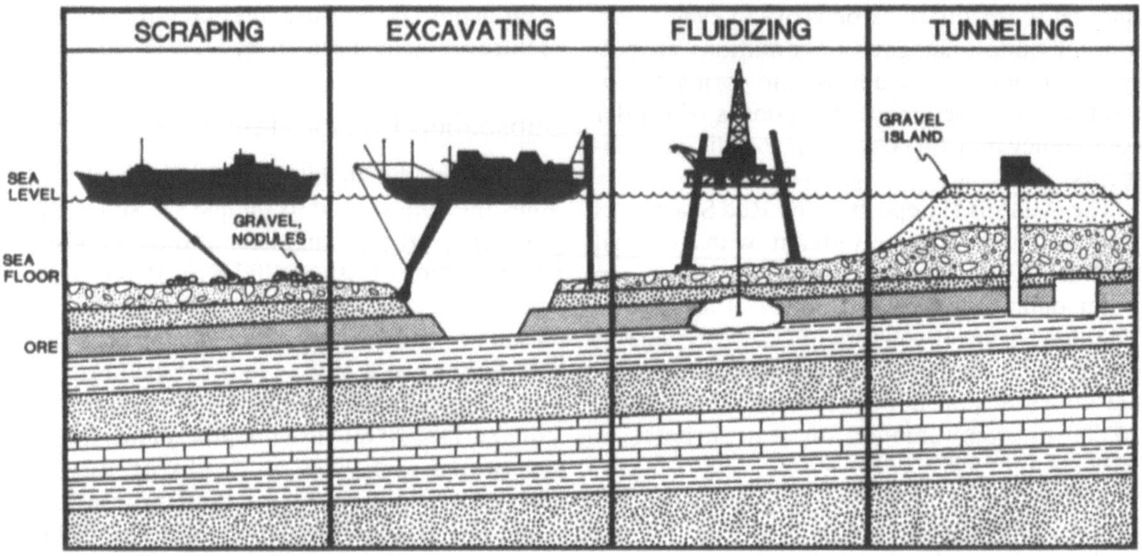

FIGURE 15.1. The four basic methods of mining (Cruickshank et al., 1987).

chromite, listing 174 examples with median and maximum tonnages of 20,000 and 100,000, respectively. No analogs of the Ni/Pt sulfides have been modeled in that source. Unless such deposits have been thrust up by massive faulting, as they have on land, it seems unlikely that they represent an attractive target at this time.

Basic Approaches to Mining

There are only four basic methods of mining: scraping the material from a surficial deposit, excavating the material from the subsurface in a pit and removal of the material as a fluid, in a slurry or solution through a borehold or pipe, tunneling into the material beneath the surface. (Fig. 15.1). Variations in these four methods are many and are based mainly on the characteristics of the ore deposits as previously noted.

Table 15.2 indicates the applicability of the methods to the various deposit types described earlier. In applying these methods to submarine metalliferous sulfides in the Gorda Ridge area, which occur in water depths of generally 2,000 to 4,000 m and between 127 and 219 km from land, it is likely that only scraping, excavating, and fluidizing methods would be applicable. The method selected will be largely a function of the nature of the deposit, including its mechanical properties and its depth of burial.

Scraping

Scraping involves the removal of a shallow layer, generally less than 1 m, from the surface being mined. There are many analogous operations on land and in shallow waters, using specially designed machines such as wheeled scrapers, bucket 'dozers, draglines, and drag suction hopper dredges, the last being used for the maintenance of shipping channels all over the world. Deep-water examples of scraping are the bucket dredges of all shapes and sizes used by marine scientists for sampling the deep seabed, the slightly more sophisticated continuous line bucket (CLB) dredge (Fig. 15.2), and the variety of active and passive seabed miners attached to hydraulic risers designed for the exploitation of manganese nodules (Fig. 15.3).

Preliminary designs have been made for a rock scraper to remove high-cobalt manganese crusts from the seabed (Fig. 15.4). The use of scraping may be confined to those deposits that are exposed at the surface of the seabed and

TABLE 15.2. Methods of mining potentially applicable to submarine metalliferous sulfides.

Nature of resource model	Nature of deposit	Depth of occurrence	Scraping	Excavating	Fluidizing Slurry	Fluidizing Solution	Tunneling
Geothermal energy	Hot rocks or interstitial fluids	0–1.5 km	na	na	na	[a]	na
Hot vent fluid	Fluids	Surficial	na	na	na	[b]	na
Metalliferous muds	Semifluid	0–30 m	[c]	[d]	[e]	na	na
Volcanogenic sediment (Mn)	Hard rock	Surficial	[c]	[d]	na	na	na
Surficial precipitates							
Individual stacks or mounds	Porous, brittle	Surficial	[c]	[d]	na	na	na
Coalesced stacks or mounds or cemented tallus	?	Surficial	[c]	[d]	na	na	na
Subseabed precipitates							
Cyprus massive sulfides	Hard rock	0–0.8 km	[c]	[d]	na	[f]	[g]
(Kuroko massive sulfides)	Hard rock	0–0.8 km	[c]	[d]	na	[f]	[g]
Stock work	Hard rock	0.8–1.5 km	[c]	[d]	na	[f]	[g]
Vein systems	Hard rock	1–1.5 km	na	na	na	[f]	[g]
Exhalative beds	Soft to hard rock	0–4 km	[c]	[d]	[e]	[f]	[g]
Deep seabed massive differentiates							
Podiform chromatic	Hard rock	8–10 km	na	na	na	[f]	[g]
Ni-Pt sulfides	Hard rock	>10 km	na	na	na	[f]	[g]

[a] Heat transfer to fluids in boreholes.
[b] Vent capture: minerals already in solution.
[c] Applicable only to materials on or above seabed.
[d] Subseabed materials exposed at or close to seabed.
[e] Unindurated or easily granulated material.
[f] Assumes technical advances in in situ hydrometallurgy.
[g] Applicable only if close to land or if new techniques of seabed entry are developed.
na: Not applicable.

are of relatively extensive lateral dimension such as muds, surficial sediments or encrustations, or the outcrops of deeper-seated beds or massive deposits exposed by faulting.

Excavating

Excavating is the extraction, from a progressively deeper work face, of the material being mined. On land, such operations take the form of open-pit mines, quarries, and dredge ponds. Offshore in shallow water, anchored or stationary dredges excavate by the use of draglines, grab buckets, bucket lines, bucket wheels, rotary cutters, or plain suction pipes. Deep-water excavation has been limited to one known example of bulk sample dredging for unconsolidated sulfide muds in the Red Sea at depths of around 2,000 m. In this case a stationary suction pipe with a vibratory head and water jets to agitate the material was lowered into the deposit to excavate a series of pits (Fig. 15.5). There is no precedent at this time for the excavation of hard rock deposits in the deep seabed, but a number of concepts have been suggested (Kaufman, 1985). The use of some form of rock fracturing followed by more conventional dredging of the broken material seems likely, but considerable advances in rock penetration and fragmentation techniques are needed at this time, particularly in this environment.

Fluidizing (Slurry Mining)

Slurry mining is the conversion to a slurry, in place, of a subsurface, solid mineral deposit and its removal to the surface as a fluid,

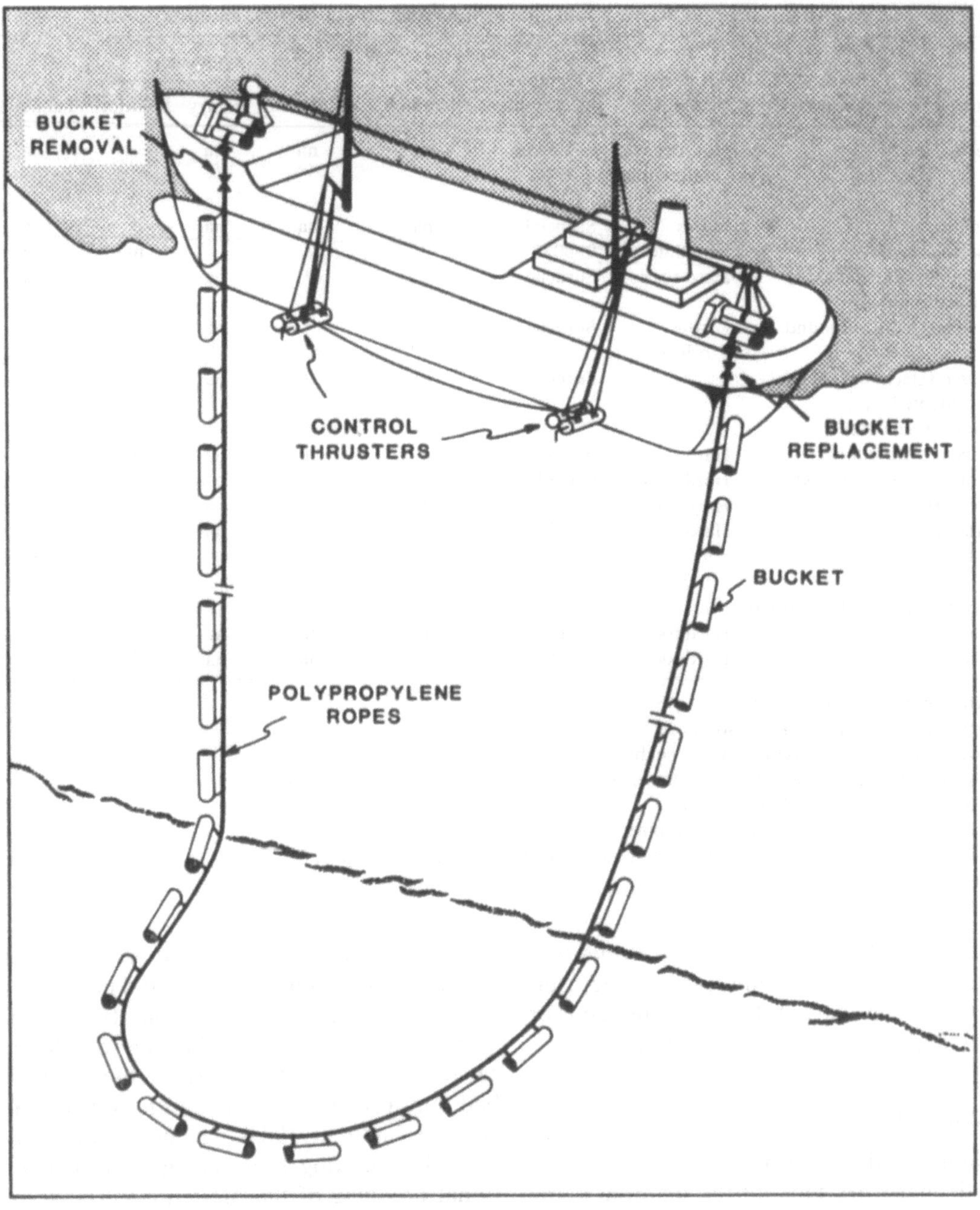

FIGURE 15.2. Continuous line bucket (CLB) dredge.

through a borehole. Examples on land of this technique are the removal of bedded phosphorites (Hrabik and Godesky, 1985) using high-pressure water jets to slurry the material, or the removal of sulfur by the Frasch process using superheated water to fluidize the ore (TGS, 1961). The method requires the drilling of many holes so that the pattern of excavations leaves a minimum of material unmined. Prerequisites are that the ore is amenable to fluidizing and that the overburden is strong enough to sustain the excavated cavity without collapse, at least until the adjacent cavities are mined out or the excavated cavity is back-filled with

FIGURE 15.3. Manganese nodule miner.

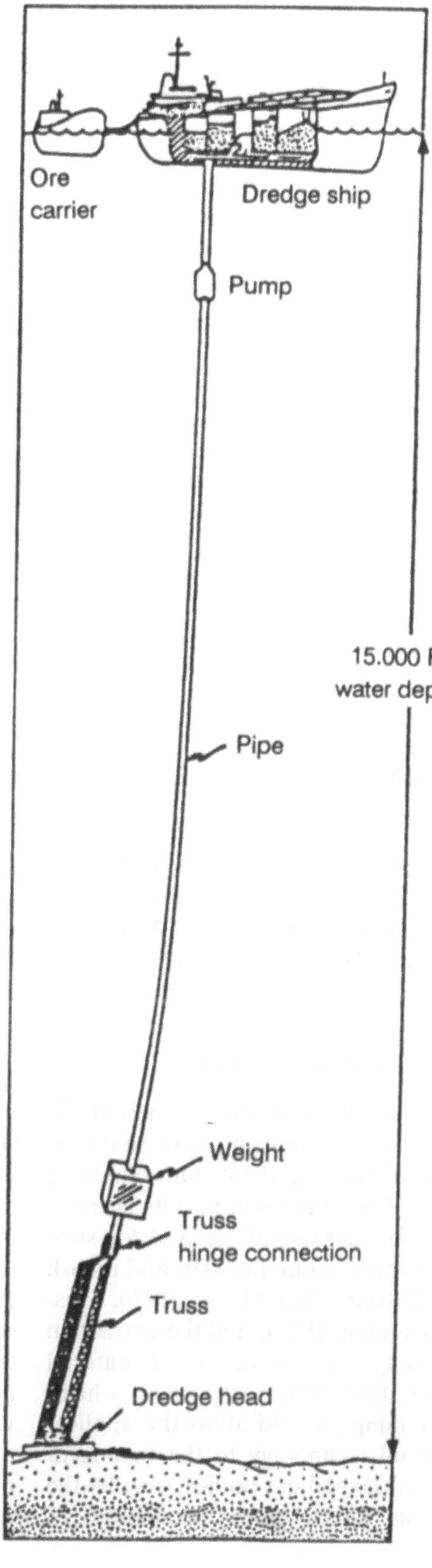

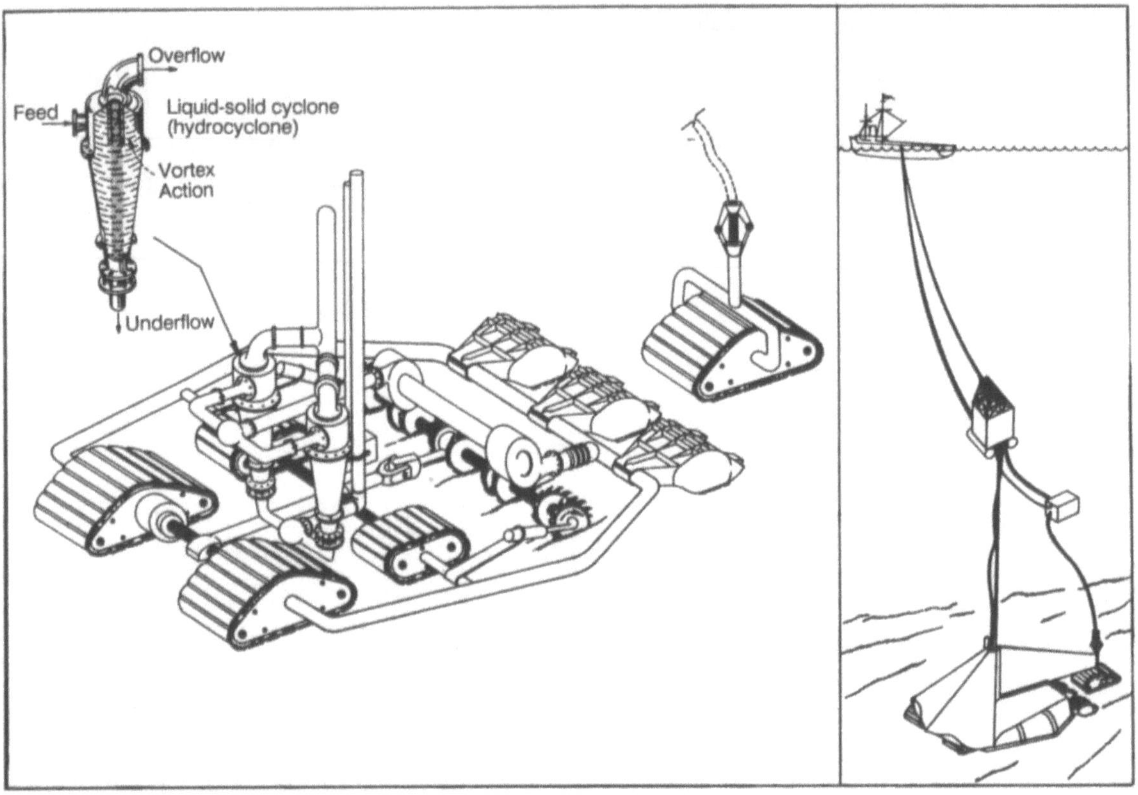

FIGURE 15.4. Cobalt crust miner (after DOI, 1987).

waste material. The only potential deposits in the Gorda Ridge area that might respond to slurry mining would be relatively soft sedimentary or exhalative beds overlain by indurated sediments or flow basalts.

Fluidizing (Solution Mining)

Solution mining involves the extraction, in place, of the valuable constituents of an ore by preferentially dissolving them and removing them from the deposit in solution. The technology is quite well developed on land for such materials as copper, uranium, salt, and potash (Hansen and Robb, 1968; E/MJ, 1970; Husband 1973; Dareing, 1971), and the extraction of gold by heap leaching of mined material is well established. This technology, which, among other things, would allow the application of deep oil technology to the mining of deep-seated subseabed ore bodies, may be the most important of the potential methods of mining for metalliferous sulfides associated with crustal plate boundaries. Areas for development that need considerable attention before such methods are generally feasible are techniques for fracturing the ore bodies in place to provide an adequate flow path for the solvents, and the development of appropriate solvents which can be used to differentially extract the desired metals from the complex sulfides that are typical of these deposits. Examples of such developments are the proposed use of contained nuclear blasts to produce porous chimneys suitable for mining (Fig. 15.6) (AEC, 1967; Hardwick, 1967; Tomblin et al., 1969; World Mining, 1969) and the further use of selective bacterial leaching, enhanced by current advances in metallurgical biotechnology (Mining Journal, 1970; Sokolova and Karavaiko, 1967; Fletcher, 1970).

Breakthroughs in either of these areas could make a considerable difference in the perception of these deposits as potential ore deposits,

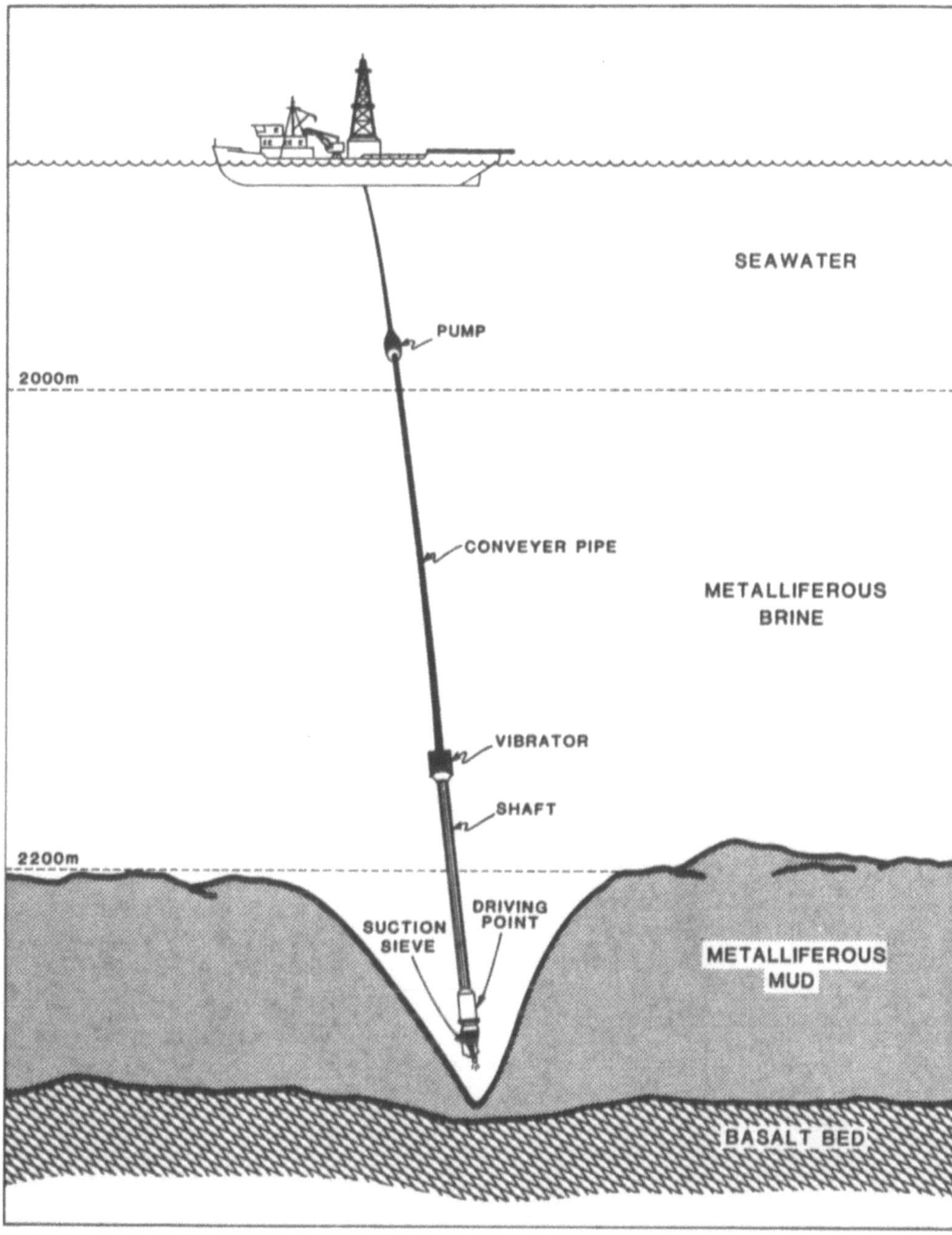

FIGURE 15.5. Anchored suction dredge for Red Sea muds at 2,200 m (after Hahlbrock, 1979).

rather than as scientific curiosities. Similar techniques involving the drilling of patterned boreholes and the circulation of fluids would apply to the extraction of geothermal energy and hot vent waters.

Conclusions

Even though there are no confirmed metalliferous sulfide deposits in the Gorda Ridge area, other than superficial samplings, the potential

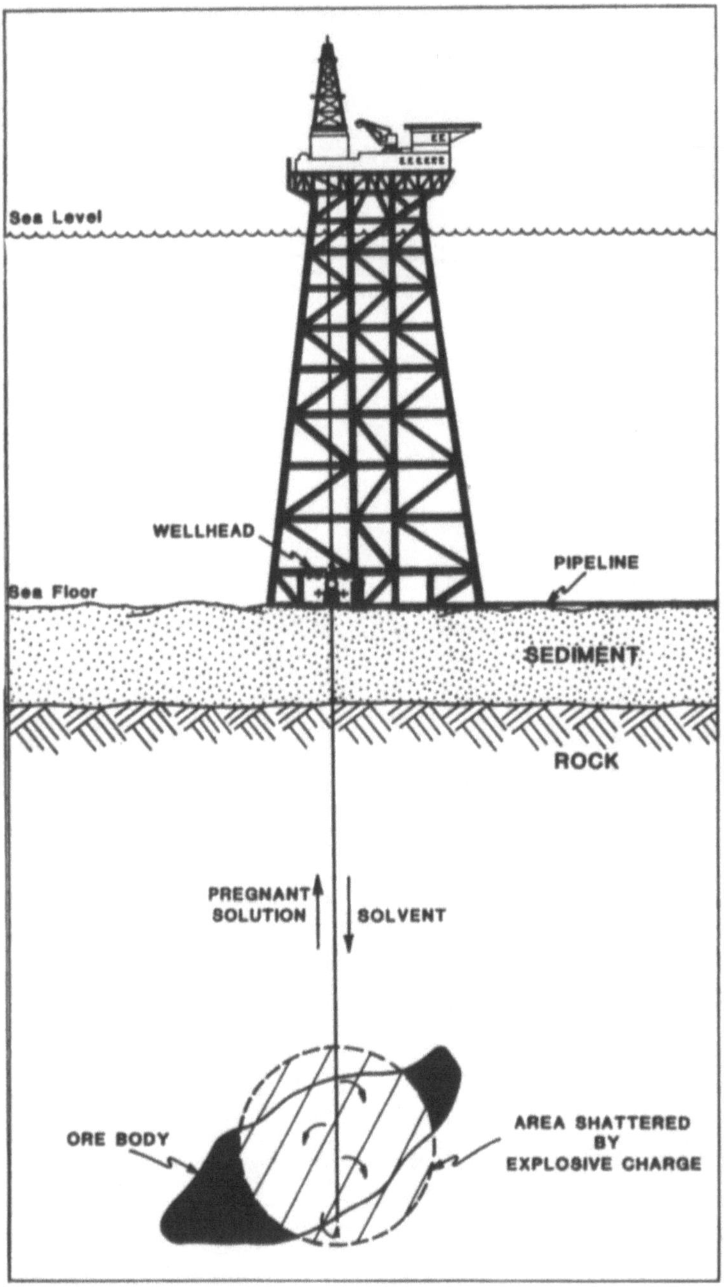

FIGURE 15.6. Contained blast for in situ leaching (after AEC, 1967).

for finding such deposits in a variety of forms for which terrestrial analogs are known appears to be good. The technology for mining such deposits is about the same level as the resource data. Nevertheless, given the incentives of confirmed deposits of appropriate grade to be potentially competitive in contemporary markets, the possibility of developing appropriate technology also appears to be good.

References

AEC (1967) The Bronco Oil Shale Study. Washington: U.S. Atomic Energy Commission, Bureau of Mines, 1CER Geo Nuclear Corp. and Lawrence Radiation Lab., PNE-1400, 64 pp.

Austin CF (1966) Undersea Geothermal Deposits—Their Selection and Potential Use. Washington: U.S. Navy, OTS Tech. paper.

Bonatti E (1978) The origin of metal deposits in the oceanic lithosphere. Sci Am Feb: 54–61.

Cruickshank M, Flanagan JP, Holt B, Padan JW (1987) Marine Mining on the Outer Continental Shelf. Washington: U.S. Department of the Interior, Minerals Management Service, OCS Report 87-0035, 66 pp.

Dareing DW (1971) Solution mining of water, brine, alkali-, or acid-soluble minerals. U.S. Patent 3,600,040.

Degens ET, Ross DA (eds) (1969) Hot Brines and Recent Heavy Metal Deposits in the Red Sea. New York: Springer-Verlag, 600 pp.

DOI (1987) Proposed Mineral Lease Sale in the Hawaiian Archipelago and Johnston Island EEZ. Washington: U.S. Department of the Interior, OCS EIS/EA, MMS 87-4.

E/MJ (1970) Cane Creek to convert to solution mining. Eng Mining J Sept: 171.

Fletcher AW (1970) Metal winning from low-grade ores by bacterial leaching. UK Inst Mining Metall Trans 79: 247–252.

Guney M, Nawab Z, Marhoun A (1974) *Atlantis II Deep*'s metal reserves and their evaluation. OTC 4780: 33–40.

Hahlbrock U (1979) Development of a system for the extraction of metalliferous mud from the Red Sea. Meerestechnik 10: 73–78.

Hansen S, Robb DD (1968) In situ copper leaching at Safford. World Mining Jan: 25–27.

Hardwick WR (1967) Fracturing a Deposit with Nuclear Explosives and Recovering Copper by the In-Situ Leaching Method. Washington: U.S. Bureau of Mines, RI6996, Nov, 48 pp.

Hrabik JA, Godesky DJ (1985) Economic Evaluation of Borehole and Conventional Mining Systems in Phosphate Deposits. Washington: Bureau of Mines I.C. 8929, 34 pp.

Husband WHW (1973) Solution mining of Potash. In: AB Cummins, IA Given (eds.) SME Mining Engineering Handbook. New York: Society of Mining Engineers of AIME pp 2135–2148.

Kaufman R (1985) Conceptional approaches for mining marine polymetallic sulfide deposits. MTS J 19: 50–56.

Maimoni A (1983) Minerals recovery from Salton Sea geothermal brines: A literature review and proposed cementation process. Geothermics 11: 239–258.

Malahoff A (1982) A comparison of the massive submarine polymetallic sulfides of the Galapagos Rift with some continental deposits. MTS J 16: 39.

Mining Journal (1970) Gold extraction developments in the U.S.S.R. Mining J May 1: 394 (microbiological methods).

MMS (1983) Draft Environmental Impact Statement, Proposed Polymetallic Sulfide Minerals Lease Offering, Gorda Ridge Area. Washington: U.S. Minerals Management Service, Dec, 580 pp.

Sokolova GA, Karavaiko GI (1967) Physiology and geo-chemical activity of Thiobacilli. Washington: U.S. Department of Agriculture, Translated from the Russian, NTIS P. 67-51275.

TGS (1961) Modern Sulphur Mining. New York: Texas Gulf Sulphur Co, 27 pp.

Tomblin JT et al. (1969) The Use of Nuclear Explosives for Sea Bed Mining. Aldermaston: U.K.: Atomic Weapons Research Establishment, report SF 1, pp 371–386.

USGS (1986) Mineral deposit models. USGS Bulletin 1693, 379 pp.

World Mining (1969) Nuclear blast to fragmentize Safford copper. World Mining October, news release.

Part IV
Benthic Ecology

16—Distributional Ecology of Benthic Megaepifauna and Fishes in Gorda Ridge Axial Valley

ANDREW G. CAREY JR., Gary L. TAGHON, DAVID L. STEIN, and PETER A. RONA

Abstract

Using bottom photographs and videotapes taken by various institutions in 1984 to 1986, we describe the taxonomic composition and patterns of distribution on two geographic scales and the relative abundance of the benthic epifauna and fishes in Gora Ridge axial valley. Gorda Ridge runs approximately northeast-southwest 200 to 300 km off northern California and southern Oregon and lies within the U.S. Exclusive Economic Zone. Suspension feeding invertebrate taxa occur in all rocky and sedimentary environments. Crinoids appear dominant at the northernmost stations, and ascidians are dominant in the central Gorda Ridge region. In the sedimented Escanaba Trough hexactinellid sponges and gorgonian soft corals are present. Suspension feeding epifauna tend to occur nonrandomly with even distributions predominant. Assemblages of invertebrates and fishes associated with rock and sediment differ in composition and abundance. Deposit feeding organisms become relatively more abundant and are interspersed with stalked suspension feeders at the southern stations of the sediment-filled southern Escanaba Trough. Many species associated with sediments are similar to those found on the surrounding abyssal plains. Biomass is unevenly distributed. The axial valley fauna is abundant and diverse, undoubtedly supported by organic materials of continental origin, including runoff from the Columbia River. In addition, scattered hydrothermal venting provides energy sources for adjacent chemosynthesis-based food webs. The enclosing valley walls and rough topography appear to constrain currents and cause turbulence that maintains the high concentrations of suspended particles observed near the valley floor.

Introduction

In 1985 we began an ecological study of the invertebrate megaepifauna and benthic fishes in Gorda Ridge axial valley. Because no vent sites with their characteristic communities (reviewed by Grassle, 1986) had been observed, we concentrated our ecological studies on fauna associated with hard substrates living within the axial valley (Carey et al., 1986, 1987). Sampling and studying benthos from deep-sea rocky substrates are notoriously difficult; consequently, few data on nonvent, hard-substrate epibenthic communities from the oceanic deep-sea and midocean ridges are available in the scientific literature (Genin et al., 1986). There are many taxonomic works on deep-sea hard-substrate fauna, but the few published ecological studies pertain to continental margins (e.g., Sartori, 1980; Conan et al., 1981; Laubier and Monniot, 1985). Submersibles, deep-towed camera systems, sea floor acoustic transponder-positioning systems, and multibeam sonars now make ecological studies of deep-sea rocky environments feasible (Clague et al., 1984; Carey et al., 1986, 1987; Morton et al., 1987; Genin et al., 1986).

Our overall objective was to describe the taxonomic composition of the fauna and relate patterns of distribution and abundance to physical and geological features of the environment. Results reported in this paper are the product of the specific objectives: (1) to determine the taxonomic composition of the Gorda

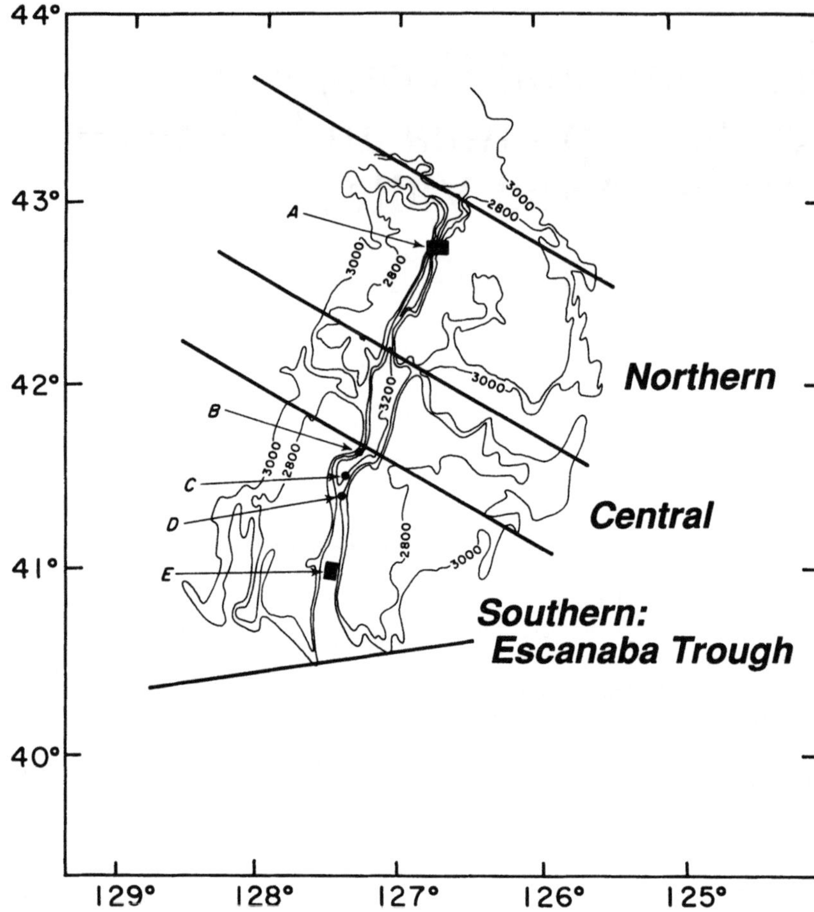

FIGURE 16.1. Geological features of the Gorda Ridge axial valley with dive-sampling-photography stations noted. Transverse geological structures divide the axial valley into three segments with differing characteristics. Area A = DSRV *Alvin* dive 1407 and R/V *S.P. Lee* 1985 stations CS 01C, CS 05, CS 09C, CS 12C, CS 15C, and CS 16C. Area B = DSRV *Alvin* dive 1406. Area C = DSRV *Alvin* dive 1405. Area D = DSRV *Alvin* dive 1406. Area E = R/V *S.P. Lee* 1985 stations 21C and 33C, and NESCA and SESCA.

Ridge megaepifauna, and (2) to define patterns of distribution and abundance of fauna along the axial valley in relation to bathymetry and substrate.

Materials and Methods

Analysis of samples, photographs, and videos obtained by the U.S. Geological Survey (USGS; Menlo Park, CA), the National Oceanic and Atmospheric Administration (NOAA; Newport, OR, and Miami, FL) and the U.S. Navy Submarine Development Group 1 (San Diego, CA) form the basis of our study. Photographic and video coverage has been extensive, providing data from three main segments of the axial valley including the east wall (Fig. 16.1). The USGS conducted rock dredging that retrieved organisms, and the submersibles DSRV *Alvin* (UNOLS) and DSV *Sea Cliff* (U.S. Navy) collected geological and biological specimens. In situ trapping by Oregon State University collected benthic fishes. The taxonomic list (Table 16.1) is derived from the 1984–1986 animal collections and bottom photographs with the aid of 16 taxonomic specialists (see Acknowledgments).

TABLE 16.1. Summary of benthic taxa observed, collected, and identified from the Gorda Ridge axial valley 1984–1987.

- Porifera
- Class Hexactinellida
  - Subclass Amphidiscophora
    - Order Amphidiscosida
      - Family Hyalonematidae
        - *Hyalonema* sp.
        - *Hyalonema* cf. *apertum* Schultz 1887
  - Subclass Hexasterophora
    - Order Lyssacinosa
      - Family Euplectidae
        - *Leroyella* sp. ?
      - Family Rossellidae
        - *Aulochone lilium* Schulze 1887 ?
        - *Staurocalyptus fasciculatus* Schulze 1899
        - ?*Staurocalyptus* sp. nr. *fasciculatus* Schulze 1899
        - *Rhabdocalyptus* sp. nr. *R. dawsoni* (Lambe 1892)
      - Family Farreidae
        - ?*Farrea occa* Bowerbank 1862
        - *Farrea aculeata* Schulze 1899
      - Family Aphrocallistidae
        - *Aphrocallistes vastus* Schulze 1887
- Class Demospongiae
    - Order Axinellida
      - Family Axinellidae
        - *Phakettia* sp.
    - Order Poecilosclerida
      - Family Cladorhizidae sp.
    - Order Hadromerida
      - Family Suberitidae
        - *Suberites* sp. ?
- Cnidaria
- Class Anthozoa
  - Subclass Alcyonaria
    - Order Gorgonacea
      - Suborder Holaxonia
        - Family Isididae
          - *Keratoisis*/*Lepidsis* sp. 1
          - *Keratoisis*/*Lepidsis* sp. 2
        - Family Primnoidae
          - *Calyptrophora* sp. ?
          - *Narella* sp. cf. *bowersi* (Nutting)
    - Order Pennatulacea
      - Suborder Subselliflorae
        - Family Virgulariidae sp.
  - Subclass Zoantharia
    - Order Antipatharia
      - Suborder Antipathida
        - Family Heterotaeniales
          - *Antipathes* sp.
          - *Bathypathes* sp.
    - Order Ceriantharia
          - *Cerianthus* sp.
    - Order Zoanthinaria (Zoanthidea)
        - Family Epizoanthidae
          - *Epizoanthus*/*Isozoanthus* sp.

Continued

TABLE 16.1. *Continued*

Order Actiniaria
Family Phymanthidae
Subtribe Mesomyaria sp.
Family Actinostolidae
Actinostola spetsbergensis ?
Family Actinoscyphiidae
Actinoscyphia sp.
Mollusca
Class Cephalopoda
Subclass Coleoidea
Order Octopoda
Family Octopodidae
Octopus sp. ?
Cirroteuthis sp.
Class Bivalvia
Subclass Heterodonta
Order Veneroida
Family Malletiidae
Malletia (*Malletia*) *truncata* Dall 1908
Annelida
Class Polychaeta
Family Sabellidae
Fabrisabella sp.
Arthropoda
Class Crustacea
Order Decapoda
Suborder Reptantia
Section Anomura
Superfamily Galatheidae
Family Galatheidae
Munidopsis sp.
Munidopsis latirostris (Henderson)
Section Caridea
Family Pasiphaeidae
Parapaisphae sulcatifrons Smith 1884
Family Oplophoridae
Hymenhodora frontalis Rathbun 1902
Superorder Peracarida
Order Mysidacea
Family Petalophthalmidae
Pelalophthalomus armiger Wilemoes-Suhm
Order Amphipoda
Family Hyperiidae
Phronima sedentaria (Forskal)
Echinodermata
Class Stelleroidea
Subclass Asteroidea
Order Platyasterida
Family Luidiidae
Luidia sp. ?
Order Spinulosida
Suborder Leptognathina
Family Echinasteridae
Echinaster sp. ?
Order Euclasterida
Family Brisingidae
Freyella cf. *insignis* Ludwig

Continued

TABLE 16.1. *Continued*

Family Pterasteridae
Hymenaster perissonotus Fisher
Family Asteriidae
Subclass Ophiuroidea
Order Ophiurida
Suborder Chilophiurina
Family Ophiuridae
Ophiocten hastatum Lyman 1878 ?
Ophiomusium multispinum Clark 1911 ?
Family Ophiacanthidae
Ophiocantha sp. a
Ophiocantha sp. b
Ophiocantha sp. c
Class Crinoidea
Subclass Articulata
Order Comatulida
Order Millericrinida
Rhizocrinus sp.
Family Bathycrinidae
Bathycrinus sp. a
Bathycrinus sp. b
Order Cyrtocrinida
Order Hyocrinida
Family Hyocrinidae
Ptilocrinus pinnatus AH Clark 1907 ?
Thalassocrinus nov. sp.
Class Holothuroidea
Order Dendrochirotida
Family Cucumaridae
Abyssocucumis abyssorum
Order Aspidochirotida
Family Synallactidae
Paelopatides confundens (Theel 1882)
Synallactes sp. a ?
Synallactes sp. b ?
Class Echinoidea
Subclass Euechinoidea
Superorder Atelostomata
Order Holasteroida
Family Urechinidae
Urechinus loveni (A Agassiz 1898) ?
Chordata
Subphylum Urochordata
Class Ascidiacea
Order Stolidobranchiata
Family Pyuridae
Culeolus sluiteri Ritter, 1913
Culeolus pyramidalis Ritter, 1907
Culeolus sp.
Bathypera ovoida (Ritter, 1907)
Family Ascidiidae
Ascidia nov. sp.
Family Molgulidae?
Pisces
Class Osteichthyes
Order Gadiformes
Family Macrouridae

Continued

TABLE 16.1. *Continued*

		Coryphaenoides armatus (Hector 1875)
		Coryphaenoides ?filifer (Gilbert 1895)
		Coryphaenoides leptolepis Gunther 1877
	Family Moridae	
		Antimora microlepis Bean 1890
Order Ophidiiformes		
	Family Ophidiidae	
		Spectrunculus grandis (Gunther 1877)
	Family Zoarcidae	
Order Scorpaeniformes		
	Family Liparididae	
		Paraliparis ?rosaceus Gilbert 1890
	Family Synodontidae	
		Bathysaurus mollis Gunther 1878

Photography and Video Recording

NOAA and the USGS took series of photographs from submersibles in 1984 and 1986. NOAA geologists made four reconnaissance dives along the Gorda Ridge rift valley during late summer of 1984 (Fig. 16.1). BENTHOS survey cameras mounted obliquely on DSRV *Alvin* and DSV *Sea Cliff* took photographs on Ektachrome 200 ASA film. During the 1986 dives of DSV *Sea Cliff*, a BENTHOS 35-mm camera took photographs and in addition, geologists and biologists took photographs with interior hand-held 35-mm cameras coupled to an exterior strobe flash unit. A hand-held automatic video camera taped segments of continuous visual sequences of the valley floor adjacent to the submersible.

The USGS (R/V *S.P. Lee*) and NOAA (R/V *Surveyor* and R/V *Discoverer*) took long series of bottom photographs with deep-towed, instrumented camera vehicles. The USGS (Branch of Marine Geology, Menlo Park, CA) photographic system (Chezar and Lee, 1985) consists of a 35-mm camera with a maximum capacity of 400 feet of film and five synchronized strobe flash units plus an in situ recording television camera. A precision depth recorder (PDR) plotted the camera system pinger trace, from which we estimated the system altitude above the bottom to the nearest meter (W.R. Normark, personal communication). The camera group at Lamont-Doherty Geological Observatory under the direction of W.B.F. Ryan designed and constructed the NOAA camera system; it has similar capabilities to the USGS system.

Animal Collections

Animal specimens collected by the USGS came from rock dredge tows in the ridge axial valley and from geological specimens collected by DSV *Sea Cliff*. USGS geologists picked the specimens from the surface of rock samples on board ship and preserved them in 10% Formalin buffered with sodium borate. In the laboratory they were sorted, identified as far as possible, and placed in 70% ethanol for later distribution to taxonomic specialists for verification or primary identification.

We made two in situ fish trap collections at stations NESCA and SESCA (see Morton et al., this volume) in Escanaba Trough during the 1986 R/V *S.P. Lee* cruise (L2-86-NC) (Fig. 16.1). Traps were cubes 1.2 m on a side, constructed with 3.8-cm vinyl-coated metal mesh. They were baited with squid, herring, and anchovy and deployed for 24 hours as free-fall vehicles. Each trap system consisted of an anchor chain link ballast weight, trap, a timer-release/pinger unit, glass floats, and a locating flag mounted on a surface float.

Photographic and Video Analysis

Photographs were examined using a Variscan aerial photogrammetric analysis device (Mark II, Westwood Division, Houston Fearless

Corp.) allowing automated projection at four magnifications (3×, 6×, 12.4×, and 31.2×). Photographs analyzed for this paper are primarily from the 1984 DSRV *Alvin* dives in the ridge valley axis (Fig. 16.1).

Animals in photographs were identified to the lowest possible taxon and counted per frame. In addition to our analysis, we sent photographs and specimens of megafaunal invertebrates to 16 taxonomic specialists for further identification and comment (see Acknowledgments). A few photographed animals could be identified to species with some confidence, although the majority were identified and counted at higher taxonomic levels during analysis of each usable frame. Many fishes, particularly macrourids, could only be identified to the family level, because of low image resolution. Others could only be identified as fishes.

Animal counts per photograph were summarized for each taxon, and faunal abundances were compared between different environmental regions of the axial valley. We counted numbers of individuals of major taxa per grid cell with a translucent rectangular grid mounted on the Variscan projection screen.

Data Analysis

For analysis of small-scale distribution patterns in the 1984 DSRV *Alvin* photographs, each photograph was treated as a similar-size quadrat. we assumed that all photographs were taken from approximately the same altitude. Examination of the photos supported our assumption; most were taken close to the bottom. Per-photograph counts of organisms were used to estimate faunal abundance. Lower taxa were combined into 10 higher taxonomic categories to provide sufficient sample size for statistical treatment (anthozoans, asteroids, crinoids, gorgonians, holothuroids, ophiuroids, pennatulids, poriferans, tunicates, and echinoids). Data were grouped into seven habitat categories depending on substrate type: soft substrate (present in photographs from dives 1404 and 1405), mixed substrates (present on dives 1404, 1405, 1406, and 1407), and hard substrate (present on dive 1407).

We used variance-to-mean ratios from per-photograph counts to determine spatial dispersion of the 10 taxa. Taylor et al. (1978) reviewed field data on a wide range of organisms and found that per-quadrat abundances best fit a model of the form:

$$S^2 = aX^b \quad (1)$$

where X is the mean abundance and S^2 is the variance. Jumars and Eckman (1983) revised this model to:

$$\log(S^2/X) = \log a + (b-1)\log X \quad (2)$$

Under the Poisson distribution, $S^2 = X$, and equation 2 reduces to:

$$0 = \log a + (b-1)\log X \quad (3)$$

which requires that $a = b = 1$ under the null hypothesis of random spatial dispersion of counts among quadrats. If the per-quadrat counts are not Poisson-distributed, S^2/X will correlate with X when these parameters are graphed in a scatterplot. Photographic data from each of the dive-habitat categories were analyzed two ways. First, we included all photographs in analyzing each taxon, including those photos where the taxon of interest was absent (referred to as "0's included"). Second, only photos containing at least one individual of the taxon of interest were selected. These smaller subsets of data were analyzed statistically only if there were five or more photos containing animals of each taxon. In both cases, we computed logarithms of the mean and variance/mean for each taxon from each of the seven habitat categories.

Log (variance/mean) = 0 indicates random dispersion, log (variance/mean) = >0 indicates aggregated dispersion, and log (variance/mean) = <0 indicates an even dispersion pattern. Departures from randomness were tested by computing the chi-square statistic (Pielou 1977); an alpha level of 0.05 was chosen.

Results

Gorda Ridge axial valley epifauna are diverse and include two species new to science: a crinoid, *Thalassocrinus* sp., and an ascidian, *Ascidia* sp. (Table 16.1). Specimens of these were collected by rock dredge from the R/V *S.P.*

Lee or by DSV *Sea Cliff*. Identifications from photographs and videotapes are generally tentative. In many cases animals observed and studied in the bottom photographs could only be identified to family or a higher taxon.

An extensive, visible benthic megafauna, consisting primarily of suspension and deposit feeders, occurs in Gorda Ridge axial valley (Carey et al., 1986, 1987). Although deposit feeders are generally associated with sedimentary substrates, they may also be prevalent on basalts with little sediment cover. Conversely, stalked suspension feeders are also found on sediments, though they are dominant on rocky surfaces.

Large-Scale Distribution Pattern

Based on available data, the taxonomic composition of rocky epifauna changes from the hexactinellid sponge and crinoid assemblage of the Northern (1) zone to a mixed assemblage in the Narrowgate area (*Alvin* dive 1406) to the ascidian-dominated south Central (2) zone of the axial valley, just north of the structural transition (Fig. 16.2). Stalked suspension feeders such as deep-sea ascidians, crinoids, glass sponges, and gorgonian soft corals form dense patches on rock surfaces.

Many deposit-feeding animals are present in the sediment-laden Escanaba Trough. Some, including holothuroids (sea cucumbers), ophiuroids (brittle starfish), asteroids (starfish), and galatheid crabs (squat lobster), are also dominant on Cascadia Abyssal Plain to the north (Carey, this volume). Stalked suspension feeders—e.g., hexactinellid glass sponges, pennatulid sea pens, and sea anemones (actiniarians)—are often interspersed among the deposit feeders.

Small-Scale Distribution Patterns

Sixty small-scale distribution patterns were identified. In 31 of these, animals clustered on the spatial scale of the photographs (Table 16.2A). The remaining patterns fit a Poisson distribution. Dividing the fauna into trophic groups, suspension feeders (anthozoans, crinoids, gorgonians, pennatulids, poriferans, and tunicates) were aggregated in 56.8% of the cases; the remaining 43.2% were random. Omnivores/deposit feeders (asteroids, holothuroids, ophiuroids, and echinoids) were aggregated 43.5% of the time; the remaining 56.5% were random.

The epifaunal dispersion patterns, when zeroes were excluded, demonstrate a shift away from aggregation toward random or even dispersion patterns (Table 16.2B). Only 13.6% of the patterns analyzed were aggregated, compared to 43.2% random and 43.2% even. Of the combinations of taxa and habitat categories common between the 0's included and 0's excluded data sets, 83.3% of the dispersion patterns moved toward more even distributions when photos with no organisms were excluded (Table 16.2C).

The majority of the distribution patterns identified do not fit a Poisson distribution. Such nonrandomness appears to be the norm in natural populations of organisms (Taylor et al., 1978). Substrate influences faunal composition and small-scale distributions (Fig. 16.3). Taxonomic composition of epifauna on soft and hard/mixed substrates (DSRV *Alvin* dive 1405) differs, particularly for ascidians (deep-sea tunicates). Short stalked ascidians (*Bathypera ovoida*) form dense populations on the basalt in this section of the rift valley.

Abundance

Axial valley fauna appears more abundant just south of the structural transition at 41° 35′ N along the southern ridge that forms the northern boundary to Escanaba Trough (Carey et al., 1986; Table 16.3). Numbers of animals per photograph are higher on *Alvin* dive 1405, particularly on the rocky and mixed substrates, than at other dive sites. Ascidians occurred on the rocks in large numbers and comprised close to 50% of the assemblage with the numbers of ophiuroids and crinoids together accounting for another 35% of the animals.

Feeding Type

Large epifaunal invertebrates in the ridge valley are primarily suspension and deposit feed-

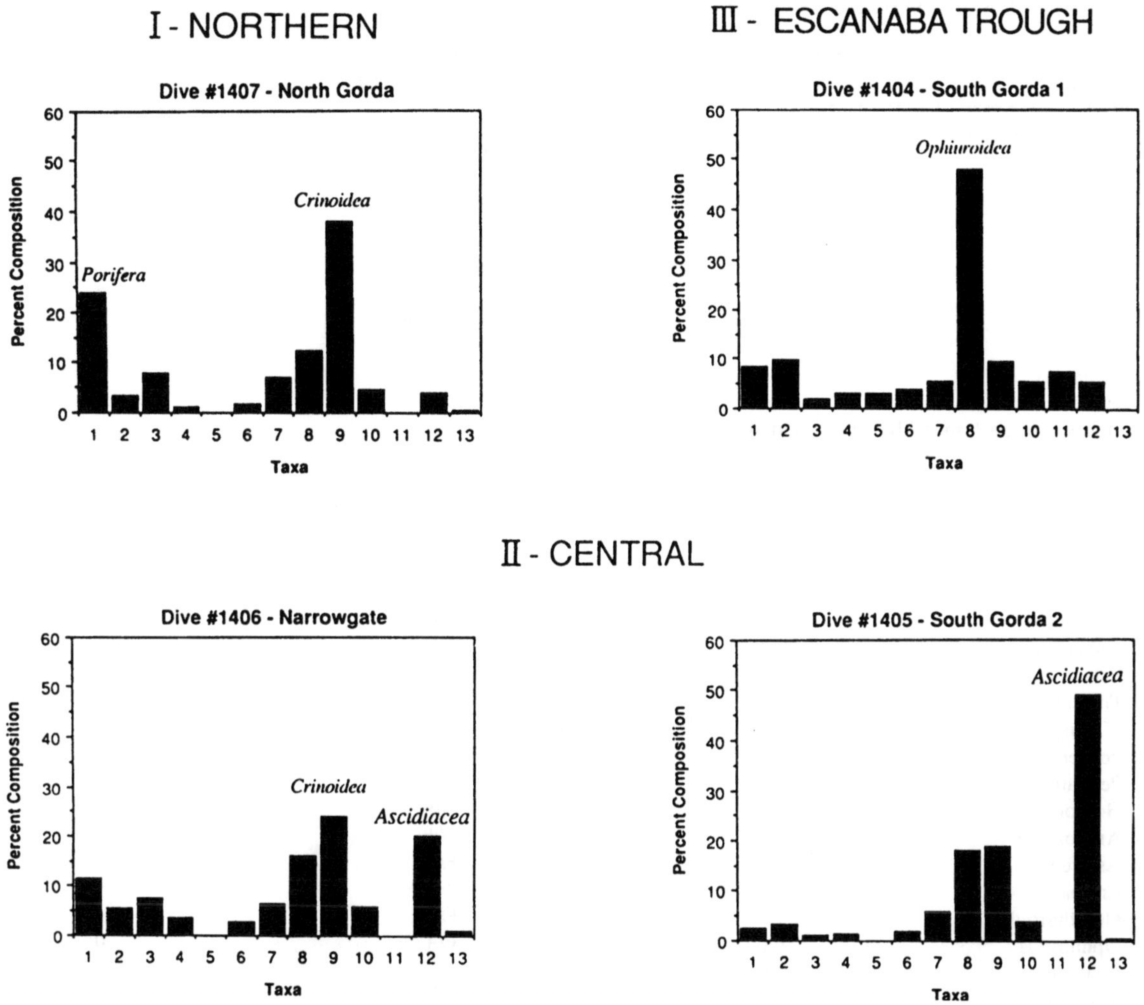

FIGURE 16.2. Percent taxonomic composition of megaepifauna down the valley axis. Note the change in dominant fauna from the northern to the southern zones of the axial valley. 1, Porifera; 2, Pennatulacea; 3, Gorgonacea; 4, Actiniaria; 5, Gastropoda; 6, Decapoda; 7, Asteroidea; 8, Ophiuroidea; 9, Crinoidea; 10, Holothuroidea; 11, Echinoidea; 12, Ascidiacea; 13, Pisces.

ers. Stalked suspension feeders—e.g., glass sponges, gorgonian corals, sea pens, soft corals, sea anemones, and crinoids—are dominant in both rocky and sedimentary substrates. A short-stalked deep sea ascidian (*Bathypera ovoida*) is a suspension feeder and often occurs in large numbers on rocks, while the slender-stalked tunicate, *Culeolus sluteri*, is found sparsely scattered in the rocky habitats. The latter species is morphologically specialized for capturing and concentrating particles from the water (Monniot and Monniot, 1978). Characteristic of the larger abyssal ascidians, they have siphons with large openings and nonciliated branchial sacs, which are unique to the deep sea. These organisms are microphagous, probably capturing small particles entering the branchial sac through the wide oral siphonal opening. A diverse group of nonstalked suspension feeders—e.g., sponges (Demospongiae), comatulids (nonstalked crinoids), and a long-armed asteroid (*Freyella* cf. *insignis*)—are often abundant on pillow lavas. An anemonelike tube dweller (Ceriantharia) with long, slender tentacles occurs within sediments frequently adjacent to rocks.

TABLE 16.2. Dispersion patterns of benthic megaepifauna.

Taxon	Soft substrate (1404/1405)	Mixed substrates (1404/1405/1406/1407)	Hard substrates (1407)
A. Dispersion patterns with zeroes included			
Poriferans	R/R	R/R/R/A	R
Pennatulids	R/A	R/A/A/A	
Gorgonians	/	R/R/A/R	A
Anthozoans	A/A	R/R/R/	A
Asteroids	R/R	R/A/R/R	R
Crinoids	A/A	R/A/A/A	A
Holothuroids	R/A	R/A/R/R	R
Ophiuroids	A/A	A/A/A/R	A
Echinoids	A/R	/ / /	
Tunicates	R/A	A/A/A/	A
B. Dispersion patterns with zeros excluded			
Poriferans	R/E	/E/E/R	E
Pennatulids	/E	E/R/E/R	
Gorgonians		/ /E/	R
Anthozoans	R/R	/ /E/	R
Asteroids	/E	/R/E/	E
Crinoids	/E	E/A/E/R	R
Holothuroids	/R	/R/E/	E
Ophiuroids	R/A	/R/R/E	A
Echinoids	R/E	/ / /	
Tunicates	/R	A/A/A/	R
C. Changes in dispersion patterns—analyzed with zeros excluded			
Poriferans	0/+	/+/+/+	+
Pennatulids	/+	+/+/+/+	
Gorgonians		/ /+/	+
Anthozoans	+/+	/ /+/	+
Asteroids	/+	/+/+/	+
Crinoids	/+	+/0/+/+	+
Holothuroids	/+	/+/+/	+
Ophiuroids	+/0	/+/+/+	0
Echinoids	+/+		
Tunicates	/+	0/0/0/	+

Numbers below substrate headings refer to DSRV *Alvin* dive numbers. A, aggregated; R, random, E, even. Blank spaces indicate insufficient sample size. +, Changes from aggregated to random, from random to even or from aggregated to even. 0, No changes in dispersion pattern. Blank space indicates insufficient sample size.

Epibenthic deposit feeding animals include the common deep-sea holothuroids (*Paelopatides confundens* and *Abyssocucumus abyssorum*) and the ophiuroids (*Ophiomusium multispinum* and *Ophiocten hastatum*?). The galatheid crab (*Munidopsis latirostris*) is a detritus-feeding omnivore always found on rock surfaces.

Geological/Ecological Zones

Based on topographic features, substrate types, and patterns of invertebrate distribution and abundance, the Gorda Ridge axial valley can be divided into three zones: (1) the Northern, (2) the Central, and (3) the Southern, or Escanaba Trough (Fig. 16.4). These environmental zones are about the same as the geological regions bounded by the two major transverse faults (Fig. 16.1), and a characteristic fauna appears to represent each zone (Fig. 16.4). The steep sediment-starved valley has stalked hard-substrate fauna, such as poriferans, alcyonarians, and crinoids. In the deep axial valley of the central zone, crinoids and ascidians are dominant. The sediment-filled Escanaba Trough supports abundant taxa including alcyonarians, ophiuroids, and holothuroids.

Visible benthic epifauna are more abundant

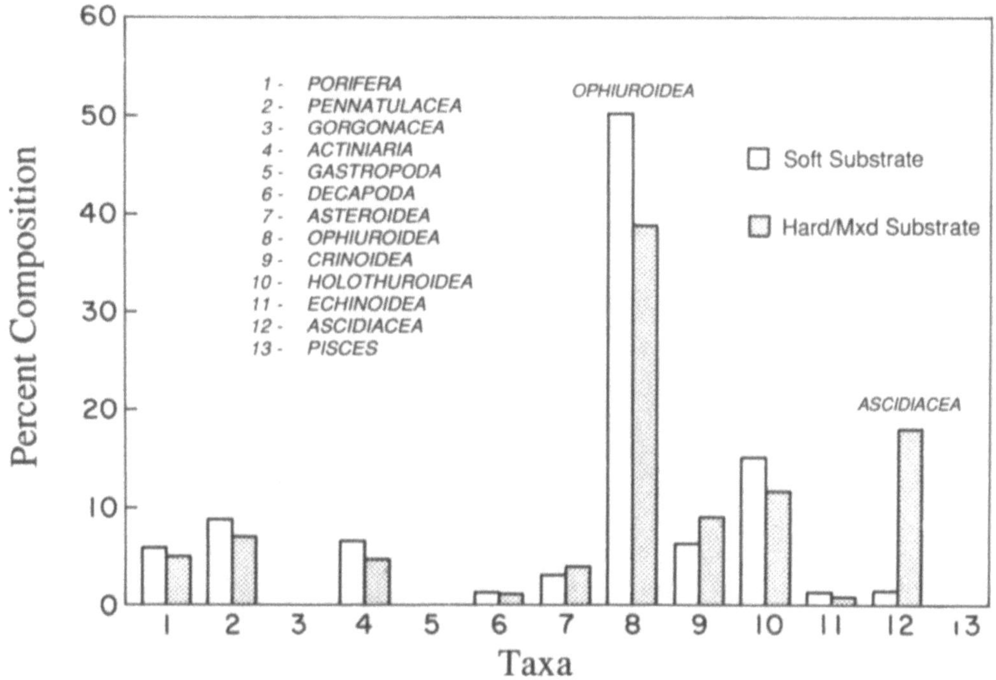

FIGURE 16.3. Percent taxonomic composition of epifauna on soft and hard/mixed substrates at DSRV *Alvin* dive site 1405 in the northern Escanaba Trough.

TABLE 16.3. Abundance of visible epifauna on the Gorda Ridge; results of analyses of 1984 DSRV *Alvin* photographs.

Substrate	Soft	Hard	Mixed	Total
Northern Axial Valley (dive 1407)				
Number of frames analyzed	(0)	(63)	(23)	(87)
Mean number animals ($\bar{X}$/frame)	—	8.1	8.6	8.2
Range ($\bar{X}$/frame)	—	1–22	3–17	1–22
Central Axial Valley (dive 1406)				
Number of frames analyzed	(0)	(30)	(45)	(75)
Mean number animals ($\bar{X}$/frame)	—	10.9	8.5	9.5
Range ($\bar{X}$/frame)	—	2–19	2–20	2–20
Central Axial Valley (dive 1405)				
Number of frames analyzed	(245)	(10)	(26)	(286)
Mean number animals ($\bar{X}$/frame)	6.5	38.5	22.4	9.0
Range ($\bar{X}$/frame)	0–25	6–63	5–72	0–72
Southern Axial Valley—Escanaba Trough (dive 1404)				
Number of frames analyzed	(16)	(1)	(8)	(25)
Mean number animals ($\bar{X}$/frame)	8.4	11.0	6.9	9.3
Range ($\bar{X}$/frame)	1–19	—	—	2–21

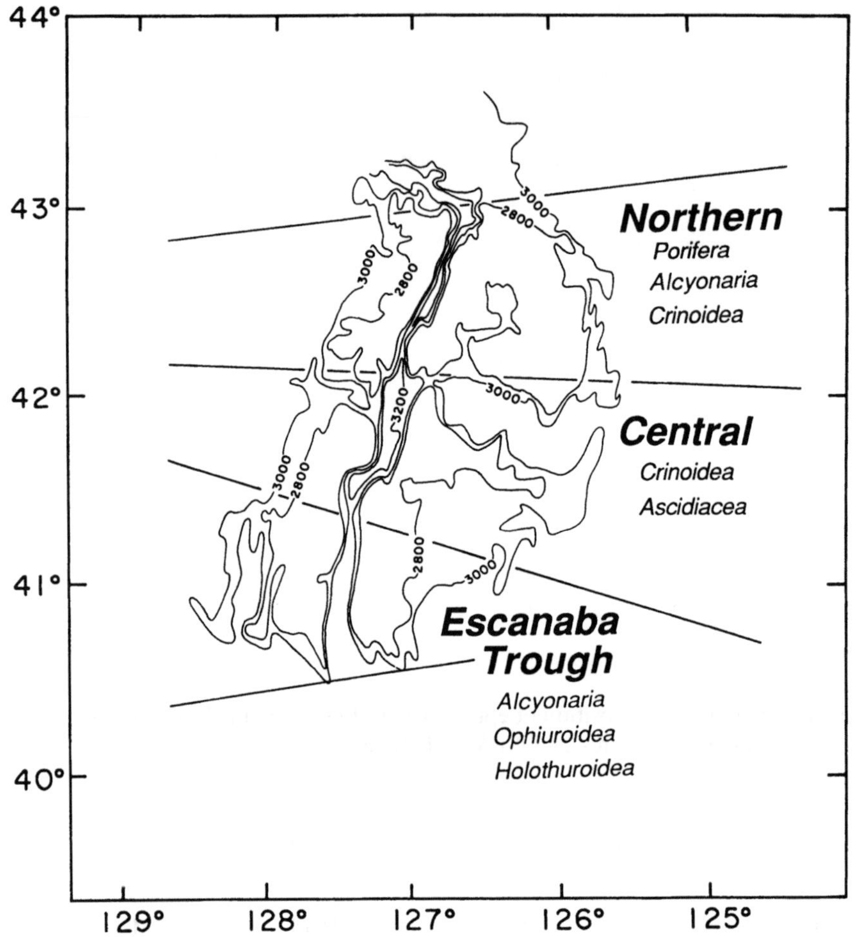

FIGURE 16.4. Gorda Ridge axial valley benthic ecological zones based on topographic and substrate characteristics and indicator animals.

on the axial valley floor in the central zone than in the northern or southern (Escanaba Trough) zones of Gorda Ridge (Table 16.3). The mean and maximum number of animals per frame are greater in the central zone (site of DSV *Alvin* dive 1405). There appear to be different standing stocks of fauna in the three environments (i.e., associated with hard and soft substrates), perhaps related to concentration of suspended particulate material overlying the sea floor.

Fishes

At least nine different fish species from six different families were tentatively identified from the 1986 Escanaba Trough fish, traps, video tapes, and photographs. These include representatives of the families Macrouridae (*Coryphaenoides armatus*, *C. ?filifer*, *C. leptolepis*), Moridae (*Antimora microlepis*), Zoarcidae (unidentifiable species), Ophidiidae (*Spectrunculus grandis*, unidentified species), Liparididae (*Paraliparis ?rosaceus*), and Synodontidae (*Bathysaurus mollis*).

The first fish trap, deployed at Station SESCA, collected 12 *C. armatus* (four males, eight females), one *S. grandis* (male), and one zoarcid (unsexed). The second trap, at Station NESCA, caught 12 *C. armatus* (eight males, four females). These trap catches suggest abundant fish populations, although we have no other fish trap data from anywhere off Washington, Oregon, or northern California for comparison. However, 2-hour 3-m beam trawl catches of similar numbers and biomass at the same depth to the north on Cascadia Plain would be considered large (Pearcy et al., 1982).

Abundances of these taxa in Escanaba Trough (Carey et al., 1986) varied greatly between camera tracks and within camera tracks. Overall order of abundance (in descending order) in still photos was (1) macrourids, (2) ophidiids, (3) zoarcids, (4) liparidids, (5) synodontids, and (6) morids.

Discussion

Large-Scale Distribution Patterns Down the Valley Axis

As Genin et al. (1986) suggest, the distribution of hard-substrate suspension-feeding deep-sea fauna, such as gorgonians (horny corals) and antipatharians (black corals), is probably influenced by the bottom current regimen. They found numerous corals associated with higher current velocities on the summits of seamounts than on the lower flanks. The seamount peaks rise to about 600 m depth from the 4,200-m-deep abyssal Pacific floor; however, different topographies were compared at similar depths. Although current measurements are not available for Gorda Ridge Valley, the topography of the valley and the amount of suspended particulate material observed by us from submersibles suggest that bottom currents may concentrate particulate organic material (POM) in certain regions, leading to an increase food supply for suspension feeders. Both the quantity and quality of POM may control species distributions; for example, crinoids are prevalent in the northern sector of the valley, and ascidians are prevalent the central sector, within similar depth ranges. The Venus flytrap-like anemone (*Actinoscyphia* sp.) appears to occur only in Escanaba Trough (See Fig. 6.17, Holmes and Zierenberg, this volume). Depth zonation of species and the amount and quality of food supply may also affect the distribution and abundance of the epibenthic hard-substrate fauna in the Gorda Ridge axial valley. Because of distribution patterns of sediment and topographic features, the Southern Ecological Zone may not extend as far north as the transverse fault that separates the lower part of the ridge into geological zones.

Small-Scale Distribution Patterns Down the Valley Axis

The higher incidence of aggregation in suspension feeders is not surprising given the assumption that these taxa select microhabitats in environments with complex substrate morphology. Suspension feeders tend to concentrate where there are abundant suspended particulate food materials and where they are exposed to bottom currents. Large- and small-scale topography probably affect current pattern and velocity as well as sediment resuspension and, therefore, the availability of POM to the sedentary epifauna.

Relative Abundance of the Fauna

Animals in Gorda Ridge axial valley appear to be more numerous than oceanic deep-sea populations at the same depth on either sedimentary or rocky environments (T. van Andel, personal communication; S. Hammond, personal communication, Carney and Carey, 1982; Haedrich et al., 1980; Pearcy et al., 1982). We hypothesize that there is a larger supply of food materials on the axial valley floor than at equivalent depths on the oceanic abyssal plains. The increased food supply would support the higher-standing stocks of benthos. Three possible sources of POM associated with the valley are (1) from continental run-off via the Columbia River plume during its southwestern summer extension (Barnes et al., 1972); (2) from the entrainment of particles and currents by the valley walls; and (3) from bacterial chemosynthetic primary production associated with active hydrothermal vents (Cavanaugh, 1985; Rona et al., 1988; Zierenberg et al., 1988).

Export of terrestrial organic materials may be an important food source for the axial valley fauna. Lyle and Prahl (personal communication) estimate that about 20% of the plant lipids associated with sediments on the ridge flanks originate in the terrestrial environment. Though lipids are but a small fraction of sedimentary organics, these data show that POM from the continent reaches the valley floor. Hydrographic data demonstrate that the large outflow of the Columbia River reaches the location of the Gorda Ridge during the summer (Conomos and Gross, 1972). The Columbia River plume is advected by wind stress to the southwest as a lens of low-salinity water (Barnes et al., 1972); it contains particulate material and terrestrial debris (Conomos and Gross, 1972; Carey, personal observation). In addition, turbidites from the Oregon and California continental margins, including terrestrial organic material, occasionally cascade down the adjacent continental slope and the floor of Gorda Basin. The Mendocino Escarpment funnels sediment flows into the southern axial valley (Vallier et al. 1973).

The extent of the influence of the chemosynthetic food web associated with active hydrothermal vents at station GR14 in the north and in Escanaba Trough to the south is unknown (Hessler and Smithey, 1983; Cavanaugh, 1985; Zierenberg et al., 1988; Rona et al., 1988). Intense chemosynthetic primary production may add to the POM in the bottom water layer (Tuttle, 1985). Bottom currents entrained along the valley axis would resuspend and transport these materials to sessile stalked suspension feeders along the valley floor and adjacent side walls.

Origin of the Fauna

Results from previous studies of the megaepifauna on Cascadia Plain, Tufts Abyssal Plain, and the Oregon continental margin (Carney and Carey, 1976; 1982; Ambler, 1980; Pearcy et al., 1982; Carey and Kyte, in preparation; Carey, this volume) suggest that much of the fauna in Escanaba Trough is derived from the surrounding abyssal plain environments. The holothuroids *Paelopatides confundens* and *Abyssocucumus abyssorum* are found in both the Cascadia Basin to depths of 3,000 m and Escanaba Trough at 3,200 m (Carney and Carey, 1976, 1982). Some species occurring in Gorda Ridge axial valley also live on abyssal plains surrounding the ridge (Carey, this volume). These include holothurians (*Paelopatides confundens* (see Fig. 6.10, Holmes and Zierenberg, this volume) and *Abyssocucumus abyssorum*), ophiuroids (*Ophiomusium multispinum*), an echinoid (*Urechinus loveni*), an asteroid (*Hymenaster perissonotus*), a galatheid crab (*Munidopsis latirostris*), and fishes (*Coryphaenoides armatus*, *C. ?filifer*, *Antimora microlepis*, *Spectrunculus grandis*, and *Paraliparis ?rosaceus* (McCauley and Carey, 1967; Carney and Carey, 1976, 1982; Ambler, 1980; Pearcy et al., 1982; Carey, this volume; Carey and Kyte, unpublished data.

Conclusions

An extensive visible epibenthic fauna exists in the central axial valley of Gorda Ridge. The animal assemblages may prove to have significantly higher-standing stocks than those on the surrounding abyssal plains. The invertebrate magafauna is composed mainly of suspension and deposit feeders. Taxonomic composition, even within feeding type, changes down the valley axis, and the changes are correlated with three geological zones of the Gorda Ridge—northern, middle, and southern (Escanaba Trough). Many species of softbottom fauna in the valley are common to the fauna on the surrounding abyssal plains. We hypothesize that the near-bottom concentration of suspended particulate organic material is enhanced in the valley by topographic entrainment of currents, offshore transport of continental runoff by the Columbia River plume in the summer, offshore advection of organic matter from nearshore primary production, and the chemosynthetic primary production associated with active hydrothermal vent fields.

Acknowledgments. We thank the Marine Geology Branch, U.S. Geological Survey, Menlo Park, CA; the National Oceanic and Atmospheric Administration's VENTS Program; and the U.S. Navy's Deep Submergence Group San Diego, CA, for their helpful field support during these biological studies.

W. Normark, J. Morton, R. Koski, R. Zierenberg, H. Chezar, R. Holmes, C. Gutmacher, and support technicians (USGS Marine Geology Branch, Menlo Park, CA) provided bottom photographs, videotapes, navigational and topographic data, and specimens acquired on the ridge since 1985. S. Hammond (NOAA, Hatfield Marine Science Center, Newport, OR) provided photographs taken from the DSRV *Alvin* during the 1984 NOAA dives on the ridge. Officer-in-Charge B. Bosshard and his DSV *Sea Cliff* pilots and crew (U.S. Navy Deep Submergence Group, San Diego, CA) are thanked for their strong and willing support during the combined 1986 USGS and Navy diving cruise.

Early photographic analyses at OSU College of Oceanography were undertaken by M. Boudrias, G. Braun, and H. Jones, E. Carey is thanked for her invaluable assistance with the photographic analyses and the manuscript.

Identification by collaborating taxonomists are gratefully acknowledged: H. Reiswig and W. Austin (Porifera); F. Bayer and D. Fautin (Anthozoa); W. A. Newman and A. B. Williams (Arthropoda); F. Bernard (Mollusca); M.A. Kyte, D.L. Pawson, D.L. Meyer, and M. Roux (Echinodermata); and C. and F. Monniot (Urochordata).

The Minerals Management Service, through the Gorda Ridge Technical Task Force and the State of Oregon Department of Geology and Mineral Industries, funded the project through contracts 63-630-8503, 63-630-8604, and 63-630-8606.

References

Ambler JW (1980) Species of *Munidopsis* (Crustacea, Galatheidae) occurring off Oregon and in adjacent waters. Fishery Bull 78:13–34.

Barnes CA, Duxbury AC, Morse B-A (1972) Circulation and selected properties of the Columbia River effluent at sea. In: Pruter AT, Alverson DL (eds) The Columbia River Estuary and Adjacent Ocean Waters. Seattle: University of Washington Press, p 41.

Carey AG Jr, Kyte MA (1989) Distributional ecology of Ophiuroidea on the Oregon continental slope and adjacent abyssal plains. (In preparation.)

Carey AG Jr, Stein DL, Taghon GL (1986) Analysis of epifauna and infaunal community structure at the Gorda Ridge. Oregon Department of Geology and Mineral Industries Open-File Report 0-86-11, 34 pp.

Carey AG Jr, Stein DL, Taghon GL (1987) Analysis of epifaunal community structure on the Gorda Ridge. Oregon Department of Geology and Mineral Industries Open-File Report 0-87-05, 48 pp.

Carney RS, Carey AG Jr (1976) Distribution of holothuroids on the northeastern Pacific (Oregon, U.S.A.) continental slope, shelf, and abyssal plain. Thallasia Yugoslavica 12:67–74.

Carney RS, Carey AG Jr (1982) Distribution and diversity of holothuroids on Cascadia Basin and Tufts Abyssal Plain. Deep-Sea Res 29:597–607.

Cavanaugh, CM (1985) Symbioses of chemoautotrophic bacteria and marine invertebrates from hydrothermal vents and reducing sediments. Biol Soc Wash Bull 6:373–388.

Chezar H, Lee J (1985) A new look at deep-sea video. Deep-Sea Res 32:1429–1436.

Clague DA, Riesen W, Quinterno P, et al. (1984) Preliminary geological, geophysical, and biological data from Gorda Ridge. U.S. Geological Survey Open-File Report 84-364.

Conan G, Roux M, Sibuet M (1981) A photographic survey of a population of the stalked crinoid *Diplocrinus* (*Annacrinus*) *wyville-thomsoni* (Echinodermata) from the bathyal slope of the Bay of Biscay. Deep-Sea Res 28:441–453.

Conomos TG, Gross MG (1972) River-ocean suspended particulate matter relations in summer. In: Pruter AT, Alverson DL (eds) Columbia River Estuary and Adjacent Ocean Waters. Seattle: University of Washington Press, p 176.

Genin A, Dayton PK, Lonsdale PF, Spiess FN (1986) Corals on seamount peaks provide evidence of current acceleration over deep-sea topography. Nature 322:59–61.

Grassle JF (1986) The ecology of deep-sea hydrothermal vent communities. Adv Marine Biol 23:301–362.

Haedrich R, Rowe G, Polloni P (1980) The megabenthic fauna in the deep-sea south of New England. J Marine Res 57:165–179.

Hessler, RR, Smithey WM Jr (1983) The distribution and community structure of megafauna at the Galapagos Rift hydrothermal vents. In: Rona PA, Bostrom K, Laubier L, Smith KL (eds) Hydrothermal Processes at Seafloor Spreading Centers,

NATO Conference Series IV. New York: Plenum Press, pp 735–770.
Jumars PA, Eckman JE (1983) Spatial structure within deep-sea benthic communities. In: Rowe GT (ed) The Sea, Vol 8. New York: Wiley Interscience, pp 399–451.
Laubier L, Monniot C (eds) (1985) Peuplements Profonds de Golfe de Gasgogne. Brest, France Institute Français de Recherche pour l'Exploitation de la Mer.
McCauley JE, Carey AG Jr (1967) Echinoidea of Oregon. J Fish Res Board Can 24:1385–1401.
Monniot C, Monniot F (1978) Recent work on the deep-sea tunicates. Oceanogr Marine Biol Annu Rev 16:181–228.
Morton JL, Holmes ML, Koski RA (1987) Volcanism and massive sulfide formation at a sedimented spreading center, Escanaba Trough, Gorda Ridge, Northeast Pacific Ocean. Geophys Res Lett 14:769–772.
Pearcy WG, Stein DL, Carney RS (1982) The deep-sea benthic fish fauna on the northeastern Pacific Ocean of Cascadia and Tufts Abyssal Plains and adjoining continental slopes. Biol Oceanogr 1:375–428.
Pielou EC (1977) Mathematical Ecology. New York: John Wiley and Sons.
Rona PA, Denlinger R, Fisk M, et al. (1988) Hydrothermal activity on the Gorda Ridge. Eos 69:1588.
Sartori R (1980) Factors affecting the distribution of ahermatypic corals on the Mediterranean seafloor: A probablistic study. Deep-Sea Res 27:655–663.
Taylor LR, Woiwod IP, Perry JN (1978) The density-dependence of spatial behaviour and the rarity of randomness. J Anim Ecol 47:383–408.
Tuttle JH (1985) The role of sulfur-oxidizing bacteria at deep-sea hydrothermal vents. Biol Soc Wash Bull 6:335–343.
Vallier TL, Harold PJ, and Girdely WA (1973) Provenances and dispersal patterns of turbidite sand in Escanaba Trough, northeastern Pacific Ocean. Marine Geology, 15, 67–87.
Zierenberg RA, Koski RA, Shanks WC III, and SLACK JF (1988) Preliminary results of ALVIN dives on active sediment-hosted massive sulfide deposits in the Escanaba Trough, southern Gorda Ridge. Eos, 69, 1488.

17—Distributional Ecology of Selected Megaepifauna on Abyssal Plains Adjacent to Gorda Ridge, Northeast Pacific Ocean

ANDREW G. CAREY JR.

Abstract

Patterns of distribution and abundance of benthic megaepifauna demonstrate mesoscale zoogeographic trends across the nearshore Cascadia Abyssal Plain and offshore Tufts Abyssal Plain. Species composition analysis of Holothuroidea and Ophiuroidea (Echinodermata) and Galatheidae (Crustacea: Decapoda) separates the fauna into sediment fan, slope-base, Cascadia Plain, Tufts Plain, Gorda Ridge, and northern California abyssal elements. Faunal distribution and abundance are correlated with distance from the continent and depth of the water column. Gradients in rates of sedimentation and food flux to the seafloor are probable causes. Unlike macrofaunal distribution and abundance patterns, the species composition, abundance, and diversity of megafauna (e.g., Holothuroidea) appear to be more controlled by depth than by distance from the continent.

Megafaunal species that live on the surrounding plains have been identified from Gorda Ridge axial valley, particularly the sediment-filled Escanaba Trough at the southern end of the ridge system. Deposit-feeding holothuroids and ophiuroids, and omnivorous galatheid crabs have been collected and observed from sedimented areas along the length of the axial valley; three species have been positively identified. Further sampling in the valley should demonstrate that many of the abundant deposit feeding species associated with abyssal plains also live in association with sediment cover in Gorda Ridge axial valley. Data suggest that the valley deposit-feeding mega-epifauna is derived from the surrounding abyssal plains from a broad range of depths.

Introduction

The majority of research on deep-sea benthic fauna emphasizes organisms from the vast abyssal plains and from sedimented continental slopes. Until the recent technological development of submersibles, bottom acoustic transponder nets, and multibeam echosounders, the rocky, rugged environments of oceanic ridges and steep continental slopes have precluded sampling and study by biological oceanographers. This paper reviews some of the research undertaken by Oregon State University over the past two decades on the benthic fauna of abyssal plains in the Northeast Pacific Ocean off Oregon and compares the results to data from benthic ecological studies of the Gorda Ridge axial valley. Collectively, the individual benthic ecological projects form a mesoscale zoogeographic study of benthic megafaunal communities in the Northeast Pacific Ocean. Research on the deep-sea macrofauna and megafauna in this region contrasts abyssal environments for comparative studies of deep-sea benthic community structure.

The continental margin of the Pacific Northwest is complex. Vancouver Island to the north, a narrow and unusually steep continental slope to the east, Juan de Fuca Ridge to the west, and the Blanco Fracture Zone and Gorda Ridge to the south enclose a shallow nearshore abyssal plain in Cascadia Basin (McManus, 1964; Kulm et al., 1984; Fisk and Howard, this volume). Subduction of the oceanic plate beneath the North American continent and the seafloor spreading processes of the oceanic ridges formed the complicated submarine topography (Kulm and Fowler, 1974). Gorda Ridge and Juan de Fuca Ridge systems are barriers to the seaward movement of continental erosional debris (Hamilton, 1967).

Sediments derived from continental sources have accumulated behind the oceanic ridges. Sediment ponding formed Cascadia Abyssal Plain and filled the trench normally associated with the subduction of an oceanic plate. Turbidity currents have flowed across and down Gorda Basin and into the southern Gorda Ridge axial valley (Moore, 1970). Turbidity currents, originating on Astoria Fan, have moved through Cascadia Channel and flowed westward through Blanco Fracture Zone to form the eastern Tufts Abyssal Plain (Duncan and Kulm, 1970). Sediments are derived from erosion of the Pacific Northwest coastline and from continental runoff from coastal rivers and streams (Duncan and Kulm, 1970; Griggs et al., 1969). The Columbia River is a significant point source for sediments. Central Cascadia Plain, although dipping slightly to the south, remains at fairly constant depths east to west. To the west of Juan de Fuca Ridge lies Tufts Abyssal Plain, which merges with the Aleutian Plain farther west in the vicinity of 151° W longitude at depths of about 5,200 m (Hamilton, 1967, 1973). On eastern Tufts Plain, depth increases from 2,000 m at area A to 3,990 m at area C over a distance of about 450 km (Fig. 17.1).

As part of a comprehensive study of deep-sea benthos from Northeast Pacific abyssal plains, Carey and his benthic ecology group collected macrofauna and megafauna from 1962 to 1983. Overall objectives of these studies were (1) to define trends in faunal distribution and abundance and community structure along an east-west transect with increasing depth and distance from the continent (e.g., Carey, 1965); (2) to determine trends in megafaunal distributions in Cascadia Basin emphasizing the relative effects of depth and distance from the continent (e.g., Carney and Carey, 1982); (3) to determine the relationship of the faunas of Cascadia and Tufts Abyssal Plains (e.g., Ambler, 1980); and (4) to study the ecology of the slope base environment (e.g., Carey, 1981).

This chapter focuses on the relationship of the megaepifauna of the Gorda Ridge axial valley and the surrounding abyssal plains. The ecological synthesis includes taxonomic composition, patterns of distribution and abundance, and the food web structure of the holothuroids (Echinodermata: Holothuroidea), ophiuroids (Echinodermata: Ophiuroidea), and galatheid crabs (Crustacea: Arthropoda: Decapoda). The abyssal plains megafauna associated with sedimentary environments is compared and contrasted to that of the Gorda Ridge axial valley fauna. These organisms are characteristic of abyssal plains and are considered here as typical of sedimentary fauna (Carney and Carey, 1976, 1982; Ambler, 1980; Kyte, unpublished; Carey and Kyte, unpublished data).

Materials and Methods

The benthic samples are from three abyssal areas: (1) Cascadia Basin, (2) eastern Tufts Abyssal Plain, and (3) south of the Mendocino Fracture Zone (Fig. 17.1). Magaepifaunal samples from Cascadia Basin are derived from a U.S. Navy Deepwater Dumpsite (DWD), centered approximately at 48° 15.0′ N, 127° 00.0′ W (Carey, 1972), the OSU Benthos Beam Trawl Station Grid (CP-1A through CP-4E) (Carey, 1972), and the Southern Cascadia Slope Base Experiment (X) (Carey, unpublished data). The Tufts Plain samples are located on the extension of the Newport Transect Line (44° 39.1′ N); a total of three sampling sites on the plain are included in this paper. This extend to 45° 03.0′ W and to a depth of 3,900 m. Collections off northern California are from south of the Mendocino Fracture Zone at station M at 40° 13.4′ N, 126° 30.0′ W and from site W-N, a potential dumpsite for the U.S. Navy Low Level Waste Ocean Disposal Program (LLWODP) centered approximately at 39° 15.9′ N, 127° 33.6′ W just west of the Delgada Fan (Fig. 17.1).

Sampling strategy for the beam trawl station

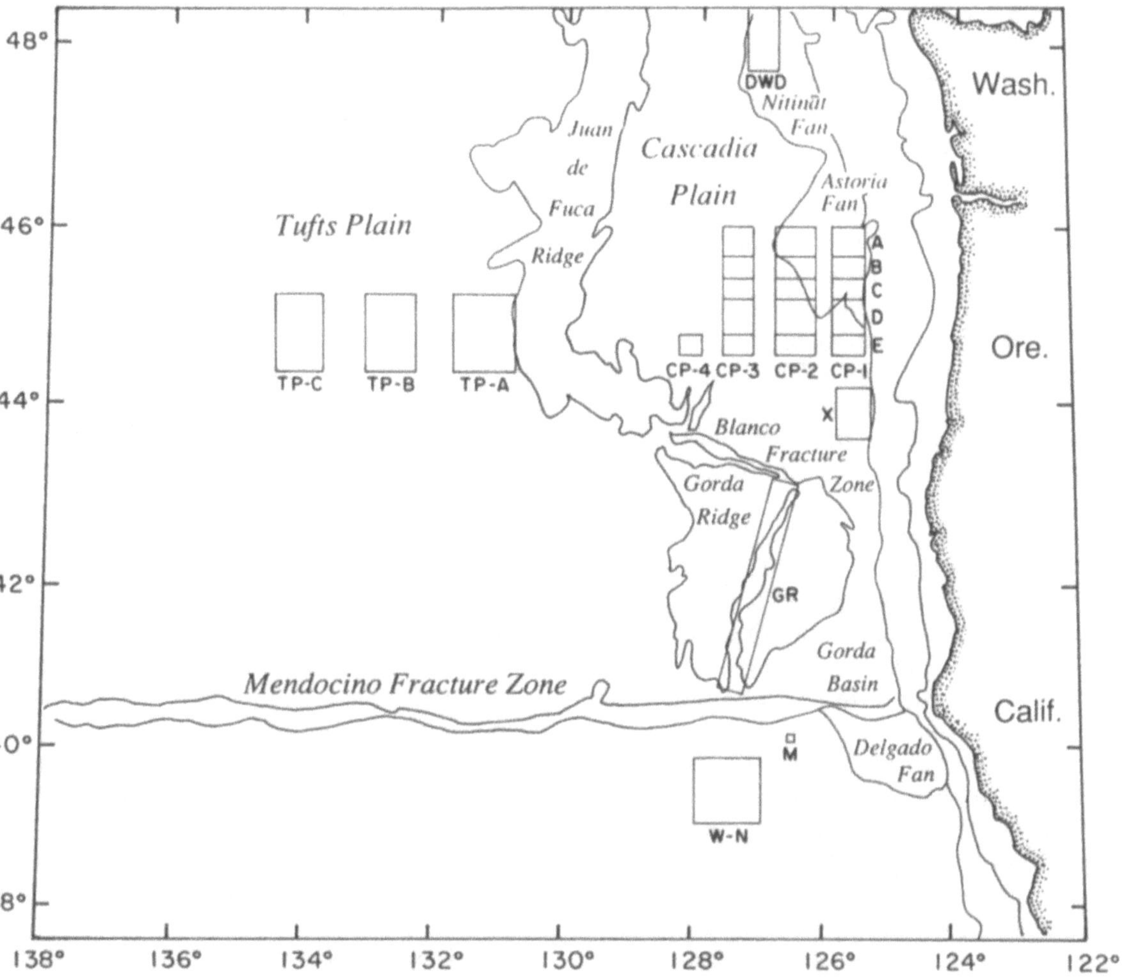

FIGURE 17.1. Location of OSU sampling sites for megaepifauna on the abyssal fans and plains in the Northeast Pacific Ocean off Washington, Oregon, and California: DWD (U.S. Navy Deepwater Dumpsite) on Nitinat Fan; stations CP-1-A through CP-4-E (OSU beam trawl station grid on Cascadia Plain); site X (continental slope base experiment on southern Cascadia Plain); TP-1 through TP-9 (extended Newport station line on eastern Tufts Plain; GR (Gorda Ridge axial valley); site M (Mendocino benthos station), and W-N (potential U.S. Navy low-level waste ocean disposal program site west of Delgada Fan).

grid emphasized characterization of the benthos in Cascadia Basin from Astoria Fan and slope base to Juan de Fuca Ridge. Stations on the north-south lines are at approximate equidistances from the continent. The OSU Benthic Ecology Program sampled the DWD stations in 1971 to provide data from Nitinat Fan at the northern end of the basin. The DWD stations average 2,030 m in depth. However, the CP-1 station line increases in depth southward from an average of 2,215 m at CP-1-A to 3,000 m at station X in southern Cascadia Basin. Although the depth increases 1,000 m along this station line, the distance from the continent does not change significantly.

Magafauna were collected with a 7.1-m semiballoon Gulf of Mexico shrimp trawl, a 3-m OSU beam trawl, or a 29.7-m commercial otter trawl (Carey and Heyamoto, 1972; Stein, 1985). Two odometer wheels, attached to the steel beam trawl frame, measured the distance traveled by the trawl across the bottom (Carney and Carey, 1980). The beam trawl net of otter trawl design contained a 1.3-cm stretch-mesh liner in the entire net. We utilized standard procedures for the beam trawl hauls, allowing direct comparison of trawl sample sizes. The paired odometers provided information on the distance towed along the seafloor and on how well the trawl tended the bottom. When available, time-depth recorders provided data for evaluation of trawl trajectory (Carney and Carey, 1980). The large commercial otter trawl had 3.8-cm stretch-mesh liner (Stein, 1985). Data are derived from 106 beam trawls (BMT), 29 benthos otter trawls (OTB), and nine large otter trawls (LOT) from 23 locations at depths greater than 2,000 m (Fig. 17.1). A minimum of three BMT samples were collected from each sampling area on Cascadia and Tufts Plains. The number of OTB samples varied from one to seven per station.

Species grouping of the holothuroids on the abyssal plains and fans was analyzed statistically by a cosine similarity matrix and factor analysis that utilized porportional data and emphasized the most abundant species (Carney, 1976; Joreskog et al., 1976; Carney and Carey, 1982). The cosine similarity indices were computed with the averaged holothuroid assemblage composition at each station.

Results

Utilizing distributional data from three epifaunal groups that differ significantly in taxonomy and ecology, we will examine the relationships among the megaepifauna for the major plains surrounding Gorda Ridge and the ridge axial valley. The depth and geographical distributions for all species of holothuroids, ophiuroids, and galatheid crabs are compared, and their ecologies are contrasted with those of the Gorda Ridge fauna.

Characteristic species groups of benthic fauna can be statistically delineated on portions of the plains. The bathymetry of Cascadia Plain, a planar surface tilted to the south and relatively level east to west in the central basin, provides insight into the relative effects of depth and distance from the continent on benthic fauna. Beyond Cascadia Basin, there are species groups that are characteristic of ocean basin abyssal environments.

Holothuroidea

The distribution and species grouping of 22 species of holothuroids (sea cucumbers) were investigated by inspection of geographic distributions of individual species and by statistical analysis of intersample similarities and of species groupings by factor analysis (Table 17.1; Fig. 17.2). The holothuroid fauna made up from 3% to 31% of the megafaunal catch (Carney and Carey, 1982).

Three main patterns are apparent: (1) present on Cascadia Abyssal Plain, but generally absent from the continental slope base environment; (2) present over the entire Cascadia Abyssal Plain and sometimes extending to eastern Tufts Plain; and (3) present only at the deeper stations on eastern Tufts Plain, south of the Mendocino Fracture Zone, and in the Gorda Ridge axial valley (Carney and Carey, 1976; Carey et al., 1987; Table 17.1; Fig. 17.2). We excluded stations on Nitinat Fan and south of the Mendocino Fracture Zone from these statistical analyses.

The Cascadia Basin beam trawl stations beyond the slope base and the bathyal sediment fans were similar in species composition. With increasing depth, southward along the slope base environment (station line CP-1), there is a shift from the shallower fan species groups to those of the Cascadia Basin floor fauna. The intersample similarity analysis of the fauna from Cascadia Basin and eastern Tufts Plain demonstrated that the distribution of the holothuroid fauna is influenced more by depth than by distance from the continent (Carney and Carey, 1982; Fig. 17.2). The faunal similarities between stations at similar depths (2,760 to 2,960 m) on Cascadia Plain are significant, in spite of varying distances from the continent. For example, the holothuroid fauna at stations CP-1-X and CP-3-A in the southeast and northwest corners of the basin have similar species composition (similarity value = 0.98; see Fig. 17.2). The depth at the

TABLE 17.1. Species list of selected abyssal megafauna (>2,000 m)—Holothuroidea, Ophiuroidea, and Galatheidae. The depth range off Washington, Oregon, and California and the distribution by environmental region are listed for each species.

Species	Depth (m)	North Cascadia Plain	East Cascadia Plain	West Cascadia Plain	South Cascadia Plain	East Tufts Plain	Gorda Ridge Valley	South Mendocino Fracture
Holothuroidea								
Pannychia moseleyi	2,030–2,500		×					
Scotoplanes globosa	2,030–3,100		×	×	×			
Peniagone cf. *dubia*	2,195–3,932		×	×	×	×		
Paelopatides confundens	2,000–4,194	×	×	×	×	×	×	×
Psychropodes longicaudata	2,450–3,932		×	×	×	×	×	×
Pseudostichopus nudus	2,100–3,600	×	×	×				×
Caphiera sulcata	2,425–3,358		×	×		×		
Benthodytes incerta	3,600–3,900	×				×		
Synallactes gilberti	2,500–3,354		×	×	×	×		
Amperima narsei	2,000–2,800		×			×		
Molpadia musculus	2,030–3,021		×	×	×			×
Molpadia granulata	2,500–3,700		×	×		×		×
Abyssocucumis albatrossi	2,600–3,932			×	×	×		×
Benthodytes sanguinolenta	2,600–3,932		×	×		×		×
Ypsilothuria bitentaculata	2,938–3,000				×			
Oneirophanta mutablis	2,900–3,932			×		×		×
Pseudostichopus villosus	2,900–3,700			×		×		
Mesothuria murrayi	2,900–3,800			×		×		
Penisagone gracilis	3,600–3,900					×		
Laetmogone wyville-thompsoni	3,700–3,700					×		×
Scotoplanes clarki	3,300–3,700					×		×
Abyssocucumis abyssorum	4,194–4,350						×	×
Ophiuroidea								
Asteronyx loveni	50–3,000	×	×	×				
Ophiomusium jollienis[a]	150–4,250							
Ophiomusium normani[a]	180–2,800							
Ophiocten hastatum	219–3,585	×	×	×	×	×		
Ophiura leptoctenia	400–3,000	×	×	×	×	×		
Ophiophthalmus cataleimmoidus[a]	400–2,000							
Amphiura diomedeae	450–3,030		×	×				
Amphiura carchara	630–3,100	×	×	×	×	×		
Ophiomusium lymani	800–2,200	×						
Ophiomusium multispinum	800 3,100		×				×?	
Ophiolimna bairdi	1,260–2,800	×	×					
Ophiura irrorata	1,600–2,100		×					
Ophiacantha bathybia	1,600–3,100		×	×	×	×		×
Amphilepis platytata	2,100–3,100	×	×	×		×		×
Ophiura bathybia	2,146–3,100	×	×	×	×	×	×	×
Ophioscolex corynetes	2,760–3,720			×		×		
Ophiacantha abyssa	3,354–4,260					×		×
Amphilepis patens[a]	3,927–4,200							
Galatheidae								
Munidopsis quadrata	950–2,189	×	×					
Munidopsis latirostris	1,900–3,021		×	×	×			
Munidopsis ciliata	2,030–2,875		×					
Munidopsis yaquiniensis	2,377–2,763			×				
Munidopsis subsquamosa	1,829–3,000			×	×			
Munidopsis bairdii	2,377–2,940		×	×	×			
Munidopsis cascadia	2,743–2,926	×	×					
Munidopsis beringana	2,800–3,990		×	×	×	×		
Munidopsis aries	2,850–3,025			×				
Munidopsis tuftsi	3,500–3,858					×		
Munidopsis verrucosus	3,932–4,194					×		×

[a] Distributional data not available.

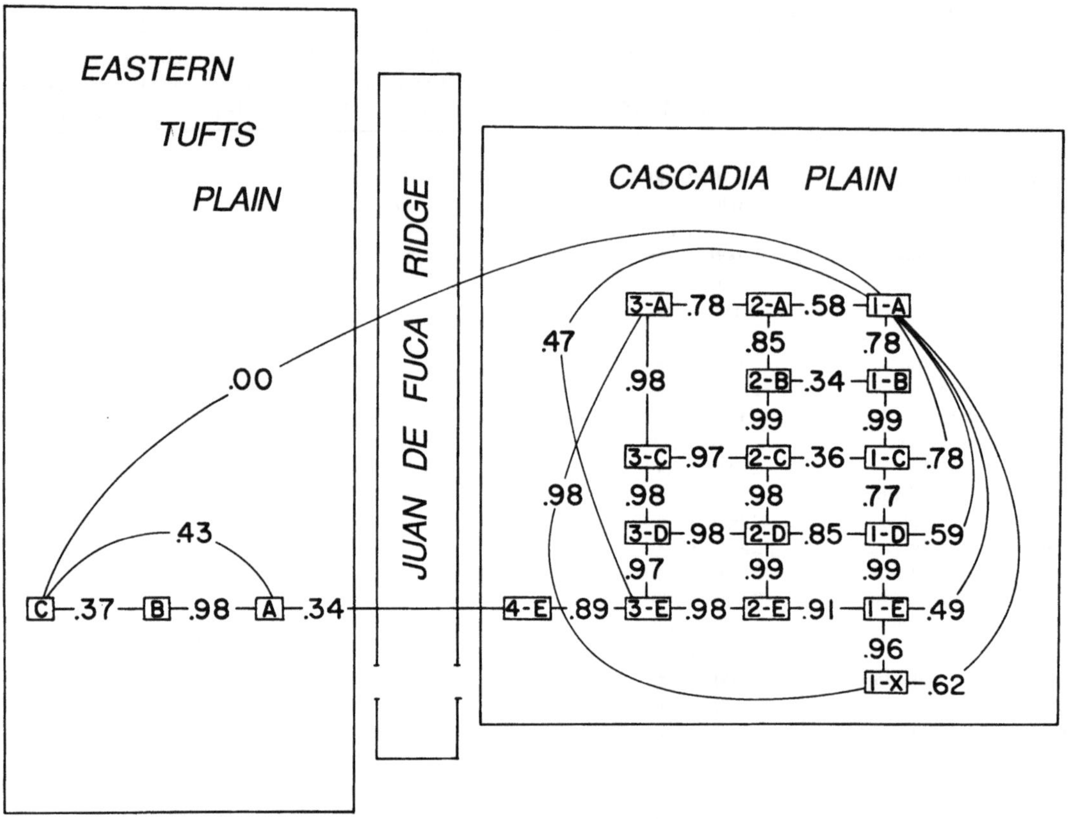

FIGURE 17.2. Diagram of the similarity relationships of the holothuroid fauna on abyssal plains adjacent to Gorda Ridge. Selected cosine similarities between stations are given on the interconnecting lines; the higher the index value, the more similar are the species groups at the stations. (Redrawn from Carney and Carey, 1982; with permission from Pergamon Press PLC)

two stations is similar (2,760 m and 2,960 m, respectively), though the difference in distance from the continent is about 170 km. On the other hand, stations CP-1-A and CP-1-E-X are less similar in species composition, although the stations on the CP-1 line are equidistant from the continent. However, the depth ranges from a mean of 2,160 m at CP-1-A to 2,820 m at CP-1-E and 2,960 m at CP-1-X. Therefore depth, and all of its correlative attributes, is a more important controlling factor on holothuroid species distributions than distance from the continent.

Nine species are associated with the Tufts stations. The species composition at the deeper stations on eastern Tufts Plain (Station TP-C) at a depth of 3,990 m is composed of totally different species from the shallowest Cascadia Basin samples (2,510 m) on the Nitinat Fan in the northeast corner (Table 17.1; Fig. 17.2; Carney and Carey, 1982). The cosine similarity index analysis demonstrated that the species groups at stations TP-A and TP-B had greater affinity to one another than to the more western and deeper station TP-C.

Eight of the 16 species of holothuroids that live on eastern Tufts Plain (from 131° 24′ W to 135° 23′ W) at depths of 3,354 to 3,724 m also occupy the western reaches of Delgada Fan, south of the Mendocino Fracture Zone at depths of 4,200 to 4,300 m (Table 17.1). Two species, *Abyssocucumis abyssorum* and *Paelopatides confundens*, have also been collected and/or observed in the Gorda Ridge axial valley. *Paelopatides confundens* is the most eurybathic holothuroid species encountered. We collected this species from all seven study locations, descending in depth from 2,000 m to

4,194 m (Table 17.1). From photographic evidence and direct observations by submersible in the axial valley, *P. confundens* is adapted to a wide range of sediment conditions from extensive soft sediments in the Escanaba Trough to a light sediment coating on pillow basalts in the northern Gorda Ridge axial valley (Carey et al., 1987). Other species found on eastern Tufts Plain and on the lower reaches of the Delgada Fan off northern California probably live in the ridge axial valley associated with sediments and sediment-coated basalts.

Factor analysis defined the following three species groups: (1) Astoria Fan, (2) Cascadia Plain extending to the western slopes of the Juan de Fuca Ridge on eastern Tufts Plain, and (3) eastern Tufts Plain (Carney, 1976; Carney and Carey, 1982).

The holothuroids are most abundant on Cascadia Plain with little change in numerical density for 185 km across the plain at a depth of about 2,800 m (Carney and Carey, 1982). Their numerical abundance remains high across Cascadia Plain to the western most stations, 1,000 km from the continent. In contrast, the numerical density of the sea cucumbers decreases about 1 order of magnitude across eastern Tufts Abyssal Plain.

Ophiuroidea

Eighteen species of ophiuroids (brittle stars) are associated with abyssal plains at depths of 2,000 to 3,100 m (Kyte, unpublished; Kyte and Carey, unpublished data; Table 17.1). The ophiuroids represent the most abundant taxonomic group in the abyssal plains megaepifauna. Eight species live only in Cascadia Basin, five species occur on both Cascadia and Tufts Plains, two species live only on Tufts Plain, and four species occur south of the Mendocino Fracture Zone.

The eight species in Cascadia Basin (*Ophiura irrorata*, *Ophiomusium lymani*, *Ophiomusium multispinum*, *Ophiolimna bairdi*, *Amphiura diomedeae*, *Amphiura carcharea*, *Ophiura leptoctenia*, and *Asteronyx loveni*) divide into characteristic distributional patterns. First, there is a group that lives on the shallow, northeastern portion on Astoria Fan. Second, others inhabit the total basin environment. Third, we define another species group from the deeper southern and southwestern segment of the basin. *Ophiura irrorata* and *Ophiomusium lymani* inhabit the shallower portion of the basin on the Astoria Fan, and *Asteronyx loveni* lives only at the deeper, southwestern part of the plain.

The five species collected from both Cascadia and Tufts Plains are *Ophiocten hastatum*, *Ophiocantha bathybia*, *Amphilepis platytata*, *Ophiura bathybia*, and *Ophioscolex corynetes*. These species exhibit overlapping distributions along the depth gradient. We collected five species—*Ophiocantha bathybia*, *Ophioscolex corynetes*, *Ophiomusium jolliensis*, *Amphilepis patens*, and *Ophiura bathybia*—from site W-N south of the Mendocino Escarpment at depths of 4,200 to 4,300 m. Two species, *Ophiocantha abyssus* and *Ophiophthalmus eropoma*, are deep abyssal forms and are found only on Tufts Abyssal Plain.

Two ophiuroid species, tentatively identified as *Ophiomusium multispinum* and *Ophiocten hastatum*, occur in the Gorda Ridge axial valley. *Ophiomusium multispinum* also inhabits the northern and eastern regions of Cascadia Plain. Other species that live on plains surrounding Gorda Ridge arc *Ophiocantha bathybia*, *Amphilepis platytata*, *Ophiura bathybia*, and *Ophiocantha abyssus* (Table 17.1).

Galatheidae

A total of 146 beam and otter trawl hauls from the abyssal plains off Oregon and Washington contained 11 species of *Munidopsis* (Decapoda: Anomura—squat lobster) (Ambler, 1980). Four categories of abyssal distributions for the galatheid species are apparent: (1) slope base and eastern Cascadia Plain; (2) Cascadia Plain; (3) the southern and deeper Cascadia Plain and eastern Tufts Plain; and (4) eastern Tufts Plain only (Table 17.1).

Of the eight species living on Cascadia Plain, three (*M. subsquamosa*, *M. cascadia*, and *M. ciliata*) occur predominantly on the eastern half of Cascadia Plain. Only two, *M. latirostris* and *M. bairdii*, occur at the slope base and extend across Cascadia Plain to the western side of the basin. We also collected *M. subsquamosa* on the abyssal plain south of the Mendocino Fracture Zone at 4,200+ m.

One species, *M. beringana*, lives on both eastern Tufts and Cascadia Abyssal Plains (Table 17.1). Two species, *M. verrucosus* and *M. tuftsi*, occur on Tufts Plain at depths of 3,200 m. *M. verrucosus* was also collected south of the Mendocino Escarpment at a depth of 4,200 to 4,300 m at station M and the LLWODP study site (Fig. 17.1; Ambler, 1980). We have not yet collected *M. verrucosus* from the Gorda Ridge axial valley (Carey et al., 1987); however, it is a lower abyssal form and inhabits abyssal environments deeper than the valley floor.

The U.S. Geological Survey collected specimens of *Munidopsis latirostris* with a rock dredge on Gorda Ridge (Carey et al., 1987). This species is broadly distributed across Cascadia Plain, occurring from the slope base to the southwestern and deepest stations on the plain at 3,041 m depth. *M. latirostris* is also the most abundant galatheid crab species sampled on the plains in the Northeast Pacific (Ambler, 1980).

Discussion

Patterns of abundance for macrofauna and megafauna differ across Northeastern Pacific abyssal plains (Carey, 1981; Carney and Carey, 1982). Distributional studies demonstrate that the macrofauna (defined here as 1.0 mm to 30.0 mm in length) community on the Newport Station Line at the base of the continental slope at 2,860 m depth at CP-1-E is greatest in biomass and numerical density. Just beyond the slope base, numbers and biomass decrease significantly and maintain a fairly constant low value across Cascadia Plain. On eastern Tufts Plain the biomass significantly decreases, while the numerical density remains comparable to that on Cascadia Plain. Consequently, the average size of individuals decreased on eastern Tufts Plain, presumably because of a depressed food supply to the seafloor. A deeper water column and greater distance from the continent influence the availability of food materials. In contrast, megafaunal numerical density, as exemplified by holothuroids, remains fairly constant across Cascadia Plain; however, it diminishes west of Juan de Fuca Ridge on eastern Tufts Plain (Carey, 1981; Carney and Carey, 1982).

The larger, more motile megaepifauna graze and scavenge on detritus and animal remains over much larger areas of the abyssal sediments than the smaller macrofauna. Also, with increasing depth and decreasing food supply the diet of asteroids becomes more generalized (Carey, 1972). The smaller macrofauna appear to react to a finer scale of physical and chemical gradients (Fauchald and Hancock, 1981; Dickinson and Carey, 1978; Carey, 1981).

Benthic fishes, similar to the macrofauna, are more abundant at the slope base environment (Pearcy et al., 1982), perhaps because of plentiful food availability. The fishes have a relatively uniform composition across Cascadia Basin (Pearcy et al., 1982), though Pearcy and co-workers delineated two species assemblages of benthic fishes associated with the lower continental slope and abyssal plains.

Many of the other megafauna—e.g., ophiuroids and galatheids—are also generalized feeders, and they extend across Cascadia Plain (Ambler, 1980; Carey and Kyte, unpublished data). Sokolva (1972) reported predominant feeding types in abyssal asteroid communities distributed over extensive areas of abyssal plains in the Pacific Ocean. She suggests that feeding type correlates with level of food input; productive environments support communities dominated by deposit feeders.

Within the megaepifauna there is a greater overlap of species distributions between abyssal environments for the holothuroids than the galatheid crabs. The distributions of galatheid species tend to cover smaller portions of the plains (Carney and Carey, 1982; Ambler, 1980; Table 17.1). In contrast to the broadly distributed surface deposit-feeding holothuroids, however, the infaunal sea cucumber deposit feeders, *Molpadia musculus* and *M. granulata*, live primarily at the slope base. Presumably, the slope base environment receives larger inputs of food derived from particles falling from the water column and from gravity flows of resuspended detrital materials and sediment particles down the continental slope interface (Carey, 1981). At the slope base utilizable organic materials are probably found buried deeper within the sediments.

Our program has sampled, observed, and identified in the ridge axial valley relatively few species of holothuroids, ophiuroids, and galatheid crabs that inhabit the surrounding abyssal plains (Table 17.1). Many of these species probably live in the sedimented portions of the valley, particularly the sediment-filled Escanaba Trough at the southern segment of Gorda Ridge. Evidence from bottom photographs and submersible observations indicate that an abundant deposit-feeding fauna is associated with sediments in the valley (Carey et al., 1987, this volume). Further sampling and study may discover in the valley habitat more of the species previously described from the surrounding plains.

On the other hand, few characteristic megaepifauna identified from the Gorda Ridge axial valley have been collected from the adjacent plains. The majority of these species are rocky fauna adapted to hard substrates and to suspension feeding in the bottom water layer. There may be rocky communities on the flanks of Gorda Ridge on rocky outcrops (Carey et al., this volume). A few species appear to be broadly adaptive, e.g., *Paelopatides confundens*, and live on a range of substrate types. Stalked suspension feeders, such as some of the hexactinellid sponges, are adapted to sediments and live in both environments. Many of the benthic fishes inhabit the valley in addition to the sedimentary plains (Pearcy et al., 1982).

Carey et al. (this volume) have not detected populations of organisms, such as the brachyuran crab *Bythograea*, that are more abundant adjacent to active hydrothermal venting (Van Dover et al., 1987). Van Dover and Hessler (this volume) collected the galatheid *Munidopsis subsquamosa* at vent sites at 21° N and 13° N along the East Pacific Rise and the Galapagos. Van Dover et al. (1987) reported that munidopsid crabs become more abundant closer to vent fields, presumably because of their omnivorous diet and their ability to scrape detrital materials from substrates. Ambler (1980) reported that *M. subsquamosa* is plentiful on the eastern side of Cascadia Abyssal Plain, although it is cosmopolitan in distribution. There are no reports that this species inhabits the Gorda Ridge rift valley, but it is likely to be common in organically rich areas near vent sites in the valley (Van Dover et al., 1987).

Although the Gorda Ridge axial valley is a semiisolated tectonic rift, the sedimentary fauna is probably derived from all the surrounding plains (Fig. 17.3). The described axial valley fauna includes deposit-feeding species from Cascadia Plain, eastern Tufts Plain, and the sedimented region south of the Mendocino Escarpment. The dense concentration of suspended particulate material and pelagobenthic zooplankton overlying the valley floor suggest plentiful food supplies for the benthos (Carey et al., this volume). Larval stages of many species on the adjacent plains could colonize the axial valley sediments.

Conclusions

Characteristic species assemblages are present on deep-sea fans, the slope base, Cascadia Plain, Tufts Plain, south of Mendocino Escarpment, and Gorda Ridge axial valley. However, motile megafaunal species are broadly distributed with depth and across environments. A complex of megafaunal species from Cascadia Abyssal Plain, Tufts Abyssal Plain, and south of the Mendocino Escarpment live in the Gorda Ridge axial valley. The low correspondence between the valley sedimentary deposit-feeding megafauna and that from the surrounding plains probably indicates that the valley fauna is undersampled. The majority of the megaepifauna shared between the axial valley and the surrounding plains are deposit feeders.

Acknowledgments. This chapter is dedicated to the memory of Bent Hansen—a fine colleague, a dedicated holothuroid systematicist, and a friend.

Many students, research assistants, colleagues, and ship's captains and crews are

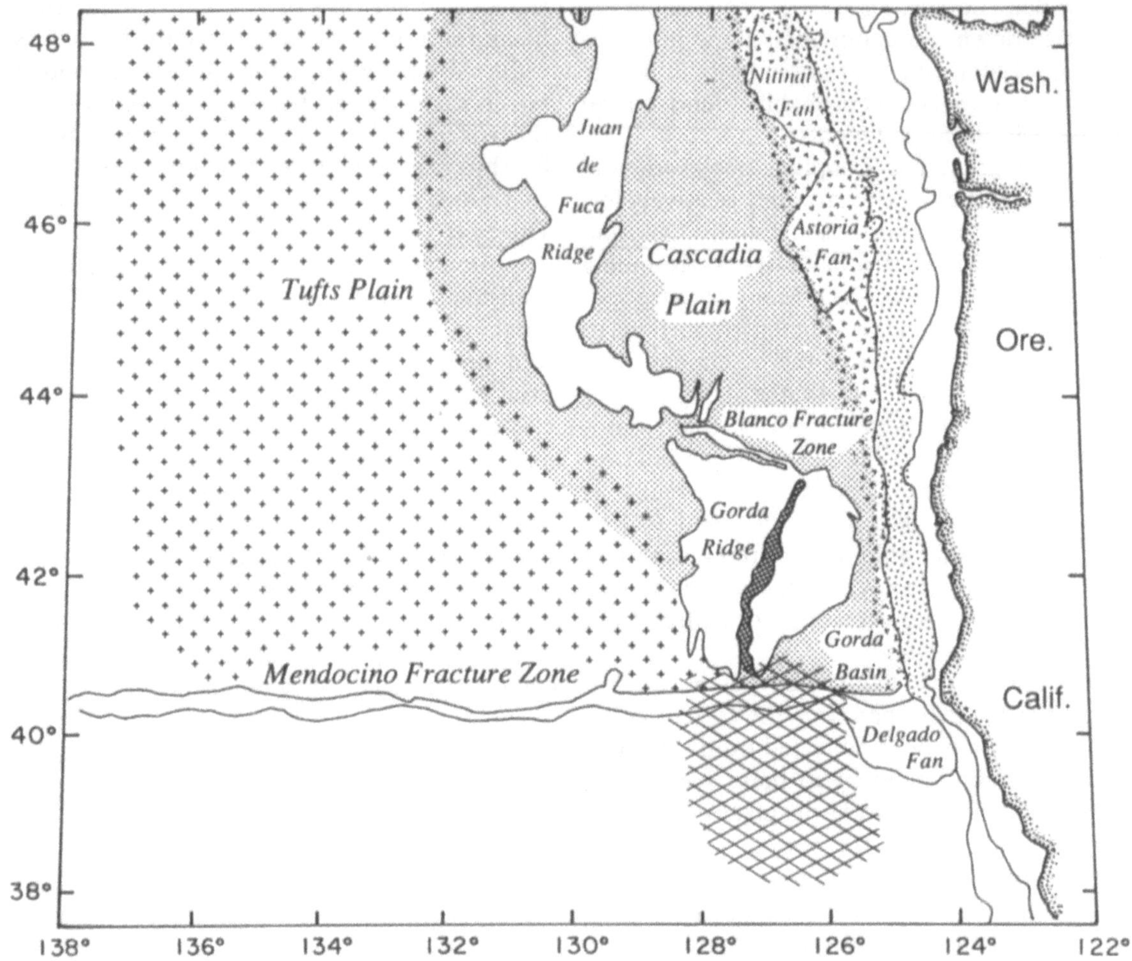

FIGURE 17.3. Synthesis of hypothesized megaepifaunal mesoscale zoogeographic patterns in the Northeast Pacific Ocean based on results from the OSU trawling program. The surrounding abyssal plains are the source for the sedimentary fauna in the Gorda Ridge axial valley (dense stipple).

thanked wholeheartedly. The results would not have been possible without their enthusiastic and expert assistance over the years. I gratefully acknowledge the taxonomic assistance of J.W. Ambler (Galatheidae), D.L. Pawson (Echinodermata), M.A. Kyte (Ophiuroidea), B. Hansen (Holothuroidae), R.S. Carney (Holothuroidae), and W.G. Pearcy and D.L. Stein (Pisces). E.M. Carey helped with the manuscript preparation, and W.G. Pearcy made valuable comments on the manuscript. The National Science Foundation, U.S. Atomic Energy Commission, Sandia Corporation, and Office of Naval Research supported the abyssal plains benthic ecological research. The Minerals Management Service through Gorda Ridge Technical Task Force and the State of Oregon Department of Geology and Mineral Industries funded the Gorda Ridge studies.

References

Ambler JW (1980) Species of *Munidopsis* (Crustacea, Galatheidae) occurring off Oregon and in adjacent waters. Fish Bull 78:13–34.

Carey AG Jr (1965) Preliminary studies on animal-sediment interrelationships off the central Oregon coast. Ocean Sci Eng 1:100–110.

Carey AG Jr (1972) Food sources of sublittoral, bathyal and abyssal asteroids in the Northeast Pacific Ocean. Ophelia 10:35–47.

Carey AG Jr (1972) Benthic ecology and radioecology. In: Cutshall N (ed) Ecological Studies of Radioactivity in the Columbia River Estuary and Adjacent Pacific Ocean. Progress Report, U.S. Atomic Energy Commission, RLO-2227-T12-30, p. 23.

Carey AG Jr (1981) A comparison of benthic infaunal abundance on two abyssal plains in the northeast Pacific Ocean. Deep-Sea Res 28A: 467–479.

Carey AG Jr, Heyamoto H (1972) Techniques and equipment for sampling benthic organisms. In: Pruter AT, Alverson DL (eds) The Columbia River Estuary and Adjacent Ocean Waters, Bioenvironmental Studies. Seattle: University of Washington Press, p 378.

Carey AG Jr, Stein DL, Taghon GL (1987) Analysis of benthic epifauna community structure on the Gorda Ridge. State of Oregon Department of Geology and Mineral Industries Open-File Report 0-87-05, 48 pp.

Carney RS (1976) Patterns of abundance and relative abundance of benthic holothurians (Echinodermata: Holothuroidea) on Cascadia Basin and Tufts Abyssal Plain in the northeast Pacific Ocean. Ph.D. Thesis, Oregon State University, Corvallis, 185 pp.

Carney RS, Carey AG Jr (1976) Distribution pattern of holothurians on the Northeastern Pacific (Oregon, U.S.A.) continental shelf, slope, and abyssal plain. Thallassia Jugoslavica 12: 67–74.

Carney RS, Carey AG Jr (1980) The effectiveness of odometer wheels for the measurement of area sampled by beam trawls. Fish Bull 78: 791–796.

Carney RS, Carey AG Jr (1982) Distribution and diversity of holothuroids (Echinodermata) on Cascadia Basin and Tufts Abyssal Plain. Deep-Sea Res 29: 597–607.

Dickinson JJ, Carey AG Jr (1978) Distribution of gammarid Amphipoda (Crustacea) on Cascadia Abyssal Plain (Oregon). Deep-Sea Res 25: 97–106.

Duncan JR, Kulm LD (1970) Mineralogy, provenance, and dispersal history of late quaternary deep-sea sands in Cascadia Basin and Blanco Fracture Zone off Oregon. J Sediment Petrol 40: 874–887.

Fauchald K, Hancock DR (1981) Deep-water polychaetes from a transect off Oregon. Monogr Allan Hancock Found 11: 73.

Griggs GB, Carey AG Jr, Kulm LD (1969) Deep-sea sedimentation and sediment-fauna interaction in Cascadia Channel and on Cascadia Abyssal Plain. Deep-Sea Res 16: 157–170.

Hamilton EL (1967) Marine geology of abyssal plains in the Gulf of Alaska. J Geophys Res 72: 4189–4213.

Hamilton EL (1973) Marine geology of the Aleutian Abyssal Plain. Marine Geol 14: 295–325.

Joreskog K, Klovan J, Reymont R (1976) Methods in Geomathematics, Vol 1: Geological Factor Analysis. Amsterdam: Elsevier, 178 pp.

Kulm LD, Fowler GA (1974) Oregon continental margin structure: A test of the imbricate thrust model. In: Burk CA, Drake CL (eds) The Geology of Continental Margins. New York: Springer-Verlag, pp 261–283.

Kulm LD et al. (1984) Western North American Continental Margin and Adjacent Ocean Floor Off Oregon and Washington. Atlas 1. Ocean Margin Drilling Program, Regional Atlas Series. Woods Hole, MA: Marine Science International, 32 sheets.

McManus DA (1964) Major bathymetric features near the coast of Oregon, Washington, and Vancouver Island. Northwest Sci 38: 65–82.

Moore GW (1970) Sea-floor spreading at the junction between Gorda Rise and mendocino Ridge. Geol Soc Am Bull 81: 2817–2824.

Pearcy WG, Stein DL, Carney RS (1982) The deep-sea benthic fish fauna of the Northeastern Pacific Ocean on Cascadia and Tufts abyssal plains and adjoining continental slopes. Biol Oceanogr 1: 375–428.

Sokolova MN (1972) Trophic structure of deep-sea macrobenthos. Marine Biol 16: 1–12.

Stein DL (1985) Towing large nets by single warp at abyssal depths: Methods and biological results. Deep-Sea Res 32: 183–200.

Van Dover CL, Franks PJS, Ballard RD (1987) Prediction of hydrothermal vent locations from distributions of brachyuran crabs. Limnol Oceanogr 32: 1006–1010.

18—Spatial Variation in Faunal Composition of Hydrothermal Vent Communities on the East Pacific Rise and Galapagos Spreading Center

CINDY LEE VAN DOVER and ROBERT R. HESSLER

Abstract

Comparisons of distributions of megafaunal species at hydrothermal vents on the East Pacfic Rise and Galapagos Spreading Center can be made at three levels: (1) within a vent field, (2) among vent fields within a cluster on a ridge segment, and (3) among ridge segments. Across 40° of latitude, megafaunal compositions of vent communities are remarkably consistent at the familial level. Along any given ridge segment, there appears to be a shared pool of species from which any vent field draws its fauna.

Introduction

In the decade since the discovery of hydrothermal vents on the deep-sea floor, hydrothermal activity and chemosynthetically based animal communities have been observed at many locations along major spreading axes including the Galapagos Spreading Center (GSC), the East Pacific Rise (EPR), the Juan de Fuca and Explorer Ridges, the Mid-Atlantic Ridge (MAR), and the Marianas Back-Arc Spreading Center. Comparisons of the communities at these diverse locations reveal both global similarities and individual differences. For example, a characteristic feature of nearly all known vent communities, regardless of geographical location, is the localized, concentrated biomass of living organic material. At the same time, a distinguishing feature of individual vent communities is the relative abundance of component species dominating that biomass.

Not surprisingly, the greatest differences in faunal composition are seen in comparisons of dominant species between geographically disjunct spreading centers. On the EPR and GSC in the eastern Pacific Ocean, tube worms, clams, and/or mussels typically dominate vent fields (reviewed in J.F. Grassle, 1985, 1986); on the Marianas Back-Arc Spreading Center in the western Pacific Ocean, a large species of snail is predominant at vent sites (Hessler et al., 1988a); at two known vent fields on the Mid-Atlantic Ridge in the Atlantic Ocean, a caridean shrimp prevails (Rona et al., 1986; Detrick et al., 1986; J.F. Grassle et al., 1986). There may even be fundamental differences in the trophic structure of vent communities: at most known vents, the dominant megafaunal invertebrate species are sessile and harbor chemoautotrophic endosymbiotic microorganisms from which the hosts presumably derive their nutrition by translocation (Cavanaugh et al., 1981; Felbeck, 1981; Cavanaugh, 1985), but the shrimp that dominate the MAR vent communities are highly mobile and are thought to graze directly on free-living microorganisms (Van Dover et al., 1988). Despite these differences in the general nature of hydrothermal vent communities on a global scale, certain elements of the vent fauna on widely separated spreading centers may be closely related in a phylogenetic sense. Some of the best examples of this are brachyuran crabs in the family Bythograeidae, caridean shrimp in the family Bresiliidae, and mussels in the family Mytilidae, all of which are known to occur at vents on the MAR, EPR, GSC, and Marianas Back-Arc Spreading Center. Vesicomyid clams and vestimentiferan tube worms occur at most of the these sites.

Before considering variation in faunal composition of vent communities on any scale, one needs to place vent environments within some context with respect to their spatial and chemical relationships. This context is best known for the East Pacific Rise and Galapagos Spreading Center, where the spreading axes are broken up at intervals into ridge segments by transform faults and lateral offsets up to tens of kilometers in length (e.g., MacDonald et al. 1988). Most hydrothermal vent fields of the EPR and GSC explored to date occur along the shallower portions of ridge segments and are predictably arranged in short linear clusters ~10 to 20 km total length) along the strike of the neovolcanic zone (e.g., Ballard et al., 1982; Hekinian et al., 1983; Francheteau and Ballard, 1983; Argorise Group, 1988).

South of the Gulf of California, ridge axes are exposed as bare basaltic lavas (i.e., sediment starved). Water depths on the order of 2,400 to 2,600 m are typical of the shallowest portion of each ridge segment, while water depths at transform faults reach 3,000 m and more. The meager areal extent of individual vent fields ranges from small pockets a few meters in diameter to fields with largest dimensions on the order of 60 to 100 m. Vent fields comprising a cluster are separated by distances of 1 to 10 km and typically range in number from 5 to 10 per cluster; distances between clusters of vent fields are less well established, but there are perhaps zero, one, or two clusters per ridge segment and therefore there may be 10s to 100s of kilometers separating clusters (Ballard et al., 1982; Hekinian et al., 1983; Francheteau and Ballard, 1983; Argorise Group, 1988). Geological relationships among vent fields within a linear cluster are not clear. One hypothesis is that such a cluster of vent fields is the consequence of a single en echelon tectonic event. Biological relationships among vent fields within a cluster seem more clear-cut, with exchange of propagules among populations more likely between vent fields within a cluster than between clusters of vent fields (J.P. Grassle, 1985).

Implicit in the concept of a vent field is the notion of a single hydrothermal system feeding each field. This has at least one important consequence that affects the distributions of animals within the vent community: hydrothermal circulation at one location within a vent field is likely to be strongly influenced by hydrothermal circulation elsewhere within the field. If, for example, conduits at one locale become clogged, either by inorganic precipitation and/or biologically mediated obstruction, flow to other conduits may be increased. Alternatively, seismic activity may open up new channels, diverting flow from other locations. Presumably, the biological community will respond to modified flow rates. This response might be expressed as changes in rates of growth, reproduction, mortality, and/or colonization. Fields of dead clams juxtaposed with vigorous populations of live clams attest to the dramatic effect modified flow rates in a low-temperature regime can have on vent communities (Hessler et al., 1985). The time scale for significant changes in flow rates appears to be on the order of years to tens of years for individual vents. Fustec et al. (1987), for example, report changes in flow rates from one chimney and the opening of an entirely new black smoker at Actinoir on the EPR (13° N). These events occurred within a 2-year time period. Campbell et al. (1988) cite other examples of changes in flow patterns of black smoker vents on a decadal scale of observation. Hydrothermal activity of a vent field may last longer (Lalou and Brichet, 1982, 1987) and is dependent on such considerations as flow rate and time required to quench a subsurface heat source (Macdonald et al., 1980; Converse et al., 1984).

While some aspects of the geochemistry of exiting hydrothermal fluids at individual vents are more similar within a vent field than between vent fields (Bowers et al., 1988), two chemically distinct end members have been observed within a single vent field of 50 m radius (Butterfield et al., 1988). Variability in the geochemistry of venting water among vent fields is expressed in the availability of reduced chemical species which in turn influences the species composition of vent communities. This is perhaps seen most dramatically in three vent fields at 13° N, where Bowers et al. (1988) report a ratio of Fe-H_2S greater than unity in end-member hydrothermal fluids. In this chemical regime, the H_2S upon which the vent fauna depends precipitates during subsurface mixing and becomes unavailable. Vent communities at these sites were absent, corresponding to the "azoic" hydrothermal fields of Ballard et al. (1984).

There are other important differences in the physical environment of vent fields to which the biological communities respond. For example, the type of venting also influences the species composition of vent communities (e.g., Fustec et al. 1987). Certain species appear to be adapted to diffuse, low-temperature flows emanating from cracks in the seafloor, while others appear to require the habitat associated with high-temperature plumes venting from massive sulfide chimneys. In addition, local hydrodynamic regimes vary with the microtopography at each vent field. On sediment-starved ridge axes, vent communities may be located on fields of bulbous pillow lavas and flat sheet flows, on talus at the base of fault scarps, or associated with the pits formed by the collapse of ponded lavas following subsurface draining. The patterns of distributions of vent and peripheral nonvent fauna probably respond to the hydrodynamics associated with these different topographic regimes as well as to the hydrodynamics of the venting water. On sedimented ridge axes, such as the northernmost extension of the EPR including Guaymas Basin, the geochemistry of the venting water (Von Damm et al., 1985) and sediment (Gieskes et al., 1982) is distinctive, adding further dimensions with which comparisons of vent fauna on a regional scale can be made.

From these spatial and chemical relationships of vent fields and their biological consequences, there emerges a three-tiered, spatial hierarchy that we will follow in our qualitative comparisons of vent fauna on the EPR and GSC, namely, patterns of distribution of fauna: (1) within a vent field, (2) among vent fields within a linear cluster of vent fields on a ridge segment, and (3) between ridge segments (Fig. 18.1). Underlying this spatial hierarchy is a gradient in the probability of dispersal and genetic exchange among populations of vent species. While we know little about these two aspects of the biology of vent fauna, they are likely to be the most important processes determining the variation in composition of vent fauna at regional and global scales. At local scales, other processes, such as recruitment, competition, and predation, may control faunal composition and relative abundance of species at vent sites.

Comparisons of faunal variation within vent communities can best be made among vent fields on the East Pacific Rise and Galapagos Spreading Center. Specific fields on these ridges have been targets of repeated investigations by biologists, geologists, and geochemists. These include Guaymas Basin (27° N); 21° N, 13° N, and 20° S on the EPR; and Galapagos vents (86° W) on the GSC. In addition, there have been several large-scale photographic surveys of ridge segments and transform faults on the EPR. Documentation of vent communities on the East Pacific Rise and the Galapagos Spreading Center ranges in quality from detailed mapping of megafaunal populations and water chemistry at Rose Garden on the GSC, to video coverage of otherwise unexplored vents and single-frame snapshots of vent animals beneath towed camera sleds on the EPR. For this reason, the comparison of vent communities that follows is restricted to a discussion of the patterns of distribution of the large, easily photographed species.

Distribution of Fauna Within a Vent Field

Low-Temperature, Diffuse Flow

Our notion of a "typical" hydrothermal vent community is derived from the historical discovery of the first active deep-sea fields on the Galapagos Spreading Center. Rose Garden is one of these fields, first visited by biologists in 1979 (Galapagos Biology Expedition Participants, 1979). Subsequent discoveries and exploration of vent communities shows that Rose Garden and Galapagos vents in general are atypical in that they lack high-temperature

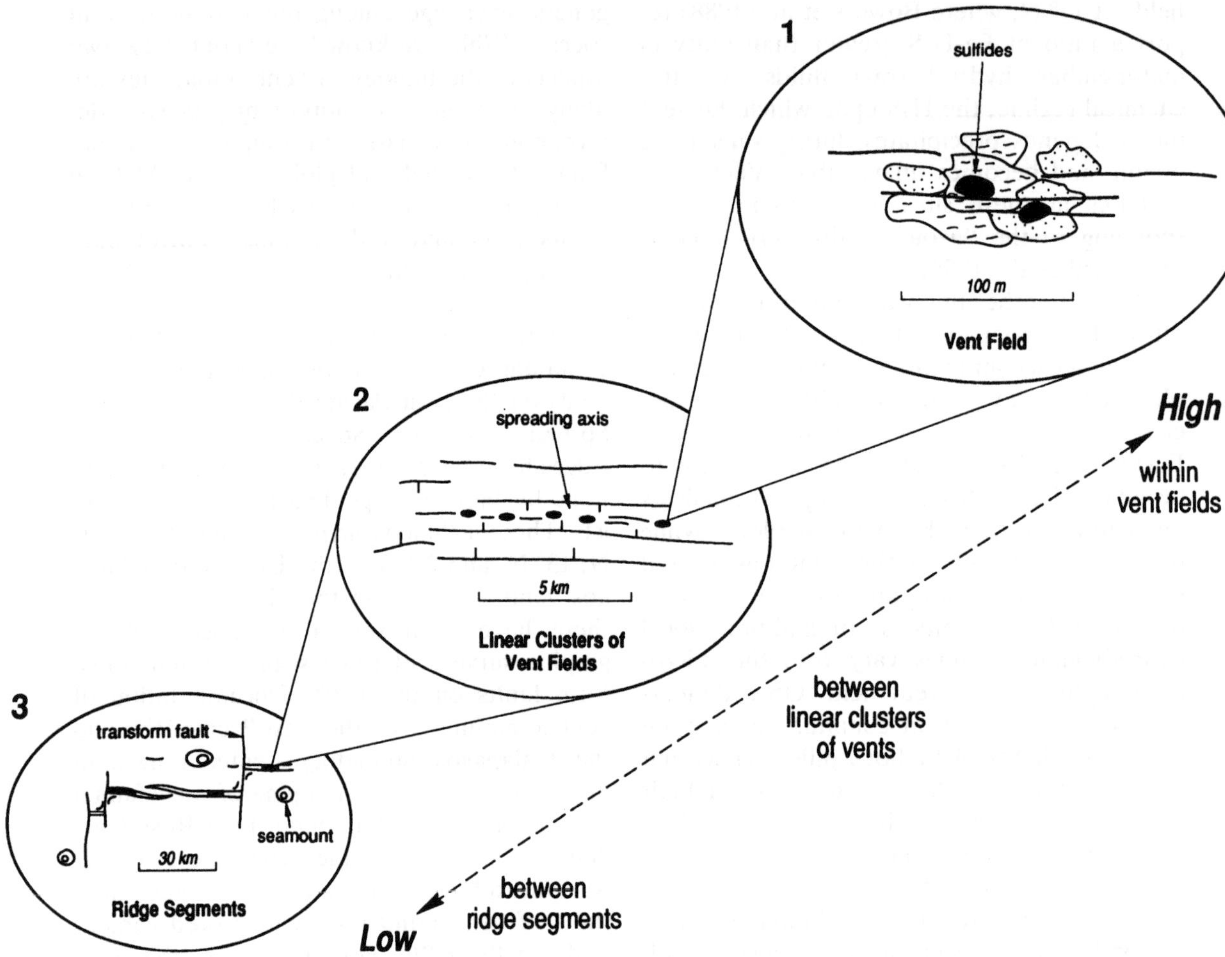

FIGURE 18.1. A three-tiered hierarchy of (1) individual vent fields, (2) clusters of vent fields on a ridge segment, and (3) ridge segments provides the context in which distribution of vent megafaunal species is discussed. Underlying this spatial hierarchy is a gradient in the probability of dispersal and genetic exchange. At global and regional scales, these processes are most likely to control the faunal composition of vent communities.

(350° C) "black-smoker" chimneys. Recent exploration of the EPR at 13° N (Fustec et al., 1987) and between 10° N and 12° N (Van Dover, personal observation), however, has shown that low-temperature vent fields are not uncommon along the EPR. Rose Garden does typify a generic low-temperature, diffuse-flow vent community, particularly in the characteristic zonation of the vent fields into three distinct habitats, each with its own suite of potentially dominant megafaunal types. Using Hessler and Smithey's (1983) original nomenclature and descriptions, these regions are designated "vent,""near-field," and "peripheral."

The warm-water "vent" region of a diffuse-flow field is typically dominated by dense clusters of megafaunal species that rely to a large extent on chemoautotrophic endosymbiotic bacteria for their nutrition. These clusters are usually constrained within sharp boundaries defined by water chemistry and flow patterns of the venting water. At Rose Garden and at other diffuse flow fields on the GSC and EPR, one or another species of clam, mussel, and/or tube worm typically dominates this region. The "vent" region is the most chemically and spatially heterogeneous environment of a diffuse-flow field. Gradients of temperature and water chemistry are strongest in this region, and complex surfaces and interstices are found within the clusters of megafauna. A bythitid fish and macrofaunal populations of limpets, polychaetes, anemones, bresiliid shrimp, lysianassid amphipods, and other small crustaceans occur within this region.

Beyond the central region are "near-field" populations of suspension-feeding and grazing megafaunal species that are presumably dependent on primary chemosynthetic production by free-living microorganisms, either directly or through short planktonic, benthic, and detrital food webs. Boundaries of these populations are sharp and again appear to be defined by gradients of water chemistry as well as by the hydrodynamic regime within the vent field. Anemones, serpulid worms (suspension feeders), gastropods, and galatheid crabs (grazers) are among the fauna that often characterize these regions.

The "peripheral" region of a diffuse-flow field has the least well defined boundaries. It is in this region that certain nonvent species may occur in greater abundance than elsewhere in the deep sea—sometimes spectacularly so, as in the case of enteropneusts at Rose Garden in 1979. Siphonophores, deep-sea shrimp, and sponges may also contribute to the peripheral populations.

Some megafaunal vent species, notably brachyuran crabs, galatheid squat lobsters, and zoarcid fish, are not restricted to a particular region of a field. They move freely between regions and even well out into nonvent terrain. Still, they tend to be most abundant in one or another specific region.

Zonation at Rose Garden is expressed as irregular bands or rings progressing outward from the central region. This pattern of zonation is most likely due to the physical characteristics of the site: venting water flows out of a single fissure and out of smaller collapse features adjacent to and within a field of sheet lava. In contrast, at another diffuse flow field (21° N, EPR), venting water flows out from cracks between lobes of pillow lava, resulting in a mosaic of "near-field" fauna on the tops of the lobes surrounded by "vent" fauna in the cracks.

High-Temperature Chimneys

As noted above, vent fields on the Galapagos Spreading Center are unusual in that the hydrothermal systems are limited to diffuse low-temperature flow. At most fields on the EPR and elsewhere, hydrothermal systems include one or more massive sulfide chimneys with high-temperature (200 to 350°C), turbulent plumes. These chimneys comprise a fourth habitat within a vent field. Physical and chemical gradients are at their most extreme and most complex within this habitat. Characteristic fauna associated with chimneys on the EPR and GSC are alvinellid polychaetes (three species); brachyuran crabs, galatheid squat lobsters, and zoarcid fish may also be present. Both low- and high-temperature microhabitats can occur in association with a sulfide chimney. Thus, low-temperature fauna (i.e., "vent" fauna above) can also be found on black-smoker chimneys. Not all chimneys are colonized.

Distribution of Fauna Among Vent Fields in a Cluster of Vents

There are three locations where exploration has been sufficiently detailed to provide data for comparisons at the level of vent fields within a linear cluster of vents.

Galapagos Vent Fields

Comparisons of megafaunal composition of seven vent fields along an 11-km length of the Galapagos spreading axis are provided in Table 18.1. Data are compiled from Corliss and Ballard (1977), Corliss et al. (1979), Crane and Ballard (1980), Hessler and Smithey (1983), Rosenblatt and Cohen (1986), and additional personal observations by R.R. Hessler. Quality of information varies among vent fields; some fields were not visited often or were more poorly photographed. Clambake 2, Dandelion Patch, and Oyster Bed were mainly documented from higher-altitude photographs taken by ANGUS, making some taxa difficult to discern. Thus it is not possible to establish

TABLE 18.1. Galapagos vent fields: comparison of megafauna.

	RG79	CB1/MB	SF	CB2	DP	OB	GoE
Warm-water vent species							
Riftia pachyptila	D	M	—		—	A	D
Calyptogena magnifica	A	M	—		—	D	dead
Bathymodiolus thermophilus	A	D	—	D	—		R
Neomphalus fretterae	—	A	—				A
?Diplacanthopoma	R	A	A				A
Near-field species							
Anemones[a]	A	—?	—			M/R	A
Serpulids[a]	A	A	M	+?	—	A	A
Bathypecten ?vulcani	M						
Turrid gastropod	R						
Peripheral species							
Thermopalia taraxaca	M	M	A		D	A/M	A
Saxipendium coronatum	M	A/M	+				
Other species							
Bythograea thermydron	A	A	R				A
Munidopsis subsquamosa	A	A	A				M
Thermarces sp.	Present (R), but vent field unknown						

Distances between Galapagos vent fields progressing east to west

Rose Garden to Clambake 1/Musselbed	8,000 m
Clambake 1/Musselbed to Small Fry	"near"
Clambake 1 to Clambake 2	1,600 m
Clambake 2 to Dandelion Patch	250 m
Dandelion Patch to Oyster Bed	250 m
Oyster Bed to Garden of Eden	650 m
Total	10,750 m

[a] More than one species present.
Key: D, dominant (biomass); A, abundant; M, moderately abundant; R, rare; +, present; —, absent; blank, unknown.
RG79, Rose Garden '79; CB1/MB, Clambake 1/Musselbed; SF, Small Fry; CB2, Clambake 2; DP, Dandelion Patch; OB, Oyster Bed; GoE, Garden of Eden.

the absence of some taxa from some vent fields. We limit ourselves to the most conspicuous species, for which the data are most likely to be complete. More detailed comparisons await further exploration. These caveats apply to subsequent sections as well.

The three best-studied fields—Rose Garden, Clambake 1/Musselbed, and Garden of Eden—share the full species list as far as one can tell, with two exceptions: (1) *Neomphalus fretterae*, a limpet that colonizes hard substrates at vent openings, is absent from Rose Garden, perhaps because all the vent openings were choked with bivalves and vestimentiferans; (2) no living *Calyptogena magnifica* were seen at Garden of Eden, but shells demonstrate past colonization.

Thus Rose Garden and Garden of Eden, which span the known length of this string of active vent fields, share a similar species pool. Yet there are obvious differences in the relative abundance of, for example, the three primary "vent" species—*Riftia pachyptila*, *Bathymodiolus thermophilus*, *Calyptogena magnifica*—among fields on this string of vents. Differences in relative abundance seem to reflect habitat differences, especially the availability of vent water, but they may be secondarily influenced by biological interactions.

A comparison of Rose Garden fauna between two visits separated by 6 years demonstrates that the relative abundance of taxa can change markedly with time (Hessler et al., 1988b). Populations of *Riftia pachyptila*, *Thermopalia taraxaca*, *Saxipendium coronatum*, anemones, and serpulids declined, while *Bathy-*

TABLE 18.2. East Pacific Rise Vent Fields, 13° N: comparison of megafauna.

	PR	PA	PM	PS	AC	PN
Warm-water vent species						
Riftia pachyptila	+	+	—	+	+	+
Tevnia jerichonana		+	dead	+		+
Calyptogena magnifica	—	—	dead	—	dead	—
Bathymodiolus thermophilus	—	+	+	+		+
Neomphalus fretterae		+	—			
?Diplacanthopoma		+	—		+	
High-temperature chimney species						
Alvinella pompejana	+	+	—		+	+
Cyanagraea praedator	+	+	—	—	+	+
Near-field species						
Anemones[a]		+	—		+	+
Serpulids[a]	+	+	+	+	+	+
Bathypecten vulcani			+	+		
Turrid gastropod		+	+	+		
Neolepas zevinae		+	+			
Ophiuroid		+		+		
Peripheral species						
Thermopalia ?taraxaca			+	+		
Other species						
Bythograea thermydron	—	+	—	+	—	+
Munidopsis subsquamosa	+	+	+	+	+	+
Thermarces sp.	+	+	+	+	+	+

Distances between 13° N vent fields progressing south to north

Pogoroux to Parigo	200 m
Parigo to Pogomort	20–30 m
Pogomort to Pogosud	50 m
Pogosud to Actinoir	300 m
Actinoir to Pogonord	480 m
Total	1,080 m

[a]More than one species present.
Key: +, present; —, absent; blank, unknown. PR, Pogoroux; PA, Parigo; PM, Pogomort; PS, Pogosud; AC, Actinoir; PN, Pogonord.

modiolus thermophilus, *Calyptogena magnifica*, *Munidopsis subsquamosa*, and turrid gastropod populations multiplied. Hessler et al. (1988b) speculate that these faunal differences are a consequence of stochastic dispersal phenomena and a decline in flow rate within the field. Thus, a single vent field can display greater differences through time than might be seen between spatially separated, contemporaneous fields.

East Pacific Rise Vent Fields, 13° N

Megafaunal compositions of vent fields along a 1-km length of the EPR at 13° N (Table 18.2) are compiled from Desbruyeres et al. (1982), Labuier and Desbruyeres (1984), Fustec et al. (1987), and personal observations by R.R. Hessler at Pogomort and Parigo. The first two papers deal with the fauna of 13° N as a whole. Fustec et al. (1987) provides descriptions of the fauna of specific vent fields; Table 18.2 lists species' presence or absence based on this work.

In 1984, 13° N fauna was composed of two distinct faunal assemblages—one found at high-temperature vents (including *Alvinella pompejana*, *Paralvinella grasslei*, *Cyanagraea praedator*, and *Thermarces andersoni*), and the other at low-temperature vents (including *Riftia pachyptila*, *Tevnia jerichonana*, *Bathymodiolus* sp., serpulid polychaetes, several gastropod species, *Bythograea thermydron*, and *Thermarces andersoni*). Most of the seven sites explored contained both habitats with some subset of associated fauna (Fustec et al., 1987), although relative abundances varied from site to site. Pogomort is the primary exception, notable especially in its lack of living *Riftia pachyptila* and *Alvinella pompejana*. This field is a site of recent, former hydrothermal activity, with massive sulfide deposits and quiescent chimneys and with residual populations of vent fauna.

Galapagos and 13° N vent fields share many of the same species. With the exception of *Tevnia jericohnana* at 13° N, the pool of warm-water vent species is identical (although living *Calyptogena magnifica* were not observed at 13° N). Exclusively high-temperature species are obviously lacking at Galapagos vents, since that habitat is not available at the known active fields; species within this category may include *Alvinella* spp. and *Cyanagraea praedator*. Ophiuroid species and a barnacle, *Neolepas zevinae*, are added to the near-field fauna at 13° N, and the enteropneust, *Saxipendium coronatum*, is absent from peripheral regions at 13° N.

Upon detailed comparison, the singularity of two species shared between Galapagos and 13° N vent fields has been brought into question. J.P. Grassle (1985) shows that Galapagos and 13° N populations of *B. thermophilus* are electrophoretically distinct, while Jones (personal communication) notes morphological dissimilarities between populations of *R. pachyptila* at these sites. There is thus the potential for significant isolation of populations on separate ridge axes.

East Pacific Rise Vent Fields, 21° N

There are at least 14 vent fields along a 30-km length of the EPR at 21° N, but the megafauna of only two of these sites is well documented (Table 18.3). Faunal composition at Clam Acres was determined from photographs and observations made on the Oasis Expedition (Hessler et al., 1985); Hanging Gardens fauna was documented on a 1985 *R/V Atlantis II* Cruise (Van Dover and Berg, personal observations).

TABLE 18.3. East Pacific Rise vent fields, 21° N: comparison of megafauna.

	Clam Acres	Hanging Gardens
Warm-water vent species		
Riftia pachyptila	M	D
Oasisia alvinae	R	A
Calyptogena magnifica	D	M
High-temperature chimney species		
Alvinella pompejana	A	
Munidopsis lentigo	M	
Near-field species		
Serpulids[a]	R	A
Coiled gastropod	A	
Neolepas zevinae	R	A
Peripheral species		
Sponges[a]	—	A
Other species		
Bythograea thermydron	M	M
Munidopsis subsquamosa	A	A
Thermarces sp.	A	

Distance between Clam Acres and Hanging Gardens: 7,000 m

[a]More than one species present.
Key: D, dominant (biomass); A, abundant; M, medium, R, rare; blank, unknown.

As at Galapagos and 13° N, comparison of the fauna at vent fields along the string of vents at 21° N reveals similarities in species composition, but differences in relative abundances. *Riftia pachyptila* dominates the vent fauna at Hanging Gardens; *Calyptogena magnifica* prevails at Clam Acres; *Oasisia alvinae*, a smaller species of tube worm, is very abundant at Hanging Gardens and all but absent at Clam Acres. *Bathymodiolus thermophilus*, one of three most characteristic vent species at Galapagos and 13° N vent fields, is absent at 21° N. Serpulids and the barnacle *Neolepas zevinae* are conspicuous near-field species at Hanging Gardens, while a coiled gastropod is the most abundant near-field species at Clam Acres. Another distinguishing feature of the Hanging Gardens site is the encroachment of peripheral, nonvent species, most notably sponges (especially *Caulophacus cyanae*), into the near-field area. In contrast, no peripheral species were noted at Clam Acres.

Hanging Gardens is in an unusual setting compared to other nearby vent fields, occurring along a wall of deep and narrowing graben. A convection cell set up within the graben by the rising plume of the black smoker at one end may carry noxious vent water away from the site while bringing in a food-enriched supply of bottom water. In this way, peripheral species could encroach upon the vent habitat and thrive in a food-rich environment. Convection cells are presumed to exist at all vent sites with active hydrothermal systems (Lonsdale, 1977; Enright et al., 1981), but the particular topographic relief at Hanging Gardens may be well suited to efficient flushing of vent effluents from the graben and entrapment of sinking chemosynthetic production from the plume.

Two hypotheses may be developed from these three comparisons of fauna among vent fields within clusters of vents: (1) Along any given ridge segment with a linear cluster of vent fields, there exists a shared pool of species from which each individual vent field draws its fauna. (2) Relative abundances of species at a vent field, and even presence or absence of a species, are most likely functions of the field itself, with determinants being such characteristics as flow rates, water chemistry, and stochastic events (Hessler et al., 1985). The relative importance of these characteristics and the role of biological interactions in determining the species composition of vent communities remain unknown.

Distribution of Fauna Among Ridge Segments

At the level of comparison of vent fauna among ridge segments on the EPR and GSC, the quality of data decreases dramatically. For this reason, we focus here on distribution of a select subset of taxonomic groups, namely vestimentiferan tube worms, bivalve mollusks, and anomuran and brachyuran decapod crustaceans. These groups typically comprise the most conspicuous elements of a vent's fauna.

As noted earlier, in many locations vent communities have been photographed but not sampled, leaving taxonomic identifications to be less than perfect. Where a vent field is distant from a well-documented vent field, we provide a generic designation from photographs with some confidence, but assume nothing about the species identification.

Data from six locations along the EPR and GSC are summarized in Table 18.4. Each of these areas is located on a different ridge segment. Sources for Galapagos, 21° N, and 13° N data have been cited above. Data from 10 to 12° N and 17 to 20° S were obtained by review of ANGUS photographs and/or ARGO video supplied by R.D. Ballard and from Hekinian et al. (1985), Renard et al. (1985), Kastens et al. (1986), McConachy et al. (1986), Bundesanstalt für Geowissenschaften und Rohstoffe (1986), Macdonald et al. (1988), and Van Dover (*Alvin-Rise* Cruise Report, 1988, unpublished). Guaymas data are from J.F. Grassle (1986). Guaymas *Riftia* has been tentatively identified as an undescribed new species by M.L. Jones (personal communication). Guaymas decapods were identified by A.B. Williams (1988, personal communication). Guaymas clams in the family Vesicomyidae were originally identified as *Calyptogena pacifica*. Analysis of recent collections of vent and seep bivalves indicates that this identification must be revised to *Vesicomya gigas* (R.D. Turner, personal communication).

Across 40° of latitude there is a remarkable consistency at the familial level in the megafaunal composition of vent communities. Vesicomyid clams and munidopsid squat lobsters, for example, are known from each of the six locations listed in Table 18.4. Riftiid tube worms are apparently absent only at the southernmost location, while bythograeid crabs do not occur at Guaymas, the northernmost extreme. Mytilid mussels are missing from the two most northern sites.

Even considerable differences in habitat, such as those encountered in the hydrocarbon-rich, sediment-hosted environment of Guaymas Basin vents, do not seem to have significant effects on the familial character of the megafaunal community: vesicomyids, munidopsids, and riftiids all occur at Guaymas; mytilids are absent, but they do not occur at 21° N either. On the other hand, species composition of Guaymas Basin megafauna (*Vesicomya gigas*, *Munidopsis alvisca*, *Riftia* sp.) is unique compared to the other locations.

There are sites along the East Pacific Rise

TABLE 18.4. Comparison of vent fauna on ridge segments.

	GY	21° N	13° N	10–12° N	GA	17–20° S
Vestimentiferan tube worms						
Riftia pachyptila		+	+	+	+	
Riftia (undescribed nov. sp.)	+					
Tevnia jericohnana			+	+		+
Oasisia alviniae		+				
Bivalve mollusks						
Vesicomya gigas (formerly identified as *Calyptogena pacifica*)	+					
Calyptogena magnifica		+	+[a]	+	+	
?*Calyptogena* sp.						+
Bathymodiolus thermophilus			+		+	
?*Bathymodiolus* sp.				+		+
Idasola sp.	+					
Anomuran and brachyuran decapod crustaceans						
Neolithodes diomedeae	+					
Munidopsis subsquamosa		+	+	+	+	
Munidopsis lentigo		+				
Munidopsis alvisca	+					
?*Munidopsis* sp.						+
Bythograea thermydron		+	+	+	+	
Bythograea microps			+			
?*Bythograea* sp.						+
Cyanagraea praedator			+	+		

[a] Shells only.
Key: GY, Guaymas; GA, Galapagos. Numeric designations refer to degrees latitude, EPR. ? Refers to uncollected taxa.

where towed camera arrays have captured images of concentrations of megafauna that are unusual in being composed of only a single, or perhaps two species. Lonsdale (1977 and in preparation) reports isolated, dense populations of anemones on the ridge crest at 3.5° S and 4° N, and Van Dover et al. (1987) mention assemblages of ophiuroids and brachyuran crabs at several ridge crest sites near 10° 45′ N. Entangled masses of enteropneusts and galatheid crabs characterize the fauna of the ridge axis between 11° 16′ and 11° 18′ N on the EPR (Van Dover, unpublished). Without more detailed surveys, it is difficult to decide if these unusual faunal assemblages comprise entire vent communities, if the camera array has simply skirted the center of a more typical vent field, or if unusual flow regimens have concentrated suspended food particles independent of any hydrothermal activity. In some instances, these sightings are correlated with temperature anomalies and therefore are demonstrably hydrothermal situations. If these observations do indeed include the core of the vent field, these unusual faunal compositions may result from environmental and/or biological circumstances unlike anything we have studied so far. There is also documentation of conventional vent taxa in relatively unconventional places—viz., vesicomyid clams in the Tamayo Fracture Zone (Gallo et al., 1984, and photographs from *Alvin* dive 977). These, too, await explanation.

Acknowledgments. We thank R.D. Ballard for access to ANGUS and ARGO images of the East Pacific Rise, and P. Lonsdale, K. Macdonald, and D. Gallo for sharing their observations of vent communities and directing us to some of the descriptions of vent fauna in the geological literature. We are especially grateful to J.F. Grassle, with whom we have had many discussions about the ecology of hydrothermal vent communities on midocean ridges. Review of the original draft of this paper by Verena Tunnicliffe and Kim Juniper greately improved the final product. This is WHOI contribution No. 6654. C.L.V.D. was supported by an NSF graduate fellowship and the WHOI/MIT Joint Program.

References

Argorise Group (1988) Geology of the East Pacific Rise axis (11°50′–10°15′N) using the ARGO and ANGUS imaging systems. Can Mineral 26:467–486.

Ballard RD, Van Andel TH, Holcomb RT (1982) The Galapagos Rift at 86°W. 5. Variations in volcanism, structure, and hydrothermal activity along a 30-km segment of the rift valley. J Geophy Res 87:1149–1161.

Ballard RD, Hekinian R, Francheteau J (1984) Geologic setting of hydrothermal activity at 12°0′N on the East Pacific Rise: A submersible study. Earth Planetary Sci Lett 69:176–186.

Bowers TS, Campbell AC, Measures CI, Spivack AJ, Khadem M, Edmond J (1988) Chemical controls on the composition of vent fluids at 13°N–11°N and 21°N, East Pacific Rise. J Geophys Res 93:4522–4536.

Bundesanstalt für Geowissenschaften und Rohstoffe (1986) Fahrtbericht GEOMETEP 4 (20.10.1985–12.01.1986), Tgb. 11336/86, Archiv 99554, Hanover, FRG, 291 pp.

Butterfield DA, McDuff RE, Lilley MD, Massoth GJ, Lupton JE (1988) Chemistry of hydrothermal fluids from the ASHES vent field: Evidence for phase separation. EOS 69:1468.

Campbell AC, Bowers TS, Measures CI, Falkner KK, Khadem M, Edmond J (1988) A time-series of vent fluid compositions from 21°N, East Pacific Rise (1979, 1981, 1985) and the Guaymas Basin, Gulf of California (1982, 1985). J Geophys Res 93:4537–4549.

Cavanaugh CM (1985) Symbioses of chemoautotrophic bacteria and marine invertebrates from hydrothermal vents and reducing sediments. In: Jones ML (ed) Hydrothermal Vents of the Eastern Pacific: An Overview. Bull Biol Soc Wash 6:373–388.

Cavanaugh CM, Gardiner SL, Jones ML, Jannasch HW, Waterbury JB (1981) Procaryotic cells in the hydrothermal vent tubeworm *Riftia pachyptila* Jones: Possible chemoautotrophic symbionts. Science 213:340–342.

Converse DR, Holland HD, Edmond J (1984) Flow rates in the axial hot springs on the East Pacific Rise (21°N): Implications for the heat budget and the formation of massive sulfide deposits. Earth Planetary Sci Lett 69:159–175.

Corliss JB, Ballard RD (1977) Oases of life in the cold abyss. Natl Geogr 152:441–453.

Corliss JB, Baross JA, Gordon LI, et al. (1979) Submarine thermal springs on the Galapagos Rift. Science 203:1073–1083.

Crane K, Ballard RD (1980) The Galapagos Rift at 86°W. 4. Structure and morphology of hydrothermal fields and their relationship to volcanic and tectonic processes of the Rift Valley. J Geophys Res 85:1443–1454.

Desbruyeres D, Crassous P, Grassle J, et al. (1982) Donnees ecologiques sur un nouveau site d'hydrothermalisme actif de la ride du Pacifique oriental. CR Acad Sci Paris (Ser III) 295:489–494.

Detrick RS et al. (1986) Mid-Atlantic bare-rock drilling and hydrothermal vents. Nature 321:14–15.

Enright JT, Newman WA, Hessler RR, McGowan JA (1981) Deep-ocean hydrothermal vent communities. Nature 289:441–445.

Felbeck H (1981) Chemoautotrophic potential of the hydrothermal vent tube worm, *Riftia pachyptila* Jones (Vestimentifera). Science 213:336–338.

Francheteau J, Ballard RD (1983) The East Pacific Rise near 21°N, 13°N and 20°S: Inferences for along-strike variability of axial processes of the Mid-Ocean Ridge. Earth Planetary Sci Lett 64:93–116.

Fustec A, Desbruyeres D, Juniper SK (1987) Deep-sea hydrothermal vent communities at 13°N on the East Pacific Rise: Microdistribution and temporal variations. Biol Oceanogr 4:121–164.

Galapagos Biology Expedition Participants (1979) Galapagos '79: Initial findings of a biology quest. Oceanus 22:2–10.

Gallo D, Kidd WSF, Fox PJ, et al. (1984) Tectonics at the intersection of the East Pacific Rise with Tamayo Transform Fault. Marine Geophy Res 6:159–185.

Gieskes JM, Elderfield H, Lawrence JR, Johnson J, Meyers B, Campbell A (1982) Geochemistry of interstitial waters and sediments, Leg 64, Gulf of California. In: Curray JR, Moore DG, et al. (eds) Initial Reports of the Deep-Sea Drilling Project, Vol 64, Pt 2. Washington: U.S. Government Printing Office, pp 675–694.

Grassle JF (1985) Hydrothermal vent animals: Distribution and biology. Science 229:713–717.

Grassle JF (1986) The ecology of deep-sea hydrothermal vent communities. Adv Marine Biol 23: 301–362.

Grassle JF, Humphris SE, Rona PA, Thompson G,

Van Dover CL (1986) Animals at Mid-Atlantic Ridge hydrothermal vents. EOS 67:1022.

Grassle JP (1985) Genetic differentiation in populations of hydrothermal vent mussels (*Bathymodiolus thermophilus*) from the Galapagos Rift and 13°N on the East Pacific Rise. In: Jones ML (ed), Hydrothermal Vents of the Eastern Pacific: An Overview. Bull Biol Soc Wash 6:429–442.

Hekinian R, Fevrier M, Avedik F, et al. (1983) East Pacific Rise near 13°N: Geology of new hydrothermal fields. Science 219:1321–1324.

Hekinian R, Francheteau J, Ballard RD (1985) Morphology and evolution of hydrothermal deposits at the axis of the East Pacific Rise. Oceanol Acta 8:147–155.

Hessler RR, Smithey WM (1983) The distribution and community structure of megafauna at the Galapagos Rift hydrothermal vents. In: Rona PA, Bostrom L, Laubier L, Smith KL (eds) Hydrothermal Processes at Seafloor Spreading Centers, NATO Conference Series IV. New York: Plenum Press, pp 735–770.

Hessler RR, Smithey WM, Keller CH (1985) Spatial and temporal variation of giant clams, tubeworms and mussels at deep-sea hydrothermal vents. In: Jones ML (ed) Hydrothermal Vents of the Eastern Pacific: An Overview. Bull Biol Soc Wash 6:411–428.

Hessler RR, Lonsdale P, Hawkins J (1988a) Patterns on the ocean floor. N Sci 117:47–51.

Hessler RR, Smithey WM, Boudrias MA, Keller CH, Lutz RA, Childress JJ (1988b) Temporal changes in megafauna at the Rose Garden hydrothermal vent. Deep-Sea Res. 35:1681–1709.

Kastens K, Ryan WBF, Fox PJ (1986) Structural and volcanic expression of a fast slipping ridge-transform-ridge-plate boundary: SeaMARC I and photographic surveys at the Clipperton Transform Fault. J Geophys Res 91:3469–3488.

Lalou C, Brichet E (1982) Ages and implications of East Pacific Rise sulfide deposits at 21° N. Nature 300:821–826.

Lalou C, Brichet E (1987) On the isotopic chronology of submarine hydrothermal deposits. Chem Geol 65:197–207.

Laubier L, Desbruyeres D (1984) Les oasis du fond des oceans. Recherche 15:1506–1517.

Lonsdale P (1977) Clustering of suspension-feeding macrobenthos near abyssal hydrothermal vents at oceanic spreading centers (preliminary communication). Deep-Sea Res 24:857–863.

Macdonald K, Becker K, Spiess FN, Ballard RD (1980) Hydrothermal heat flux of the "black smoker" vents on the East Pacific Rise. Earth Planetary Sci Lett 48:1–7.

Macdonald K, Sempere J-C, Fox PJ (1984) East Pacific Rise from Siqueiros to Orozco Fracture Zones: Along-strike continuity of axial neovolcanic zone and structure and evolution of overlapping spreading centers. J Geophys Res 89:6049–6069.

Macdonald K, Haymon RM, Miller SP, Sempere J-C, Fox PJ (1988) Deep-tow and Sea Beam studies of dueling propagating ridges on the East Pacific Rise near 20° 40′ S. J Geophys Res 93: 2875–2898.

McConachy TF, Ballard RD, Mottl MJ, Von Herzen RP (1986) The geological form and setting of a hydrothermal vent field at 10° 56′ N, East Pacific Rise: A detailed study using *Angus* and *Alvin.* Geology 14:295–298.

Renard V, Hekinian R, Francheteau J, Ballard RD, Backer H (1985) Submersible observations at the axis of the ultra-fast-spreading East Pacific Rise (17° 30′ to 21° 30′ S). Earth Planetary Sci Lett 75:339–353.

Rona PA, Klinkhammer G, Nelson TA, Trefry JH, Elderfield H (1986) Black smokers, massive sulfides and vent biota at the Mid-Atlantic Ridge. Nature 321:33–37.

Rosenblatt R, Cohen D (1986) Fishes living in deep-sea thermal vents in the tropical Eastern Pacific, with descriptions of a new genus and two new species of eelpouts (Zoarcidade). Trans San Diego Nat Hist Soc 21:71–79.

Van Dover CL, Franks PJS, Ballard RD (1987) Prediction of hydrothermal vent locations from distributions of brachyuran crabs. Limnol Oceanogr 32:1006–1010.

Van Dover CL, Fry B, Grassle JF, Humphris SE, Rona PA (1988) Feeding biology of Mid-Atlantic Ridge hydrothermal vent shrimp: Functional morphology, gut content analyses and stable isotopic compositions. Marine Biol 98:209–216.

Von Damm KL, Edmond JM, Measures CI, Grant B (1985) Chemistry of submarine hydrothermal solutions at Guaymas Basin, Gulf of California. Geochim Cosmochim Acta 49:2221–2237.

Williams AB (1988) New marine decapod crustaceans from waters influenced by hydrothermal discharge, brine and hydrocarbon seepage. Fish Bull 86:263–287.

19—Regional-Scale Features of Northeast Pacific, East Pacific Rise, and Gulf of Aden Vent Communities

S. KIM JUNIPER, VERENA TUNNICLIFFE, and DANIEL DESBRUYÈRES

Abstract

Increasing exploration of sea floor spreading centers is revealing regional-scale differences in the nature of hydrothermal venting. Fluid physicochemical properties, mineral deposition, vent field geology, and recent volcanic history can vary considerably between regions. Comparison of northeast Pacific vent biocenoses with sites at 13° N and 17° S on the EPR and in the westernmost Gulf of Aden reveals differences in community structure and ecological strategies that can be related to interregional differences in venting characteristics. The most distinctive environmental influence on northeast Pacific sites is the extensive mineral deposition, which affects fluid flows, alters the substratum, and poses fouling and detoxication problems for vent organisms. This is likely to be an important habitat feature at active vents on Gorda Ridge, where mineral deposition is also very heavy. Biogeographical constraints are also likely to be an important in determining the nature of Gorda Ridge vent communities.

Introduction

Hydrothermal activity is now known to occur at seafloor spreading centers worldwide. The profusion of discoveries that followed the finding of vents on the Galapagos Rift has revealed venting to have many forms. Exit temperatures range from a few degrees above ambient to 350 to 400°C. Hydrothermal fluids have been observed issuing from basalt and sediment substrata and from sulfide and oxide edifices. Chemically, fluids range from slighty reducing seawater to strongly reducing, acidic mineral solutions. These differences in the nature of hydrothermal activity involve fundamental properties of vent habitat, and their role in shaping vent biological communities is of interest to ecologists. An examination of the recent literature reveals that over and above local variations, hydrothermal activity can have a distinct regional character. Examples are the phenomenon of venting through organic-rich pelagic sediment in Guaymas Basin (Simoneit, 1985; Lonsdale et al., 1980), the hot brine deeps of the Red Sea (Degens and Ross, 1969), the absence of high-temperature emissions on the Galapagos Rift (Fustec et al., 1987), the extensive sulfide mineral deposition in the northeast Pacific (Scott, 1986; Tunnicliffe et al., 1986), and the diffuse venting of iron-oxide gel in the westernmost Gulf of Aden (Choukroune et al., 1985).

Except for the recent comparison of sediment infauna from Guaymas Basin, Galapagos Mounds, and EPR by Grassle et al. (1985), regional differences in vent fauna-habitat relationships have yet to be explored. In this chapter we present four examples of vent megafauna-habitat associations that appear unique to particular regions of seafloor spreading, with particular emphasis on the northeast Pacific. We examine new data for the westernmost Gulf of Aden and 17° S on EPR and review existing literature. Geological processes underlying these habitat differences are considered, and likely fauna-habitat relationships on Gorda Ridge are discussed.

Four Contrasting Regions of Hydrothermal Activity

Westernmost Gulf of Aden: Widespread Low-Temperature Venting

The opening of the Gulf of Aden–Red Sea between transform faults results in the focusing of seafloor volcanism and hydrothermal activity at a few discontinuous loci rather than along a linear spreading ridge. The Discovery, Chain, and Atlantis II deeps of the Red Sea are some of the better-known examples of this effect. No biological communities have yet been found in association with hydrothermal activity in the Red Sea. In 1984 an area of recent seafloor accretion in the westernmost Gulf of Aden was explored by the French submersible *Cyana* (Fig. 19.1). Extensive areas of low-temperature hydrothermal venting were observed in one of the explored zones. Diffuse emissions of water producing a hydroxide gel occurred on the southern wall of a basin centered around a volcanic cone (Fig. 19.1) (Choukroune et al., 1986). No high-temperature venting or sulfide deposits were found (Choukroune et al., 1986), and there were no faunal aggregations around vent openings (P. Choukroune, personal communication). However, large areas of unusually high densities of benthic organisms were observed during most dives in this area. Photographic and video images and ambient water temperature data (obtained by a hull-mounted thermistor on *Cyana*) were analyzed to investigate the relationship between faunal distribution and hydrothermal activity. These analyses are summarized in Figure 19.1. A very dense epibenthic megafaunal assemblage occurred discontinuously on basalt substrata on the walls of the basin, mainly between 1,400 and 1,600 m (Fig. 19.1), in close correlation with areas of bottom-water temperature anomalies (Figs. 19.1, 19.2). This megafauna, which consisted of unidentified shrimp, anemones, and galatheid crabs, covered surfaces of pillow basalts and basalt talus (Fig. 19.3A, B). In the only area of temperature anomaly where this faunal assemblage was absent (Fig. 19.1), the substratum was soft sediment. The most unusual aspect of the fauna was the behavior of the shrimp, which was usually the most numerous member of this assemblage. The shrimp appeared to be suspension-feeding with the posterior abdomen attached to the basalt substratum and the thoracic appendages held in the overlying water (Fig. 19.3A, B). Not one was seen to crawl or swim away from this position or change orientation, either by submersible observers or in later examination of video records. Frequently shrimp were regularly spaced over individual basalt pillows giving a bouquetlike appearance (Fig. 19.3A).

Such faunal abundances are very unusual for deep waters, and their correlation with temperature anomalies suggests a trophic dependence on hydrothermal activity. The diffuse warm-water venting observed throughout this area is probably the source of the temperature anomalies, although no near-bottom "plumes" of this scale have been reported. No data are available on the properties of plume or vent waters in the study area. Emitting fluids were depositing iron hydroxide material rather than sulfides (Choukroune et al., 1986), indicating they were transporting reduced iron but little or no H_2S.

17° 26′ S EPR: Nascent Hydrothermal Activity

The East Pacific Rise south of the Garret fracture zone and north of Easter Island has an ultrafast spreading rate (16 cm/year) (Herron, 1972; Rea and Blakely, 1975). Two major stages in ridge evolution can be observed in this region of the EPR: a stage dominated by volcanism leading to ridge formation, and a faulting-dominated stage, where an axial graben is well developed (Renard et al., 1985). The ridge at 17° 26′ S represents the greatest contrast with previously studied segments of the EPR. Recent volcanism is evident on the unrifted summit of the ridge, which is composed of fresh basalts and is almost devoid of fractures and faulting (Renard et al., 1985). Submersible exploration revealed hydrothermal activity to be confined to 5-m-diameter collapse pits formed by drainage of ponded lava.

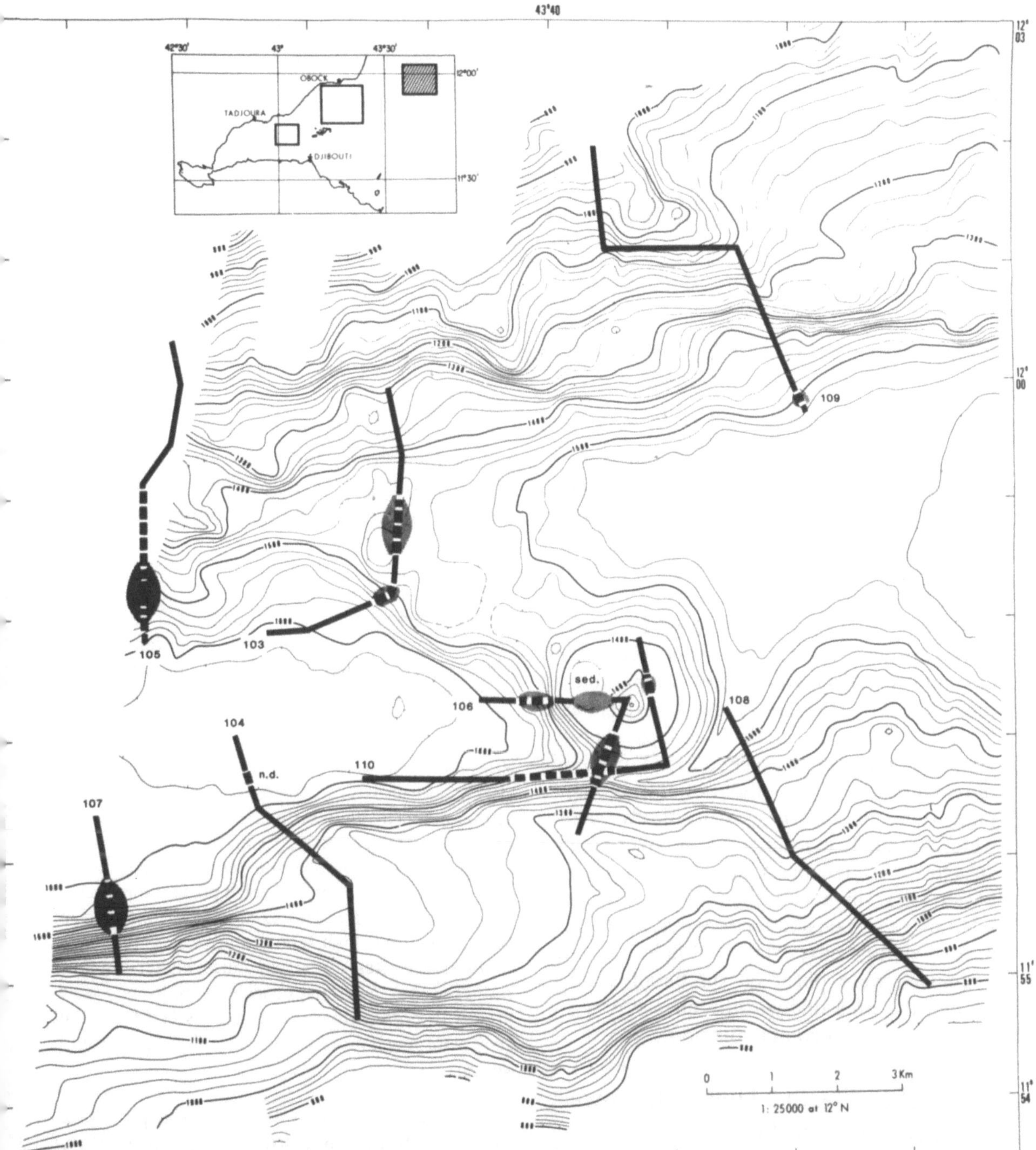

FIGURE 19.1. Bathymetry of hydrothermal site in westernmost Gulf of Aden (shaded area in insert) showing submersible dive tracks. Gray-shaded zones along dive tracks indicate bottom-water temperature anomalies registered by external thermistor on submersible *Cyana*. Broken lines along dive track represent areas of unusual benthic faunal aggregations as recorded in bottom photographs and video. Beginning of each dive track marked by 1984 dive number (103–110).

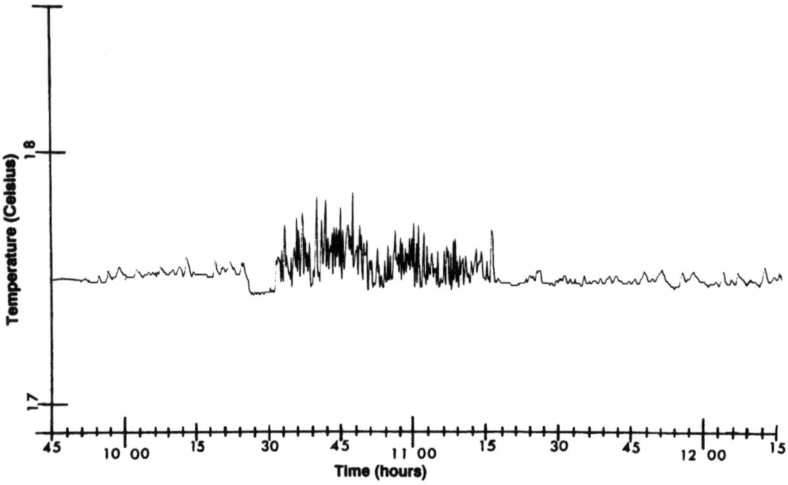

FIGURE 19.2. Example of bottom-temperature anomaly recorded by submersible *Cyana* during exploration of hydrothermal area. Data are from dive 84-105.

Venting was almost entirely in the form of diffuse, warm-water emissions. The associated fauna was dominated by suspension feeders. Shrimp and amphipods swarmed within and above actively venting pits (Fig. 19.3C, D). Serpulid polychaetes densely colonized the rim and inner walls of the pits (Fig. 19.3C, D). Vestimentiferans were occasionally present but did not show any abundance comparable to those known elsewhere on the EPR. Isolated individuals of *Riftia* and small clumps of *Tevnia* were sometimes seen attached to the underside of remaining sections of the roof of the collapse pits, where dilution of venting fluids by seawater was likely restrained by the overhanging structure. A single black smoker was seen on a basalt pillar in a collapsed pit structure. In addition to suspension feeders that aggregated around venting pits, extensive "fields" of large anemones were observed on basalt surfaces around the area of hydrothermal activity (V. Renard, personal communication).

13° N EPR: Distinct High- and Low-Temperature Habitats

Fustec's (1985) extensive survey of more than 20 hydrothermal fields on the EPR near 13° N revealed that venting occurred either in the form of white and black smokers (200 to 350°C) located on scarps of collapsed lava lakes or fissures or as low-temperature emissions (generally <30°C) on plane terrains of sheet or pillow lava flows. Fustec et al. (1987) recognized two distinct vent faunal assemblages at 13° N adapted to high- and low-temperature emissions

> High temperature emission were exclusively exploited by alvinellids and a polynoid. [Vestimentifera] occasionally colonized high-temperature edifices but only where local fluid temperatures were reduced. The low-temperature vents were the domain of *Riftia pachyptila*, the Jericho worms, the serpulids, the mytilids, and the gastropods. A distinction between high and low temperature sites could also be made for associated predatory fauna. . . . The undescribed octopod and *Bythograea thermydron* were restricted to low-temperature vents, while *Cyanograea praedator* was restricted to high-temperature vents, where it presumably preyed on the alvinellids. The zoarcids were found at both low- and high-temperature sites, but low-temperature sites supported much higher numbers of these fishes. Scavenger species were found associated with both primary consumer assemblages, but they were much more numerous at low temperature vents.

Most vent fields in the 13° N region contained both high- and low-temperature habitat and their associated faunae. Fustec et al. (1987) further classified the primary consumer species at 13° N into groups according to their tolerance of, or requirement for, various types of fluids. *Alvinella pompejana* was considered as a "strict eulithophile" with a clear dependence on element-rich fluids. *Paralvinella gras-*

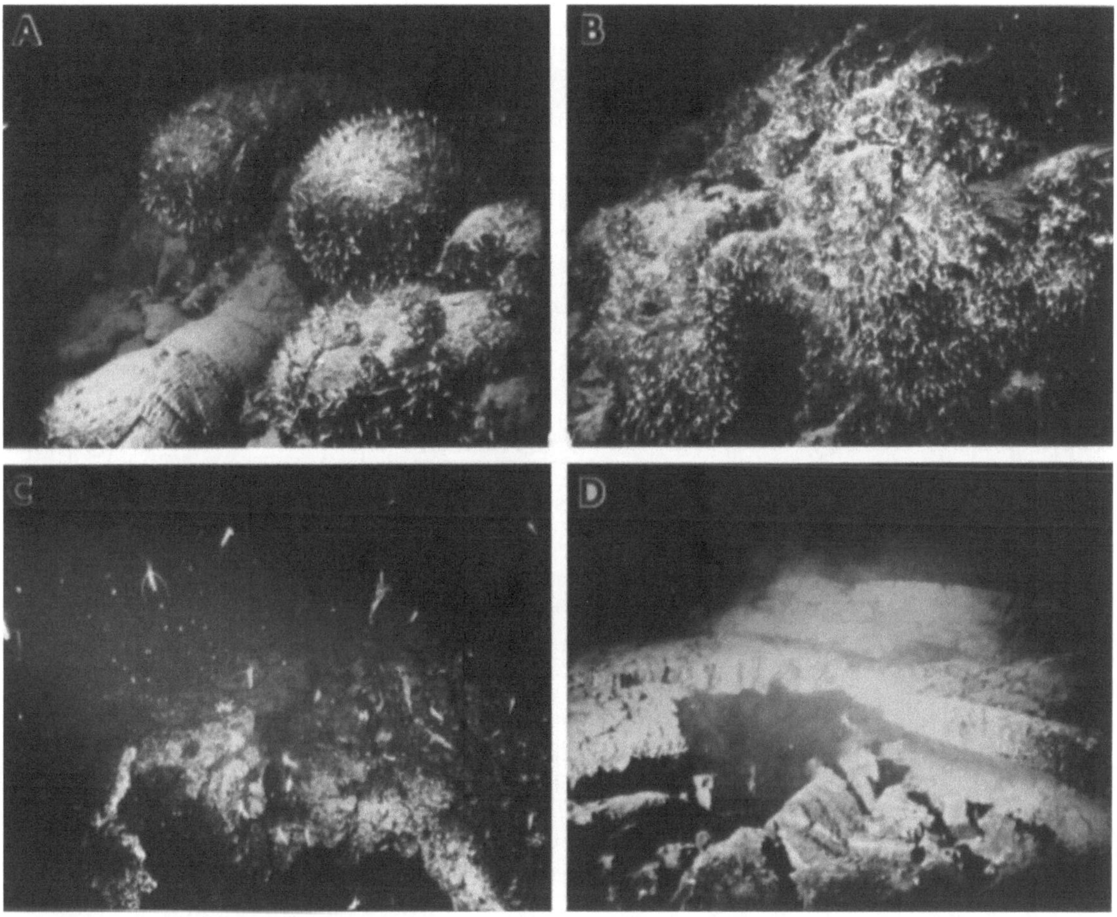

FIGURE 19.3. (A, B) Benthic faunal assemblage associated with bottom-water temperature anomalies. Photo A shows a characteristic "bouquet" of shrimp attached in an upright posture to lobate and pillow lavas. Small galatheid crabs are also visible. Actinarians are less abundant in these photos than at other sites. (Photos IFREMER Cyana.) (C, D) Characteristic collapse pit vents at 17° 26′ S on the East Pacific Rise. Abundant shrimp and amphipods feed in the particulate plume, while the rim of the collapse structure is colonized by serpulid polychaetes and occasional Vestimentifera (no Vestimentifera visible here). (Photos IFREMER Cyana.)

slei was described as a "tolerant eulithophile" present in the high-temperature biotope but less abundant there than in areas where fluids were cooler and less element-rich. "Oligolithophilic" primary consumers showed various levels of dependence or tolerance of low-temperature emissions. *Riftia pachyptila* was the most lithophilic member of this group, with waters at plume level containing 1% to 10% primary fluid. At the other extreme of this group from the low-temperature habitat were the mytilids, termed "tolerantly oligolithophilic," whose mixotrophic feeding mode allowed them to resist interruptions of hydrothermal activity.

Juan de Fuca/Explorer Ridges: Dominance of Mineral Deposition

To date, four hydrothermal areas on the Juan de Fuca/Explorer Ridge system have been explored by submersible (Fig. 19.4). Mineral deposition is very extensive, and deposits form the substrata of most vent fields (Fig. 19.5). As

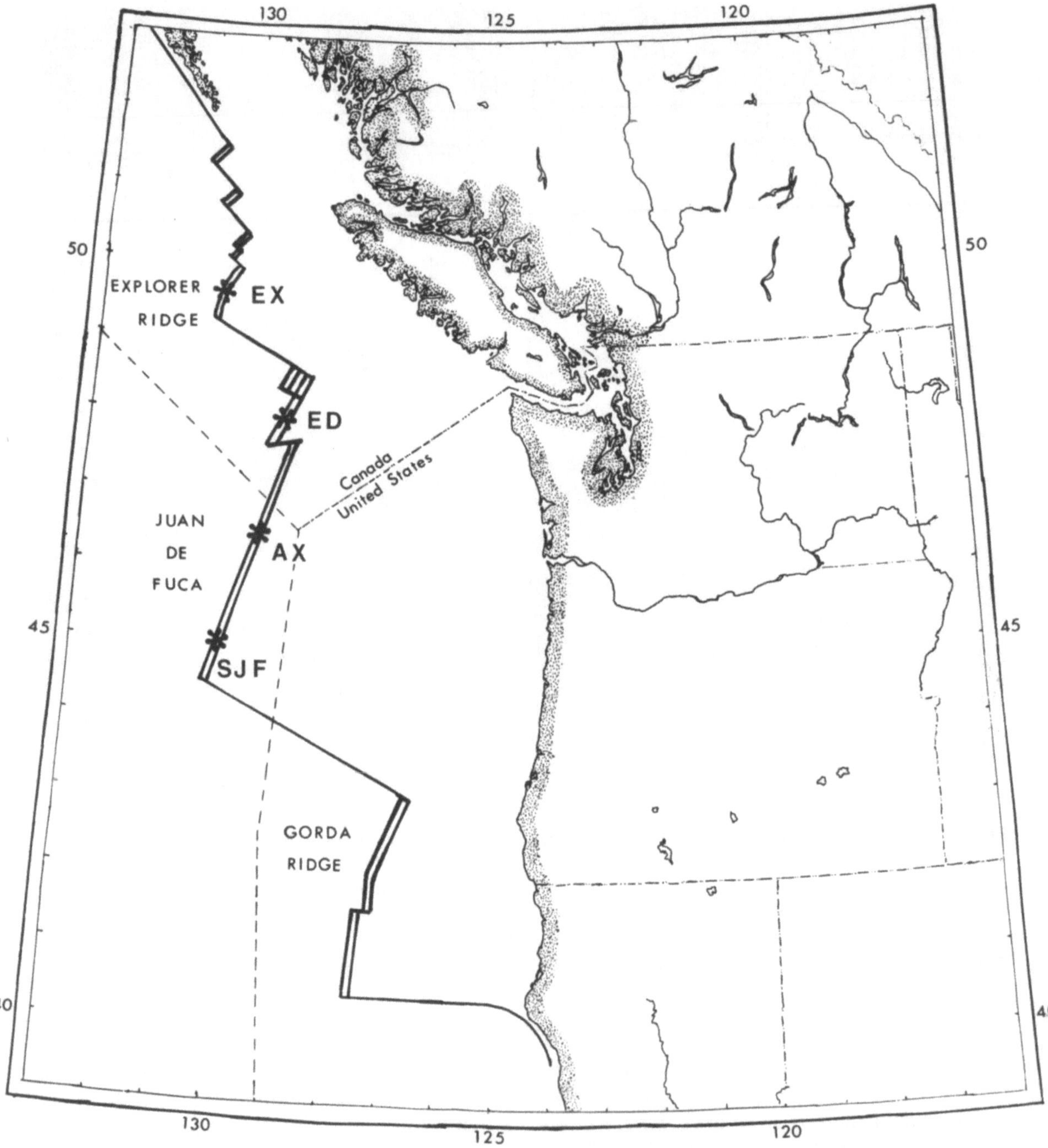

FIGURE 19.4. Map of northeast Pacific spreading centers showing active hydrothermal areas that have been explored, to date, by submersible: EX, Explorer Ridge; ED, Endeavor Segment, Juan de Fuca Ridge; AX, Axial Seamount; SJF, Southern Juan de Fuca site.

FIGURE 19.5. Artist's conception of a large active sulfide mound typical of mature hydrothermal vent fields in the northeast Pacific. Drawing represents the "Magic Mountain" vent field on Explorer Ridge. These mounds are frequently 100 to 200 m in diameter and 10s of meters in height.

a result, hydrothermal fluids generally issue from chimney structures or diffuse from deposits that cover the seafloor between chimneys. Organisms exploiting hydrothermal fluids arc thus required to colonize these substrata and live intimately with the processes of mineral deposition (Fig. 19.6). Fluid emissions cover the known temperature range for hydrothermal activity, but the marked polarization and physical separation of high- and low-temperature venting found at 13° N is rare in the northeast Pacific.

This difference between the two regions can be attributed to two phenomena. First, a large proportion of fluid emissions at Juan de Fuca and Explorer Ridge sites fall into an intermediate range of 20 to 200°C. Such temperatures are frequently measured in diffuse flows that occur on the sides and bases of sulfide chimneys and over extensive areas of the sulfide mounds that form the substratum of most vent fields. Second, at northeast Pacific sites, venting and leaking chimneys and areas of diffuse venting are often so closely spaced that benthic habitat influenced by one emission overlaps and merges with that of another. This is likely produced by lateral diffusion of fluids as they pass through the porous deposits. Together with the diversity of fluid temperatures this extensive venting produces a complex mosaic of chemical and thermal gradients where relationships of faunal distribution to environmental properties are difficult to unravel. Past submersible technology has permitted only a few point measurements of temperature and fluid chemistry. Except for dependence on the presence of HS^-, Tunnicliffe et al. (1986) found no apparent relationship between relative abundances of vent faunal species and physicochemical properties of adjacent vents in an extensive sampling of sulfide mound vent fields on Explorer Ridge.

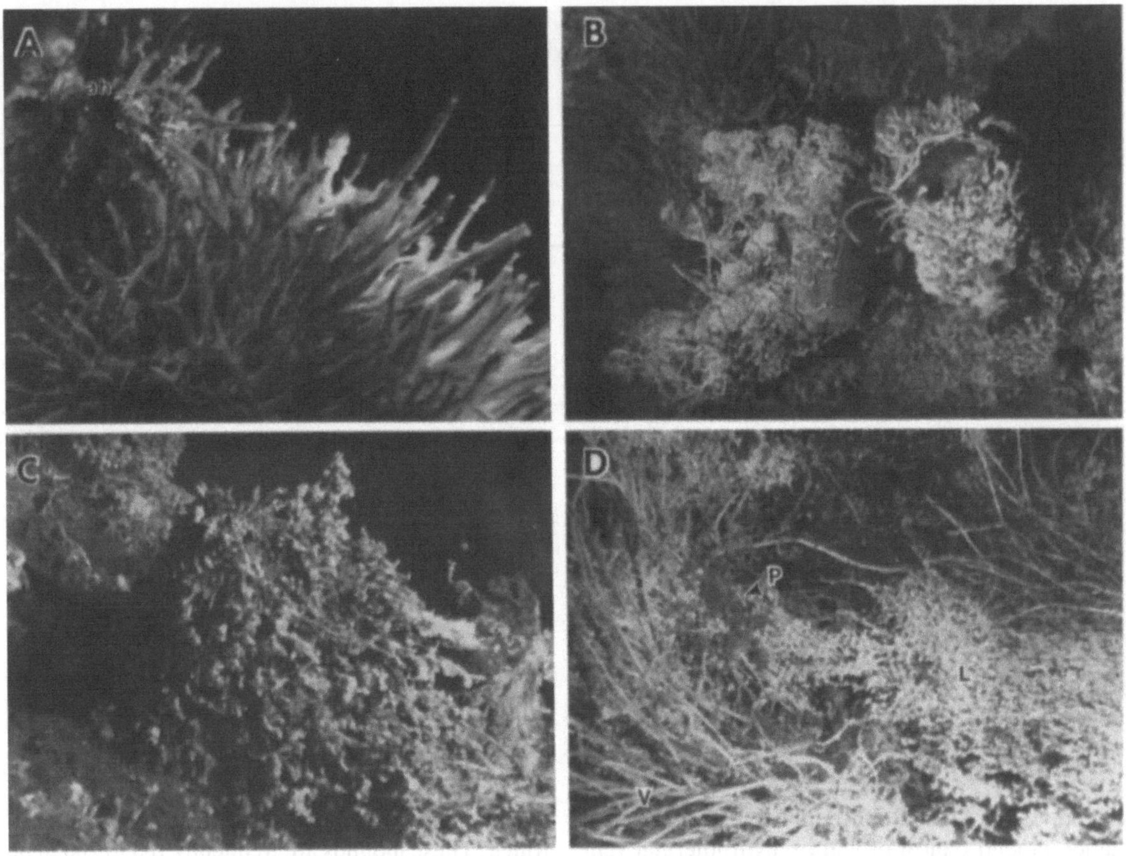

FIGURE 19.6. Examples of northeast Pacific hydrothermal vent habitat. (A) An anhydrite chimney (ah) growing up through a group of vestimentiferans that colonize a sulfide mound. (B, C) Colonization of sulfide substrata within a growing deposit. The dense, multiple species assemblage covers sulfide blocks with vent fluids discharging from multiple exits. (D) Three major components of the northeast Pacific vent fauna showing some spatial zonation. V, Vestimentifera; L, vent limpets; P, *Paralvinella palmiformis*.

Understanding of faunal/environmental relationships in northeast Pacific vent fields is improving as the numbers of in situ observations increase and more samples are analyzed. For the major megafaunal species, trends in habitat control of microdistribution are now discernible, particularly in zones of high-temperature activity or very weak hydrothermal influence. It is also becoming apparent that at least some members of the northeast Pacific vent megafauna occupy a broader range of habitat than their counterparts on the EPR. This may be an adaptation to the diversity and complexity of physical and chemical conditions within hydrothermal fields. These points emerge from observations of the most frequently encountered faunal assemblage, which consists of alvinellid polychaetes, vestimentiferans, and the vent gastropods and polynoids (Fig. 19.6C).

This assemblage colonizes active sulfide chimneys and mounds, areas of diffuse venting, and sites where warm water vents directly from fissures in seafloor basalt. In the latter case vestimentiferans usually form thickets similar to those commonly found at 13° N, and the other members of the assemblage use the vestimentiferan tubes as a substratum. Where this assemblage colonizes active sulfide edifices, the vestimentiferans are attached to the sulfide, and the other organisms utilize both the sulfide and the vestimentiferans tubes as a substratum.

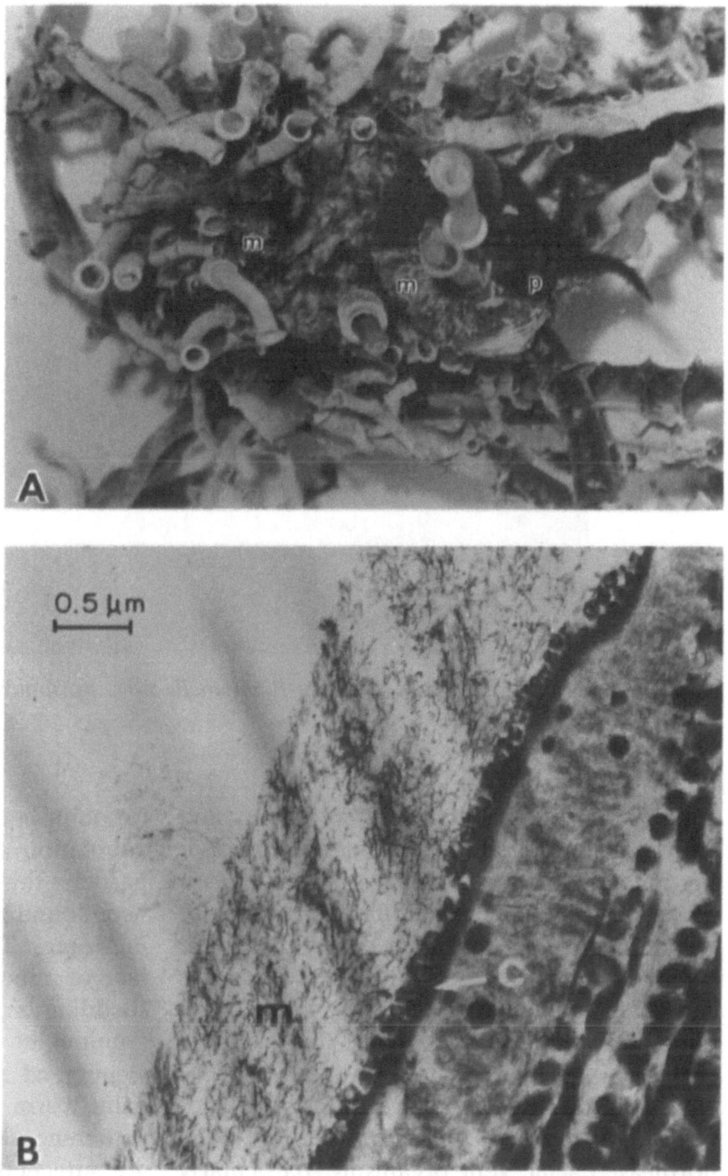

FIGURE 19.7. (A) Palm worm, *Paralvinella palmiformis* (p in photo), in vent faunal grab sample collected by submersible. Gray, mineral-laden mucus (m) is visible surrounding the midbody segments of the worm, and shed mucus (m) can be seen adhering to vestimentiferan tubes. (B) TEM micrograph of transverse section through the epidermal region of a palm worm. A lightly mineralized layer of mucus (m) overlies the cuticular surface (c).

The "palm worm" *Paravinella palmiformis*, an alvinellid polychaete, and several polynoid species are more tolerant of high-temperature emissions than the other members of the assemblage (Fig. 19.7A). The palm worm has a very flexible life-style and can be observed living within particulate tubes in close proximity to high-temperature vents (Fig. 19.8)*, nestling in crevices in sulfide substrata, and free-living attached to vestimentiferan tubes in lower-temperature flows. The vestimentiferans can be represented by one or more of at least five species, and environmental control of species distribution is as of yet not fully understood. The vent limpets appear to be less litho- or thermophilic than the alvinellids and the Vestimentifera.

By way of summary, in a theoretical gradient away from a point of high-temperature emission, the palm worm, often living in particulate tubes, and polynoids are encountered first, after which there is a zone where free-living

*The tube-dwelling palm worm now appears to be a separate species, *Paralvinella* nov. sp.

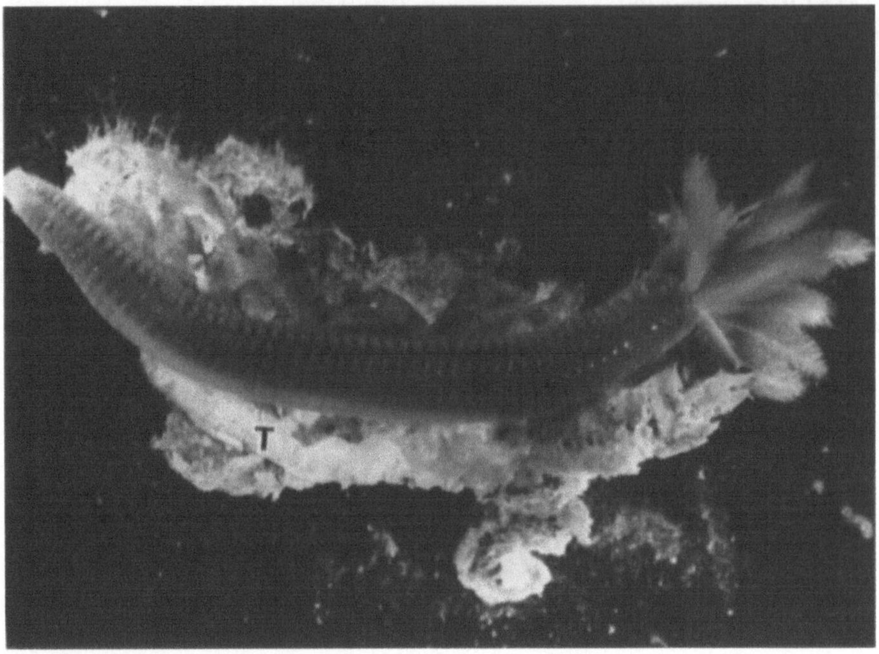

FIGURE 19.8. Tube-dwelling *Paralvinella* nov. sp. removed from its mucoparticulate tube (T).

palm worms, polynoids, Vestimentifera, and limpets occur together. Vestimentiferan species found in minimal flows, in thickets, and on active chimneys appear to differ. For example, weak flows issuing from cracks in lava flows are exploited by one species which lines the cracks in sparse numbers. Exact physicochemical tolerances and requirements for the members of this assemblage remain to be defined.

Mineralization may be an important antagonistic environmental force in the life of many northeast Pacific vent organisms, particularly the sessile species, which are unable to move away from zones where mineralization becomes particularly heavy. Many examples of vestimentiferan tubes heavily mineralized from the base almost to the level of the plume have been observed, and the presence of unoccupied, completely mineralized tubes suggests that this process may be fatal since the organisms have no escape other than to grow away from the source of mineralization. The frequency of vestimentiferan retraction (into their tubes) appears to relate to deposition of particulates that may clog branchial filaments (unpublished observation). One interesting adaptation to the constant rain of mineral particles that typifies this vent environment has been found in the palm worm (*Paralvinella palmiformis*). The worm cleans its epidermal surface of mineral accumulation by secreting and shedding large quantities of mucus (Fig. 19.7) (Juniper et al., 1986; Juniper, 1988). In the vicinity of smoker vents, the tube-dwelling palm worm uses mucous secretions to aggregate mineral particles into habitable tubes that are attached to the sulfide substratum (Fig. 19.8).

Mineral deposition at vents in the northeast Pacific is constantly modifying and transforming the environment through the growth and collapse of chimneys, the accumulation of metalliferous sediments, the blockage of vent flows, and the emergence of new vents. In addition to year-scale habitat changes, which have been observed at 13° N (Fustec et al., 1987), time-lapse camera studies of Juan de Fuca vents are revealing that mineralization can significantly affect faunal habitat within a matter of days or even hours. Johnson and

Tunnicliffe (1985) have recorded repeated growth and collapse of an anhydrite chimney within a palm worm/Vestimentifera/limpet assemblage that proved fatal for many individual organisms (Fig. 19.6A). Similar rapid chimney growth has also been observed at 13° N (Fustec, 1985), but the fauna there does not live as intimately with mineral deposition as they do in the northeast Pacific, where the habitat is effectively formed by a growing deposit.

Discussion

There appear to be important differences in structure and ecological strategies of vent communities that can be attributed to or associated with regional-scale differences in vent habitat. The four regions have very different tectonic and volcanic characteristics—a major reason for their exploration and continued study by earth scientists. These geological processes influence the nature of hydrothermal venting and thus in turn the associated megafauna.

In some cases habitat and faunal characteristics of a region appear attributable to physicochemical properties of vent fluids: the absence of smoker habitat and a high-temperature faunal assemblage at Galapagos is one such example. Properties of venting fluids are to a large degree determined by the nature of fault-produced conduits that permit seawater to circulate in the newly formed crust at spreading centers. Study of hydrothermal fluids from the northern EPR, Galapagos, and Juan de Fuca/Explorer sites has shown that exiting vent waters originate as hot, highly reducing "primary fluids" formed in a subbottom seawater-rock reaction zone (Edmond et al., 1982). During their ascent to the seafloor these fluids are modified by conductive cooling and by mixing with cold, oxidizing bottom seawater that has penetrated the upper part of the oceanic crust (Bowers et al., 1985; Edmond et al., 1982). The latter process has the effects of cooling the primary fluid and reducing the mineral content by dilution and oxydation-induced precipitation. Fluids following the most isolated conduits to the seafloor surface have the highest exit temperatures and the greatest mineral content. Leakier conduits allow considerable dilution of primary fluid by subsurface cold seawater and deposition of the mineral load in the stockwork (subsurface fracture network), producing sulfide-depleted, low-temperature emissions. Some sites are dominated by a single form of venting, and their underlying conduit systems can be simply described in terms of the above model. The Gulf of Aden and 17° S EPR vent fields appear to be fed by fluids that have been extensively cooled and oxidized below the seafloor, with the Gulf of Aden fluids being the most affected by this process. In other cases vent fields appear to be fed by a complex system of both closed and leaky conduits. This is the situation in the majority of vent fields at 13° N EPR and in the northeast Pacific, where both high- and low-temperature vents can occur in very close proximity to one another.

Jannasch and Mottl (1985) have recently proposed that the zone of subsurface oxidation of vent fluids by seawater would be an environment highly favorable to chemosynthetic bacterial productivity. This can result in the emission of low-temperature, oxidized fluids that contain significant quantities of microbial-produced suspended organic matter (Karl et al., 1980; Juniper, unpublished data). Such fluids would be attractive to suspension-feeding organisms in the deep sea, where organic matter is very scarce. This mechanism provides an explanation for the high density of suspension feeders at the Gulf of Aden site and at least at partial explanation the nature of the vent megafauna at 17° S EPR. The apparent depletion of sulfide, the major energy source for hydrothermal vent communities (Childress et al., 1987), in the Gulf of Aden vent fluids may account for the absence of faunal aggregations at vent openings.

The known vent megafauna at 17° S consist mainly of the least lithophilic members of the EPR warm-water assemblage, with the notable absence of bivalves. The most abundant organisms—shrimp, amphipods, and serpulids—are generally found at the periphery of warm-water vents at 13° N: serpulids on basalt surfaces, shrimp and amphipods swarming in the particulate plume issuing from *Riftia* thickets (Fustec et al., 1987). Subsurface mixing and oxidation and organic matter production by microorganisms are certainly the most likely reason for the large numbers of the suspension feeders at 17° S. The presence of a few members of *Riftia* and *Tevnia* at these vents suggests that H_2S is available for symbiosis.

The structure of the vent community at 17° S may also the reflect the recent resumption of hydrothermal activity following a major episode of volcanism in this area. Hekinian and Fouquet (1985) have proposed that hydrothermal activity on a given area of spreading ridge goes through a cycle of waxing and waning following a major volcanic event. Hydrothermal activity increases in intensity as the newly formed crust is fractured by cooling and tectonic activity, which increases contact between seawater and the deep-lying heat source. Eventually, as the heat source is quenched, hydrothermal circulation and venting activity decline. While this cycle cannot be observed during any human lifetime, various stages can be seen along the length of oceanic ridges. Portions of the southern EPR at 18° 30′ S and 21° 30′ S are in a more advanced stage of postvolcanic evolution. There, tectonic activity has formed an axial graben, hydrothermal activity is varied, and vent communities resembling those on the northern EPR can be found (Renard et al., 1985).

The recent episode of volcanism at 17° S would have certainly been catastrophic for local populations of vent-specific organisms. At the present time it appears that both hydrothermal activity and vent communities are reestablishing at 17° S. The newness of the volcanic terrain at 17° S is emphasized in the description by Renard et al. (1985), where they remark that even hairline fractures are very rare on the glassy surface of the seafloor. The time scale for recolonization of newly activated regions by vent organisms is unknown. It will certainly vary with respect to nearness of mother populations and the quality and abundance of vent habitat. It is reasonable to assume that vent-specific species will be slower to move into a new area than more mobile omnivores such as shrimp and amphipods. Thus the scarcity of *Riftia* and other vent-specific species at 17° S may simply represent an early stage of repopulation of the area by vent organisms. Another possible consequence of the present form of hydrothermal activity at 17° S is that the restriction of emissions to collapse pits may be unfavorable to the development of bivalve populations and large thickets of *Riftia* that occur on plane terrains at 13° N.

There is no theoretical basis for relating the major differences in vent habitat between 13° N and the northeast Pacific to ridge crest volcanic and tectonic processes. Hydrothermal activity is widespread and well developed in both regions, yet mineral deposition is much more extensive in the northeast Pacific. And, although the same range of fluid temperatures is encountered in both regions, intermediate temperature emissions appear to be much less common at 13° N. One fundamental and unambiguous difference between these two regions can be found in their respective rates of seafloor spreading. The EPR at 13° N has a fast spreading rate (near 10 cm/year), while Juan de Fuca and Explorer Ridges are classified as intermediate spreading systems (near 6 cm/year). It is interesting to note that venting through actively forming sulfide deposits also occurs at 21° N, which, like the Juan de Fuca/Explorer Ridges, has an intermediate spreading rate. However, causal links between spreading rate, hydrothermal activity, and mineral deposition are still too poorly understood to permit any generalization. As more new areas of ridge crest are studied, correlations between these processes may become more apparent.

Perspectives for Gorda Ridge

Gorda Ridge has the slowest spreading rate of the northeast Pacific systems (4 cm/year) and has shown inconsistent spreading in the past (Atwater and Mudie, 1973). The recent finding of large mineral deposits on Gorda Ridge suggests that, as at Juan de Fuca and Explorer vents, mineral deposition will have a very strong influence on vent fauna. The examples of Juan de Fuca/Explorer indicate that vent organisms will be abundant at actively forming deposits on Gorda, which, if so, will be important in assessing the impact of commercial exploitation of sulfides. Mining of active deposits would involve direct disturbance of vent faunal habitat.

Fluid-related influences on vent habitat on the Gorda Ridge will not be known until active sites are visited, although the mineral deposits at inactive sites indicate that high-temperature venting will certainly be significant. If active sites are found in the sedimented Escanaba Trough, it will be interesting to compare faunal communities between this area and northern and central Gorda, where hydrothermal activity is basalt-hosted. The presence of sediment provides habitat for infaunal organisms that are present only in microenvironments in hydrothermal areas where the substratum is basalt (Grassle et al., 1985; Dinet et al., 1988).

In addition to habitat factors, structure and composition of Gorda Ridge vent communities will be strongly influenced by biogeographic constraints. The northeast Pacific spreading ridges, once contiguous with the East Pacific Rise system, have been separated from the latter for the past 27 million years. The tectonic history of these ridges has been invoked to explain the finding that the Juan de Fuca/Explorer vent fauna is taxonomically distinct from the fauna of the EPR (Tunnicliffe et al., 1985; Tunnicliffe, 1988). Differences at the species, genus, and even family levels and absence of taxa known from the EPR suggest that the northeast Pacific vent fauna has evolved in isolation from the EPR fauna since the two spreading systems were separated. Noted habitat differences between northeast Pacific and EPR vents may also have contributed to the divergence of their faunae. On the basis of this apparent biogeographic effect we would predict that the Gorda Ridge vent fauna would resemble (taxonomically) that of the Juan de Fuca and Explorer ridges (Tunnicliffe, 1988). Nonetheless, we note that the Gorda Ridge was the last of the northeast Pacific spreading centers to remain contiguous with the EPR and that Gorda Ridge has been separated from the Juan de Fuca Ridge by the now 350-km-long Blanco fracture zone for about 8 million years (Engebretson et al., 1985). It will therefore be interesting to discover whether or not these geographic conditions together with environmental influences have produced any faunistic features that distinguishes Gorda Ridge from Juan de Fuca/Explorer and the EPR.

Acknowledgments. We thank P. Choukroune and V. Renard for access to data from, respectively, the Gulf of Aden and 17° S EPR sites. Tom Gore is thanked for preparation of photo plates. This work was supported by the University of Victoria, the Natural Sciences and Engineering Research Council Canada, IFREMER, and the French Ministère des Rélations Extérieures.

References

Atwater T, Mudie JD (1973) Detailed near-bottom geophysical study of the Gorda Rise. J Geophys Res 78: 8665–8686.

Bowers TS, Von Damm KL, Edmond JM (1985) Chemical evolution of mid-ocean ridge hot springs. Geochim Cosmochim Acta 49: 2239–2252.

Childress JJ, Felbeck H, Somero GN (1987) Symbiosis in the deep sea. Sci Am 255(5): 114–120.

Choukroune P, Auvray B, Francheteau J, et al. (1986) Tectonics of the westernmost Gulf of Aden and the Gulf of Tadkoura from submersible observations. Nature 319: 396–399.

Degens ET, Ross DA (eds) (1969) Hot Brines and Recent Heavy Metal Deposits in the Red Sea. New York; Springer-Verlag.

Dinet A, Grassle JF, Tunnicliffe V (1988) Premières observations sur la meiofaune des sites hydrothermaux de la dorsale Est-Pacifique (Guaymas, 21° N) et de l'Explorer Ridge. Oceanol Acta 8: 7–14.

Edmond JM, Von Damm KL, McDuff RE, Measures CI (1982) Chemistry of hot springs on the East Pacific Rise and their effluent dispersal. Nature 297: 187–191.

Engebretson DC, Cox A, Gordon RG (1985) Relative motions between oceanic and continental plates in the Pacific basin. Geol Soc Am Special Paper 206, 59 pp.

Fustec A, Desbruyères D, Juniper SK (1987) Deep-sea hydrothermal vent communities at 13° N on the East Pacific Rise: Microdistribution and temporal variations. Biol Oceanogr 4: 121–164.

Grassle JF, Brown-Leger LS, Morse-Porteous L, Petrecea R, Williams I (1985) Deep-sea fauna of sediments in the vicinity of hydrothermal vents. In: Jones ML (ed) Hydrothermal Vents of the Eastern Pacific: An Overview. Bull Biol Soc Wash 6:443–452.

Hekinian R, Fouquet Y (1985) Volcanism and metallogenesis of axial and off-axial structures on the East Pacific Rise near 13° N. Econ Geol 80:221–249.

Herron EM (1972) Sea-floor spreading and the Cenozoic history of the East Central Pacific. Geol Soc Am Bull 83:1671–1691.

Jannasch HW, Mottl MJ (1985) Geomicrobiology of deep-sea hydrothermal vents. Science 229:717–725.

Johnson HP, Tunnicliffe V (1985) Time-series measurements of hydrothermal activity on northern Juan de Fuca Ridge. Geophys Res Lett 12:685–688.

Juniper SK, Thompson JAJ, Calvert SE (1986) Accumulation of minerals and trace elements in biogenic mucus at hydrothermal vents. Deep-Sea Res 33:339–347.

Juniper SK (1988) Géochimie et écologie d'un microenvironnement hydrothermal: Les sécrétions de mucus de *Paralvinella palmiformis*. Oceanol Acta 8:167–172.

Karl DM, Wirsen CO, Jannasch HW (1980) Deep-sea primary production at the Galapagos hydrothermal vents. Science 207: 1345–1347.

Lonsdale PF, Bischoff JL, Burns VM, Kastner M, Sweeney RE (1980) A high temperature hydrothermal deposit in the seabed at a Gulf of California spreading center. Earth Planetary Sci Lett 49:8–20.

Rea DK, Blakely RJ (1975) Short-wavelength magnetic anomalies in a region of rapid seafloor spreading. Nature 255:126–128.

Renard V, Hekinian R, Francheteau J, Ballard RD, Backer H (1985) Submersible observations at the ultra-fast-spreading East Pacific Rise (17° 30′ to 21° 30′ S). Earth Planetary Sci Lett 75:339–353.

Scott SD (1986) Seafloor polymetallic sulfides: Scientific curiosities or mines of the future? In: Teleki P et al. (eds) Proceedings of NATO Advanced Research Workshop on "Marine Minerals: Resource Assessment Strategies." Dordrecht: D. Reidel Publishing.

Simoneit BRT (1985) Hydrothermal petroleum: Genesis, migration and deposition in Guaymas Basin, Gulf of California. Can J Earth Sci 22:1919–1929.

Tunnicliffe V (1988) Biogeography and evolution of hydrothermal vent fauna in the eastern Pacific Ocean. Proc R Soc Lond B 233:342–366.

Tunnicliffe V, Juniper SK, De Burgh ME (1985) Biological communities of the Juan de Fuca hydrothermal vents. In: Jones ML (ed) Hydrothermal Vents of the Eastern Pacific: An Overview. Biol Soc Wash Bull 6:453–464.

Tunnicliffe V, Botros M, De Burgh ME, et al. (1986) Hydrothermal vents of Explorer Ridge, northeast Pacific. Deep-Sea Res 33:401–412.

20—Notes on Fauna from Several Deep-Sea Hydrothermal Vent and Cold Seep Soft-Sediment Communities

ROSEMARIE F. PETRECCA and J. FREDERICK GRASSLE

Introduction

Since our initial summary of the macrofauna of hydrothermal soft sediments of Guaymas Basin and the Galapagos Mounds (Grassle et al., 1985), we have obtained cores from other hydrothermal vent and cold seep sites. In June 1988 we sampled a newly discovered area of hydrothermal venting in soft-sediment at about 3,200 m depth in the Escanaba Trough region of the Gorda Ridge (Van Dover et al., this volume). We also obtained samples from soft-sediment hydrothermal environments of the Mid-Atlantic Ridge, the Florida Escarpment cold seep (Hecker, 1985), and the Laurentian Fan cold seep (Mayer et al., 1988). Additional results from the 1985 Guaymas Basin Expedition are also summarized.

Results

Gorda Ridge

In the Escanaba Trough, a single scoop sample was collected from a 50-m-diameter patch of sediments venting low-temperature hydrothermal fluid. The sediments at the Escanaba Trough site smelled slightly of petroleum but, unlike some of the Guaymas Basin hydrothermal sediments, were not saturated with petroleum. A vent clam, *Calyptogena phaseoliformis*, was the largest and most abundant animal in this area. This species was previously known only from soft-sediment seeps near Japan (Metvier et al., 1986; Juniper and Sibuet, 1987). A polychaete genus found at most vents, *Amphisamytha*, was represented by a new species in the soft-sediment (V. Solis-Weiss, personal communication) New species of spionid and orbiniid polychaetes were also abundant. The fauna from the hydrothermal sediment did not have any species in common with the fauna from the surrounding nonvent, soft-sediment community. Except for *C. phaseoliformis*, the Escanaba Trough soft-sediment vent sample did not share any species with any other soft-sediment vent or seep sites.

Guaymas Basin

The geological setting and megafauna of the Guaymas Basin hydrothermal vent communities were described by Grassle et al. (1985) and Grassle (1986). Some of the results from the 34 225-cm^2 cores taken on the 1985 Guaymas Basin Expedition are shown in Figure 20.1, together with results published and in preparation from our laboratory on the number of species per 900 cm^2 at four other deep-sea sites. The community structure of most deep-sea communities studied shows a linear relationship between number of species and number of individuals per sample. Cores from sediments with a flux of hydrothermal fluid have a reduced number of species and higher densities of individuals. A similar phenomenon was found in previous samples from Guaymas Basin and the Galapagos Mounds (Grassle et al., 1985). Species diversity as measured by the

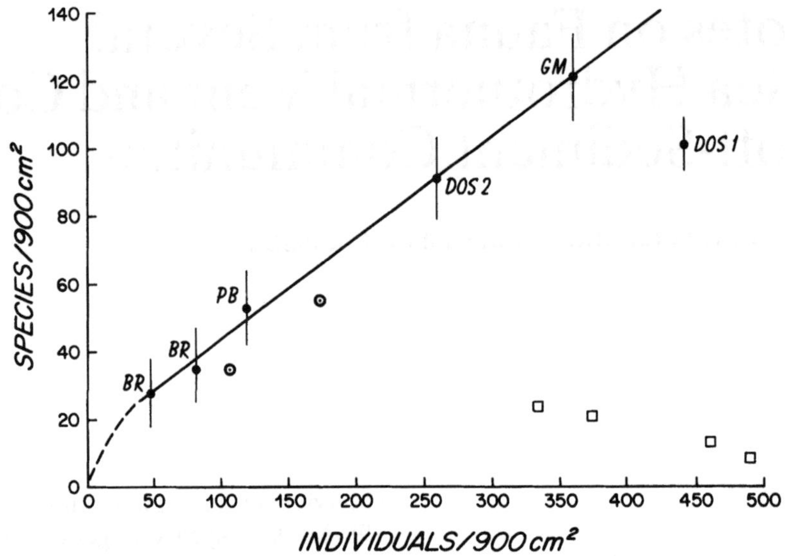

FIGURE 20.1. Soft-sediment macrofaunal species and individuals per 900 cm² from five sites within a depth range of 2,700 to 5,000 m compared with results from a hydrothermal area at 2,000 m depth (Guaymas Basin, Gulf of California) and 1,800 m depth south of New England (DOS 1). BR refers to two sites on the northeast Bermuda Rise (4,600 m and 4,900 m depth) (Grassle, unpublished); PB is Panama Basin (3,900 m depth) (R.F. Petrecca and J.F. Grassle, unpublished). DOS 2 is the 3,600-m depth station south of New England, and GM is Galapagos Mounds (2,700 m depth) with no hydrothermal activity. The two circled dots are groups of four combined Guaymas Basin box cores (900 cm² in aggregate) from areas with no hydrothermal activity. The squares represent data analyzed in the same way from Guaymas Basin areas where there is hydrothermal activity.

species/individuals relationship is progressively lower in patches with evidence of increased hydrothermal flux (e.g., increased temperature, organics, sulfides, or minerals).

The most characteristic species found in mats of filamentous bacteria (chiefly *Beggiatoa* spp.) overlying soft sediments at Guaymas Basin vents include the following polychaetes: *Amphisamytha galapagensis* (2,267/m²), an ampharetid polychaete, which has the most widespread distribution of any vent animal (see below); *Ophryotrocha akessoni* (3,222/m²), a dorvilleid polychaete, known only from the Galapagos Rift and East Pacific Rise; and *Orseis grasslei* (2,844/m²), a hesionid polychaete, known only from Guaymas Basin. *Ophryotrocha akessoni* and a new species of gastropod (gastropod 2) dominated sediments saturated with petroleum hydrocarbons, both in the presence and absence of bacterial mat. Two species in the polychaete genus *Paralvinella* (up to 1,867/m²) and, in one core, the vestimentiferan tube worm, *Riftia pachyptila* (267/m²) were found in sediments where the temperatures were 5° C at 8 cm and 55° C at 16 cm depth. A dorvilleid polychaete, *Exallopus jumarsi* (2,667/m²), and an undescribed bivalve, *Nuculana* sp. (up to 6,756 m²), are known only from the Guaymas Basin and occur most abundantly adjacent to fields of the Guaymas vent clam, *Vesicomya gigas* (this species, previously identified as *Calyptogena pacifica*, is being reclassified by R.D. Turner, personal communication). The clam fields are lower in temperature and are the most distinct and least heterogenous community associated with hydrothermal activity. The biomass of *V. gigas* in cores with as many as 2,600 clams/m² is about 10 kg/m² shell-free wet weight.

Mid-Atlantic Ridge

Biologists have not yet visited the Mid-Atlantic Ridge vent areas (Grassle et al., 1986). Collections made by geologists and geochemists at vent sites north and south of the Kane Fracture Zone have shown that the dominant organism is a bresiliid shrimp in a new genus, *Rimicaris* (Williams and Rona, 1986; Grassle et al., 1986; Van Dover et al., 1988). A shrimp belonging to another bresiliid genus, *Alvinocaris*, was found at other vent sites including the Mid-Atlantic Ridge, but *Rimicaris* had not been observed at any other vent site until its collection on the Marianas Back-Arc Spreading Center (Hessler et al., 1988).

Vestimentiferans have not been oberved on the Mid-Atlantic Ridge. Some of the tubes visible in photographs from the area belong to a large (>1 cm diameter) maldanid polychaete worm. This polychaete has not been collected, but the characteristic appearance of the tail indicates that it belongs to the genus *Nicomache*. This genus is now known from all of the well-studied vent areas including Galapagos Rift, East Pacific Rise, Guaymas Basin, Juan de Fuca Ridge, and the Florida Escarpment seep. The only soft-sediment core for faunal analysis was taken within a hexagonal grid pattern characteristic of the trace fossil *Paleodictyon* (Rona, 1980). Careful dissection of the core yielded some irregular burrow traces and a few small, poorly preserved polychaetes belonging to the families Capitellidae, Paraonidae, and Ampharetidae.

Florida Escarpment

The megafauna from this site has been described by Paull et al. (1984) and Hecker (1985). Occurring at this seep site are two species of the galatheid squat lobster, *Munidopsis*, a genus known from Pacific vent sites and throughout the deep sea and a new species of the vesicomyid clam, *Calyptogena*. The Florida seep limpet is similar to the "transparent" limpet known from the Galapagos Rift and East Pacific Rise (McLean, 1985). The vestimentiferan *Escarpia laminata* belongs to the same genus as the species collected from a strike-slip fault site off San Diego (Jones, 1985). A mytilid, which forms dense beds at the Florida seep site, is related to the vent mussels in the genus *Bathymodiolus* but may be sufficiently different to warrant a new genus (R.D. Turner, personal communication). Associated with the mussel is a polychaete, *Branchipolynoe*, belonging to the same genus as those living with *Bathymodiolus* on the East Pacific Rise and Galapagos Spreading Center (M.H. Pettibone, personal communication).

Three 225-cm^2 box cores from the soft sediments from this seep area were collected and analyzed. Two polychaete species—the ampharetid, *Amphisamytha galapagensis*; and the maldanid, *Nicomache arwidssoni*—previously known only from hydrothermal vent areas in the Pacific, occur at the Florida Escarpment site. Another polychaete species, the hesionid, *Hesiospina vestimentifera*, described from the Galapagos Rift, East Pacific Rise, and Juan de Fuca Ridge, was also found in Florida Escarpment collections (B. Hecker and J.A. Blake, personal communication).

Laurentian Fan

During the summer of 1986, when the *Alvin* was used to explore the deposits of the 1929 Grand Banks earthquake and turbidity current, large, dense communities of benthic animals were unexpectedly discovered (Mayer et al., 1988). From a dredge haul we made in 1987 and specimens collected from the *Alvin*, some species have been identified. The fauna includes living and dead vesicomyid, thyasirid, and solemyid clams (*Calyptogena* sp., *Thyasira insignis*, and *Solemya grandis*); gastropods (*Beringus ossiani frielei*); pogonophorans (*Sclerolinum* sp., *Oligobrachia* sp., and two species of *Spirobrachia*); and galatheid crabs (*Munidopsis* sp.). A polychaete, probably belonging to a new family (J.A. Blake, personal communiation), was found living in association with *Thyasira insignis*. None of these species live exclusively in vent or seep areas.

Discussion

In the dozen years since the discovery of hydrothermal vent communities at the Galapagos Spreading Center, other chemosynthetically based deep-sea communities have been observed. These communities are geographically widespread and diverse in nature, ranging from hard-substrate, hydrothermal vents, to organic-rich, soft-sediment vent and seep areas with associated petroleum hydrocarbons, to methane-rich cold seeps. Despite the geographical distance and environmental differences among sites, there are similarities in the taxonomic composition among communities. In some cases, morphologically indistinguishable species are known to occur in both cold seep areas and in Atlantic and Pacific hydrothermal vent areas. The best example of this is the ampharetid polychaete, *Amphisamytha galapagensis*, which occurs in soft-sediment habitats at the Galapagos Rift, East Pacific Rise, Guaymas Basin, and Juan de Fuca vents in the Pacific and at the Florida Escarpment in the Atlantic. In other cases, closely related species occur in several types of habitats. For example, bivalves in the genera *Calyptogena* and *Bathymodiolus* and squat lobsters in the genus *Munidopsis* are known from Pacific and Atlantic hydrothermal vents and from cold seeps.

Two of the most interesting sites, the Mid-Atlantic Ridge and the Laurentian Fan, have not yet been visited by biologists. From the sparse collections made thus far, these areas appear to have quite distinct faunae in comparison to other vent and seep areas. A more extensive sampling effort is needed in order to draw firm conclusions from comparable sets of data.

The lack of similarity of the soft-sediment fauna from the Gorda Ridge to that at other sites is consistent with the hypothesis of Tunnicliffe (1988) that the interruption of the ridge system by the North American continent about 35 Ma BP has resulted in the evolution of a distinct fauna on the northern ridges of the eastern Pacific Ocean. The one species with an extremely widespread distribution, *Amphisamytha galapagensis*, is a good candidate for further study of relatedness among populations, using biochemical characters and morphometric analyses of large numbers of individuals. Techniques for comparing DNA and RNA sequences should also prove useful for studying the divergence among populations and species of hydrothermal-vent and cold-seep organisms.

Acknowledgments. We thank Ruth Turner for identifications of bivalves and James Blake for identifications of polychaetes. We also thank Judith Grassle, Paul Snelgrove, and Cindy Van Dover for reviewing this paper, and Margaret Dimmock for typing the text. This paper was supported by National Science Foundation grant OCE-8311201 and is contribution No. 6708 from the Woods Hole Oceanographic Institution.

References

Grassle JF (1986) The ecology of deep-sea hydrothermal vent communities. Adv Marine Biol 23: 301–362.

Grassle JF, Brown-Leger LS, Morse-Porteous LS, Petrecca RF, Williams IP (1985) Deep-sea fauna of sediments in the vicinity of hydrothermal vents. Bull Biol Soc Wash 6: 445–456.

Grassle JF, Humphris SE, Rona PA, Thompson G, Van Dover CL (1986). Animals at Mid-Atlantic Ridge hydrothermal vents. EOS 67(44): 1022.

Hecker B (1985) Fauna from a cold sulfur-seep in the Gulf of Mexico: Comparison with hydrothermal vent communities and evolutionary implications. Bull Biol Soc Wash 6: 465–473.

Hessler R, Lonsdale P, Hawkins J (1988) Patterns on the ocean floor. New Sci 117: 47–51.

Jones ML (1985) On the Vestimentifera, new phylum: Six new species and other taxa, from hydrothermal vents and elsewhere. In: Jones ML (ed) The Hydrothermal Vents of the Eastern Pacific: An Overview. Bull Biol Soc Wash 6: 117–158.

Juniper SK, Sibuet M (1987) Cold seep benthic communities in Japan subduction zones: Spatial organization, trophic strategies and evidence for temporal evolution. Marine Ecol Prog Ser 40: 115–126.

Mayer LA, Shor AN, Clarke JH, Piper DJW (1988) Dense biological communities at 3850 m on the Laurentian Fan and their relationship to the deposits of the 1929 Grand Banks earthquake. Deep-Sea Res 35: 1235–1246.

McLean JH (1985) Preliminary report on the limpets at hydrothermal vents. Bull Biol Soc Wash 6: 159–166.

Metvier B, Okutani T, Ohta S (1986) *Calyptogena* (*Ectenagena*) *phaseoliformis* n. sp., an unusual vesicomyid bivalve collected by the submersible

Nautile from abyssal depths of the Japan and Kurile Trenches. Venus 45:75–86.
Paull CK, Hecker B, Commeau R, et al. (1984) Biological communities at the Florida Escarpment resemble hydrothermal vent taxa. Science 226:965–967.
Rona PA (1980) TAG hydrothermal field: Mid-Atlantic Ridge crest at latitude 26° N. J Geol Soc 137:385–402.
Tunnicliffe V (1988) Biogeography and evolution of hydrothermal-vent fauna in the eastern Pacific Ocean. Proceedings of the Royal Society of London B 233:347–366.
Van Dover CL, Fry B, Grassle JF, Humphris S, Rona PA (1988) Feeding biology of the shrimp *Rimicaris exoculata* at hydrothermal vents on the Mid-Atlantic Ridge. Marine Biol 98:209–216.
Williams AB, Rona P (1986) Two new caridean shrimps (Bresiliidae) from a hydrothermal field on the Mid-Atlantic Ridge. J Crustacean Biol 6(3):446–462.

21—Hydrothermal Vent Fauna of Escanaba Trough (Gorda Ridge)

CINDY LEE VAN DOVER, J. FREDERICK GRASSLE, and MICHEL BOUDRIAS

Biota of the NESCA Site, Escanaba Trough

Hydrothermal vent communities on Gorda Ridge were discovered and documented during 1988 field programs using the deep-diving submersibles *Alvin* and *Sea Cliff*. Venting was associated with large sulfide mounds that comprise the NESCA site in the Escanaba Trough of Gorda Ridge (41° 00′ N; 127° 29′ W; 3,200 to 3,250 m). Details of the geological setting are presented elsewhere in this volume.

Within the study area, we identified five principal habitats: (1) active hydrothermal venting associated with massive sulfide deposits (temperatures ranged from a few degrees above ambient to a maximum of 220°C); (2) a low-temperature site associated with soft sediment deposits at the base of sulfide mounds; (3) "inactive" sulfide mounds; (4) nonvent soft sediment; and (5) nonvent basalt lava.

The fauna of active hydrothermal vents associated with sulfide deposits was lush, dominated by slender vestimentiferan tube worms in the family Ridgeiidae. Tube worm tubes were heavily fouled by paralvinellid and ampharetid polychaetes. We noted local differences in the composition of macrofaunal species associated with different tube worm clumps: at some sites, the distal ends of the tube worm tubes were ornamented with one or two white-bodied anemones with pink-fringed tentacles. Some clumps of tube worms were colonized by abundant lepetodrilid limpets and small, coiled gastropods; in other clumps, these mollusks were rare or absent. At the vent site designated 6X, three arthropod taxa were abundant within the tube worm clumps. Pycnogonids of all size classes, including gravid females and juveniles, were found attached to tube worm tubes; red copepods were found in mucous and washings of tubeworm clumps; large tanaid crustaceans encased in thin mud tubes also fouled tube worm tubes. The tanaids belong to the family Neotanaidae and represent, to our knowledge, the first recorded collection of a vent-associated tanaid species. Tanaids of all size classes were collected, from small, ~3-mm juveniles to large, ~1.5-cm adult males. Polynoid polychaetes were common but not abundant in all clumps of tube worms. Copious mucous secretions by paralvinellid polychaetes at the base of tubeworm clumps appeared to form a matrix within which filamentous bacteria thrived. Populations of paralvinellid worms were also observed in tubes within sulfides adjacent to active vents where tube worms were absent. Disklike sponges (1–2 cm diameter) were abundant on sulfides surrounding areas of hydrothermal venting. A sample of fossilized worm tubes embedded in iron oxides near a 110°C vent was colonized by dense populations of small ampharetid polychaetes and aplacophoran mollusks. Small vestimentiferan tube worms, anemones, limpets, pycnogonids, and folliculinid protozoans also colonized this substrate. Galatheid crabs reached their greatest densities in the vicinity of active venting on sulfide mounds.

One site of low-temperature venting in soft sediment was observed at the base of a sulfide mound. The area (<10 m^2) was colonized by elongate vesicomyid clams identified as *Calyptogena phaseoliformis* by R.D. Turner. Live clams were oriented half-buried, with their long axes at about 45° angles with respect to the sediment surface. The exposed anterior ends of live clams were often colonized by anemones. A number of empty valves lying on the surface of the sediment were noted, but no obvious explanation for mortality was observed. Qualitative samples of the sediment at this site contained large populations of undescribed species of orbiniid, ampharetid, and spionid polychaetes.

"Inactive" sulfide mounds were often colonized by remarkably dense concentrations of suspension-feeding deep-sea fauna. This fauna included aggregations of large, solitary tunicates plus brisingid seastars, crinoids, sponges, anemones, and brachiopods. While these organisms are common elements of the local nonvent fauna (Carey, this volume), their unusual abundance on sulfide mounds suggests that there may be some low-level venting of hydrothermal fluids at these "inactive" sites that supports chemosynthetic production within the overlying water column, or that the acid-labile sulfides are mobilized by microbial activity to support primary prodution. Alternatively, the topographic relief of the sulfide mounds may modify the local flow regime, concentrating suspended particulates on which the biota feed.

Nonvent soft sediments were heavily bioturbated. Echinoderms, including asteroids, ophiuroids, holothurians, and urchins, were conspicuous elements of the megafauna, as were xenophyophores and anemones. Occasional pennatulaceans and galatheid squat lobsters were also observed. The infauna of the sediments was dominated by several small species of polychaetes and an isopod.

Fauna of nonvent basalt substrates included sponges, brachiopods, and sabellid polychaetes.

Faunal Affinities

Specific identifications of most of the Gorda Ridge biota remain to be confirmed by taxonomic specialists. Nevertheless, we can consider the general nature of the fauna and relate it to the fauna of vent communities described from other oceanic spreading centers. Not unexpectedly, the fauna of NESCA hydrothermal vents most closely resembles that of Juan de Fuca and Explorer Ridge vent communities. The Northeast Pacific assemblage of Tunnicliffe (1988) can be extended to embrace the Gorda Ridge fauna. Ridgeiid vestimentiferans, lepetodrilid limpets, paralvinellid, polynoid and ampharetid polychaetes, pycnogonid arthropods, and the small coiled gastropod are important components of this assemblage. Common and abundant faunal types present at vents on Juan de Fuca and Explorer Ridges but so far characterized as absent or rare at the NESCA site on Gorda Ridge include maldanid polychaetes and ostracods. The majid crab, *Macrooregonia macrochira*, while not strictly a vent-associated species, was not observed at NESCA, though it is common at Juan de Fuca and Explorer vents. Tanaid crustaceans occur at the NESCA site but have not been noted at Juan de Fuca or Explorer Ridge vents.

The fauna of the low-temperature soft-sediment site at NESCA, dominated by *Calyptogena phaseoliformis*, is not part of Tunnicliffe's Northeast Pacific assemblage. The same species of clam is known from cold-water seeps in soft sediment off Japan (Metivier et al., 1986; Juniper and Sibuet, 1987) and was recently identified from sites in Monterey Canyon off California (R.D. Turner, personal communication). Soft-sediment vent fauna is best known from Guaymas Basin (Gulf of California), where the clam *Vesicomya gigas* (previously identified as *Calyptogena pacifica* and reclassified by R.D. Turner on the basis of additional material) is abundant (Grassle et al., 1985). Guaymas sediments are infused with petroleum hydrocarbons formed from high-temperature cracking of recent organic material (Simoneit and Lonsdale, 1982); an odor of petroleum was detected in Gorda vent sediment, but it was neither as strong nor as pervasive as that of Guaymas material. The single, qualitative sample of Gorda vent sediment shares no common infaunal taxa with the samples of vent sediment from Guaymas Basin, and the extensive bacterial mats associated with the sediment surface at Guaymas are absent at the Gorda soft-sediment vent site.

Acknowledgments. Exploration of Gorda Ridge was funded by the Gorda Ridge Technical Task Force and the National Science Foundation. We thank Greg McMurray for his organizational efforts and his enthusiastic support. We are also indebted to the pilots and crews of the submersibles *Alvin* and *Sea Cliff* and of their support vessels *Atlantis II* and *Laney Chouest*.

References

Grassle JF, Brown-Leger S, Morse-Porteous L, Petrecca R, Williams I (1985) Deep-sea fauna of sediments in the vicinity of hydrothermal vents. Biol Soc Wash Bull 6:443–452.

Juniper SK, Sibuet M (1987) Cold seep benthic communities in Japan subduction zones: Spatial organization, trophic strategies and evidence for temporal evolution. Marine Ecol Prog Ser 40:115–126.

Metivier B, Okutani T, Ohta S (1986) *Calyptogena* (*Ectenagena*) *phaseoliformis* n. sp., an unusual vesicomyid bivalve collected by the submersible *Nautile* from abyssal depths of the Japan and Kurile Trenches. Venus 45:75–86.

Simoneit B, Lonsdale P (1982) Hydrothermal petroleum in mineralized mounds at the seabed of Guaymas Basin. Nature 295:198–202.

Tunnicliffe V (1988) Biogeography and evolution of hydrothermal-vent fauna in the eastern Pacific Ocean. Proc R Soc Lond B 233:347–366.

Part V
Conclusions

22—Results of the Gorda Ridge Symposium

By Gregory R. McMurray

Background

On March 10, 1983, President Ronald Reagan signed the Exclusive Economic Zone Proclamation, giving the United States sovereign rights to all resources within 200 nautical miles of its shorelines. Later that month, on March 28, the Minerals Management Service issued in the *Federal Register* a "Scoping Notice to Prepare an Environmental Impact Statement" for a proposed lease sale for polymetallic sulfides on the Gorda Ridge. Polymetallic sulfide deposits at other locations contain important concentrations of metals including copper, zinc, lead, silver, and gold. A Draft Environmental Impact Statement (DEIS) for the 40-million-acre lease sale was released in December 1983 (Zippin, 1983). The lease area is shown in Figure 22.1.

In the wake of negative public reaction to the proposed lease sale and DEIS, Secretary of the Interior William Clark and Oregon Governor Victor Atiyeh agreed to the creation of a federal-state working group. The primary purpose of the group was to perform a technical analysis of the economic, environmental, and engineering aspects of a possible leasing program. There were six other points in the agreement: (1) a reduction of any future lease area from 40 million acres to no more than 4 million acres; (2) indefinite postponement of the lease sale; (3) deletion of the nearshore portion of the possible lease area covering the physical continental shelf and slope; (4) a commitment to a "tiered' EIS process; (5) sharing of all nonproprietary data with the public; and (6) removal from consideration that part of the area offshore of a state that would choose to withdraw from the working group. The state of California later made a similar arrangement and joined the working group in June 1984.

The working group was officially named the Gorda Ridge Technical Task Force, and its scope was developed during meetings in May and August of 1984. The major limitation in scope was that the task force would not deal directly with policy issues but would provide only technical counsel to the secretary and governors. Working guidelines institutionalizing the structure and function of the task force were developed and adopted. The Department of the Interior and the states of Oregon and California were each given a membership of five, with a single vote through a chairperson. Twenty nonvoting advisers were named from federal agencies and academia to complete the technical breadth of the task force. The members and advisers were each assigned to one of three subgroups organized by discipline: physical/geotechnical, living resources, and socioeconomic/coastal impacts.

At the August 1984 meeting, the task force subgroups submitted reports summarizing their initial research recommendations. These subgroup reports set the tone for the ensuing task force research programs which resulted in this volume. The physical/geotechnical subgroup recommended three general activities: (1) a shipboard gathering of oceanographic and resource data; (2) definition of the resource base and generation of the appropriate development scenarios; and (3) valuation of the re-

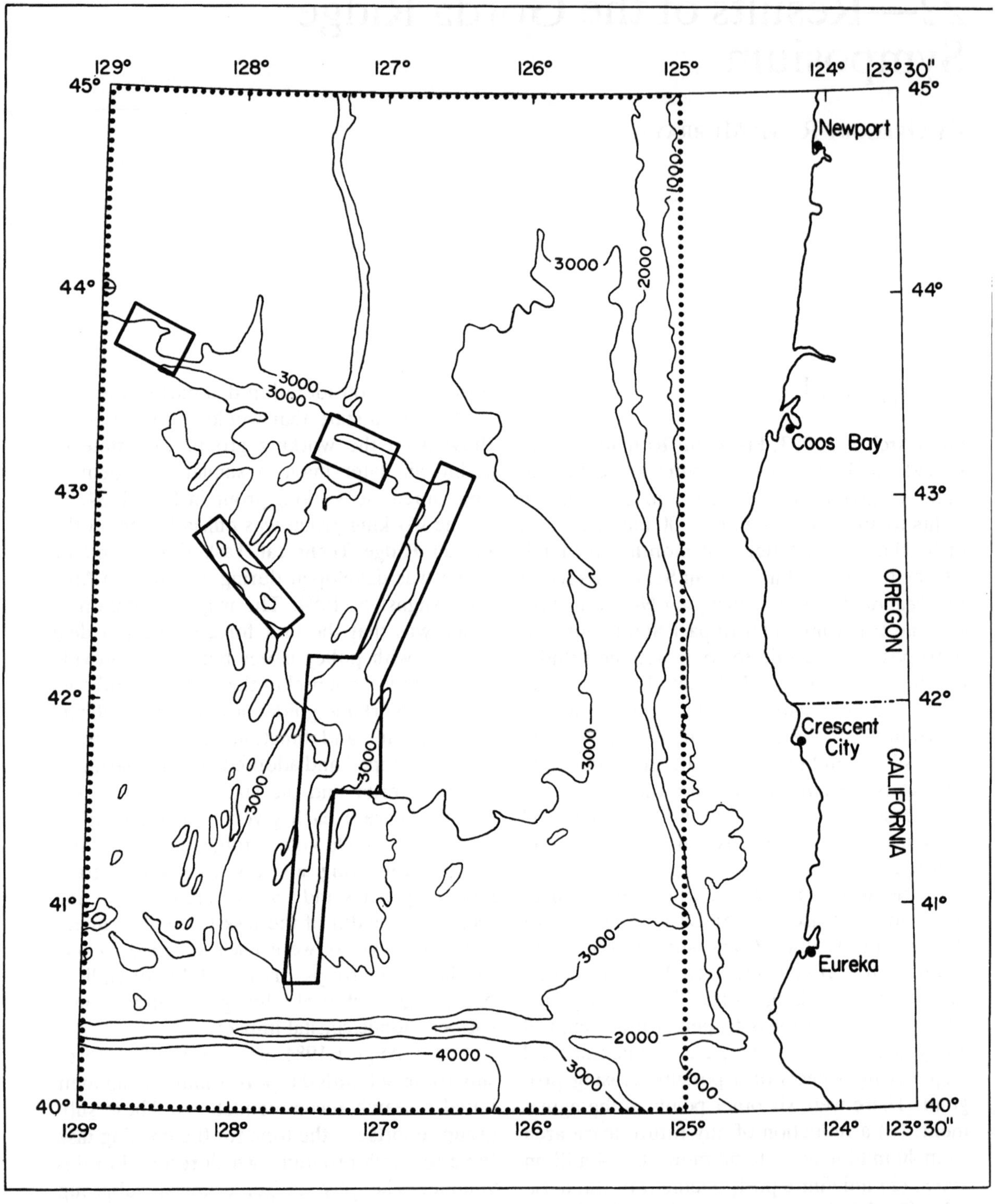

FIGURE 22.1. Map of Gorda Ridge and surrounding area showing original lease area of 40 million acres (dotted line) and task force study area of 2.65 million acres (solid lines). (From McMurray, 1986.)

source. The field studies were also to include water column studies, sediment studies, near-bottom side-scan sonar coverage, deep-sea photography, bottom sampling, and water column current velocity measurements. The living resources subgroup also recommended three general activities: (1) a review and synthesis of existing literature; (2) a symposium with discussion of the benthos of midocean ridges, methodologies for research, and vulnerability of the ridge biota to mining-related impacts; and (3) preliminary field studies. The preliminary field studies were to include benthic photo surveys and sea bird and marine mammal observations. The socioeconomic/coastal impacts subgroup recommended two studies: an analysis of public costs associated with infrastructure development, and a land use analysis for port development.

The research program pursued by the task force generally followed the original subgroup recommendations. A study area of approximately 2.65 million acres was defined (Fig. 22.1). The major objective of the program was to establish the presence or absence of potentially significant polymetallic sulfide deposits on Gorda Ridge. The secondary objective was to acquire as much baseline data as possible in support of a possible future EIS. To meet these objectives, the task force used funds from MMS to complement and augment the existing National Oceanic and Atmospheric Administration (NOAA) and U.S. Geological Survey (USGS) programs with tasks contracted to academic researchers. All of the shipboard studies recommended by the physical/geotechnical subgroup were initiated, with the exception of the current studies, and all of the living resources subgroup recommendations were begun or accomplished. The socioeconomic/coastal impacts subgroup recommendations were deferred until the resource studies could be completed.

The Gorda Ridge Symposium

The success of the 1985 and 1986 resource reconnaissance program on Gorda Ridge was great enough that the task force decided to hold a public conference to present the results. The Gorda Ridge Symposium was intended to use the presentation and discussion of data and related technical information to provide future direction for the task force. The papers presented at the symposium were used as a backdrop for five subsequent workshops for peer discussions of topics germane to possible future leasing of Gorda Ridge. Most importantly, the workshops were charged with making recommendations for future task force activities.

The Gorda Ridge Symposium was held in Portland, Oregon, May 11–13, 1987. Of the 94 symposium registrants, 42 represented government agencies, 27 represented academia, 22 represented industry, and 3 represented environmental interest groups. The industry attendees were equally divided between the mineral industries and the ocean services industries. Such limited attendance by the exploration and mining industries was part of a continuing expression of disinterest in a Gorda Ridge lease sale for polymetallic sulfide minerals.

Workshop Results

Exploration and Mining Technologies Workshop
Chaired by Michael J. Cruickshank and Sylvia A. Earle

Participants in this workshop were asked to respond in writing to three queries with respect to Gorda Ridge: (1) rank the first three needs for exploration technology; (2) rank the first three needs for mining technology; and (3) identify any other scientific goals for which new technologies are needed.

Exploration Technology

There was a consensus that tools are needed to provide a three-dimensional characterization of the target mineral deposits. In order of priority, the four most needed technologies are drilling tools, electromagnetic geophysical techniques, better (and more) remotely operated vehicles (ROVs), and a seabed navigation survey. Of a lower priority were extended submersible capability, improved high-frequency side-scan sonar, tools for measuring the engineering properties of rocks in situ, improved sample recovery, improved high-frequency seismic systems, visual aids to sampling, and deep-towed geological and geophysical capabilities.

Mining Technology

The group agreed that, in advance of any knowledge of the three-dimensional characteristics of any ore body in Gorda Ridge, specific tool development was not yet a need. General technology needs included ore removal systems, improved leaching techniques, robotics, at-sea processing, rock fragmentation below 3,000 m water depth, and seabed heavy work vehicles.

Other Scientific Goals

The Technologies Workshop also discussed the technology advances needed in a regional approach to characterize and understand the hydrothermal systems associated with Gorda Ridge. Such a regional approach would require study of the mechanisms for plate accretion, mineral deposit formation, and chemical, bacterial, and thermal budgets in the rocks, sediments, and waters on and off axis. Specific technological developments include at-sea or in situ chemical analysis, heat flow measurement at sediment-free locations, techniques for identifying and inventorying biota, and techniques for measuring bacterial activity.

Issues Relating to Technology Availability

The group identified a number of issues that may greatly affect the availability of technology or state-of-the-art data. These issues include classification of precision bathymetric data by the Department of Defense, cooperative efforts between government agencies and industry, allocation of Gorda Ridge as an Ocean Drilling Program site, and cooperation between Canada and the United States in exploration and technology development programs.

Comparative Vent Ecology Workshop *Chaired by Craig R. Smith and William G. Pearcy*

Two sets of research needs and recommendations were formulated by the workshop: scientific needs for the study of vent ecology in general, and research priorities specifically for Gorda Ridge.

General Vent Ecology

The most pressing needs in the study of hydrothermal vent ecology in areas such as Juan de Fuca and Gorda Ridges, Guaymas Basin, and the East Pacific Rise include three topics. First is the definition of spheres of influence. This involves addressing the question "To what distance vent systems, particularly their productivity, impact the ecology of the surrounding seafloor?" Answering this question requires (1) detailed mapping of ridge segments; (2) location and flow rate characterization of all vents within a field; (3) determination of the densities and trophic structure of the nonvent faunas of the surrounding seafloor; and (4) construction of carbon budgets distinguishing vent from nonvent organic carbon for communities of various distance from hydrothermal vents. This last objective is probably best addressed through use of stable isotope analysis combined with sediment trap deployments.

The second need is the assessment of faunal dispersal rates within and between vent fields. This involves addressing questions such as "What are the source populations for new recruits at vents?" and "What are the levels of gene flow between vents?" These questions are of relevance to understanding colonization processes and faunal evolution and to assessing environmental impacts at hydrothermal vents. Some specific topics worth addressing include (1) the dynamics and faunal-dispersal effects of vent plumes, (2) the velocity and direction of water currents in vent fields, and (3) levels of genetic similarity between vent sites.

The third need is the elucidation of biological succession at vents. The main question to be addressed here is "What are the nature and rates of biological changes at vents, and what are the physical/chemical correlates?" Specific data collection needs to address this question include (1) time-series observations at month-long time scales and greater of vent flow rates and chemistry, community structure, and faunal recruitment; (2) better documentation of life histories to identify important time scales; (3) further biochemical documentation of species niches; and (4) in situ experimentation to address the nature of biological and physical/biological interactions.

Gorda Ridge Topics

The priorities for the study of hydrothermal vent faunae on Gorda Ridge are (1) to locate active hydrothermal vents; (2) to describe the general faunal and microbial communities at a number of sites which will require submersibles, detailed mapping, and organism collection; (3) to determine the levels of faunal and microbial similarity (taxonomic, genetic, and physiological) to other vent systems; and (4) to elucidate the trophic structure of and construct carbon budgets for the communities.

The workshop also outlined the minimum amount of information that should accompany organism description and/or collection at any Gorda Ridge site. These include temperature where the organism was observed or collected, a photograph and verbal description of the microhabitat, and a transponder fix for the observation/collection site and survey photographs of the general area, with good benchmarks and altitude data. The workshop also recommended that a general pictorial atlas of vent organisms be assembled for use of non-biologists in describing the fauna at hydrothermal vent sites.

Resource Assessment Workshop *Chaired by James F. Franklin and Joseph L. Ritchey*

The main objective of the Resource Assessment Workshop was to formulate scientific recommendations for completing the resource reconnaissance, the initial step of a resource assessment, on Gorda Ridge. The resource reconnaissance requires information on critical geological characteristics and a sufficiently detailed regional geology. The regional aspects include five factors: stratigraphic control, structural control, "marginal" indicators, heat source, and alteration.

Subsequent to the 1986 Gorda Ridge dive program, only general observations about the Escanaba Trough deposits can be made (see Morton et al., this volume; Holmes and Zierenberg, this volume). The areal extent of the deposits is unknown, but it is greater than 100 to 200 m in diameter for a single lateral dimension. The thickness is also unknown, but it is greater than 15 m in one case and apparently greater than 30 m in the case of a morphologically "young" dome (see Holmes and Zierenberg, this volume). Finally, the grade, too, is unknown. The distribution of valuable metals in existing Gorda Ridge samples is bimodal (see Morton et al., this volume).

The group recommended three areas of study to complete the resource reconnaissance. First, mapping and surficial sampling of the domes at the NESCA and SESCA sites should be completed. This should include analysis of cobalt, rhenium, tin, and platinum group metals and will require improved submersible sampling gear and positioning for sample locations. Second, a drilling program should be conducted to obtain a three-dimensional picture of the NESCA and SESCA sites. The group noted that the USGS and the Geological Survey of Canada have made future plans to employ a sea floor drill (see Ryall, this volume) in Escanaba Trough on Gorda Ridge and at Middle Valley on Explorer Ridge, and that proposals are being submitted to the Ocean Drilling Program for a site in Escanaba Trough. Last, the group recommended study of analogous deposits, including Middle Valley, Bathurst, the Iberian Pyrite Belt, and the Besshi (see Koski, this volume).

Potential Ecological Impacts Workshop
Chaired by J. Frederick Grassle and Gary L. Taghon

This workshop was designed as a follow-up to the Comparative Vent Ecology Workshop held on the previous day, with a focus on applied research for establishing baseline conditions and assessing potential ecological impacts. An attempt was made to limit the scope to the Gorda Ridge deposits and to address the needs of a future environmental impact statement for development of the ridge for polymetallic sulfides. Research recommendations were made in three areas: general baseline data collection, toxicity and impact assessment studies, and cooperative, or "piggyback," studies.

The group recommended six general baseline data collection needs: (1) a circulation study in the vicinity of the ridge axis; (2) broad-scale biological characterization of both hard- and soft-bottom nonvent benthic habitats; (3) studies of the migration routes of tuna and salmon; (4) life history studies of vent fauna and/or similar animals; (5) establishing the baseline levels of metals in the vent benthos; and (6) modeling studies of hydrothermal vent plumes. The needs for toxicity and impact studies include detailed data on the chemistry of sulfide deposits, recolonization studies, mining and processing scenarios, and studies on biostimulation of vent species. Suggested cooperative or piggyback studies include biological data collection during the remainder of the resource reconnaissance with an increased level of biological data collection when or if hydrothermal vents are discovered, and recolonization studies subsequent to Ocean Drilling Program activity on the ridge.

Leasing and Development Workshop
Chaired by Ray G. Hunter and Robert F. Dill

The major objective of the Leasing and Development Workshop was to provide for a dialogue between the Minerals Management Service (MMS), as the federal government leasing agency, and representatives of the mining industry. Because industry attendance to the symposium was so limited, the workshop resulted in a position statement by MMS regarding its approach to leasing, and statements for and against this approach by industry representatives.

The MMS position was presented by Reid T. Stone, then program director for the Office of Strategic and International Minerals. MMS derives its leasing authority for seafloor hard minerals under the Outer Continental Shelf Lands Act (OCSLA). Regulations for hard-minerals leasing and development are currently being promulgated by MMS.[1] A decision whether or not to lease Hawaiian cobalt-enriched manganese crusts will not be made until a decision document has been prepared for the Secretary of the Interior. MMS will release a draft EIS jointly with the state of Hawaii. MMS is not ready at the time of this symposium to proceed with an EIS for a Gorda Ridge lease sale.

A statement of support for hard-minerals leasing under OCSLA was presented by Robert F. Dill, representing Geomarex (Geomarex has applied for a leasing permit for aragonite in the Gulf of Mexico). On-land costs for reclamation after phosphate mining have increased dramatically in recent years, so marine mining for phosphates may be economic at the present time. New legislation covering hard-minerals leasing on the outer continental will take at least 5 years; this is too long to suit Geomarex's needs in this case.

A statement opposing hard-minerals leasing under OCSLA was made by TS Ary from a mining industry perspective. There are three major problems with the existing regime. First, the OCSLA lease term is too short; it does not fit the typical mine life of this type of project. Second, bonus bidding[2] is not appropriate for hard minerals in a case where up to $500 million could be spent to establish a venture on

[1] The regulations for hard-minerals activities under OCSLA are now in place (Department of the Interior, 1988, 1989a, b).

[2] The OCSLA provides for lease offerings by a competitive cash bonus bidding system in which sealed bids are submitted to MMS.

Gorda Ridge. Third, the financial risk factor on this type of leasing is too high to be attractive to industry. Industry wants a first-come first-served "concession" system, with safeguards to keep speculators out. Industry can afford to wait the 5 years or more that it might take until new legislation can be put into practice to regulate hard-minerals leasing on the outer continental shelf.

Events Subsequent to the Gorda Ridge Symposium

The Gorda Ridge Symposium resulted in two major actions during the period 1987–1988. First, the symposium results were summarized in a report from the cochairs of the Task Force to the Secretary of the Interior and the governors of Oregon and California. The report included the following paragraph:

> On May 11–13, 1987, the Gorda Ridge Symposium was held in Portland, Oregon, to present the results of the scientific studies to industry and the public and to provide a forum for discussion of the future significance of the polymetallic sulfide deposits. Industry representatives attending the symposium expressed the opinion that it would be premature to consider leasing the deposits at this time because information available on the quality, composition and extent of the deposits is insufficient to permit meaningful estimates of their future economic potential. They also emphasized that economic conditions in the industry would have to improve significantly before industry could begin exploring or funding the research required to develop the drilling and mining technology required to exploit the deposits. Nevertheless, industry representatives strongly encouraged continued study of the Gorda Ridge because they believe the deposits constitute a resource that could well be developed sometime in the future when onshore resources are depleted or otherwise unavailable.

This lack of immediate interest by industry in Gorda Ridge caused MMS to publish a *Federal Register* notice in March 1988 withdrawing the draft EIS, thus ending consideration of a Gorda Ridge lease sale in the immediate future.

Second, the task force met immediately following the symposium and discussed recommendations for completing the resource reconnaissance phase of exploration on Gorda Ridge. MMS responded positively to the idea of a 1988 dive program on the ridge, and the program was planned during the fall and winter of 1987–1988. The U.S. Navy agreed to make the deep submersible vehicle *Sea Cliff* available to the task force for a 30-day period during the summer of 1988, and a two-leg cruise was planned, with the first leg to conduct 14 dives on the SESCA and NESCA sites in Escanaba Trough, and the second leg to conduct 10 dives on the northern Gorda Ridge, in the vicinity of GR14.

1988 Gorda Ridge Dive Program Results

There were actually two dive programs conducted on Gorda Ridge during 1988. John Edmond, of the Massachussetts Institute of Technology (MIT), led an *Alvin* dive expedition during June 1988. The MIT group, with personnel from Woods Hole Oceanographic Institution (WHOI) and the USGS, made 10 dives on the Escanaba Trough on southern Gorda Ridge with the intent of sampling active hydrothermal vents. Vents were discovered at the NESCA site, and the attendant vent benthos was photographed and sampled (see Van Dover, Grassle and Boudrias, this volume).

The task force *Sea Cliff* dive program began in August 1988. The first leg of the expedition was able to complete only four of the 14 scheduled dives before foul weather precluded diving for the remainder of the leg. The first dive was able to reoccupy the vent field discovered by the *Alvin* expedition, and subsequent dives were made on both NESCA and SESCA. The combined data from the *Alvin* and *Sea Cliff* dives appear to have substantially advanced the state of knowledge about the Escanaba Trough deposits.

The second leg of the task force *Sea Cliff* dive program succeeded in conducting five of the allotted 10 dives on the northern Gorda Ridge. The first dive discovered a large active vent field on the east wall of the axial valley, as predicted by Clague and Rona (this volume). The vent field had some "black smoker' activity and attendant vent fauna.

References

Department of the Interior (1988) Geological and geophysical exploration of the outer continental shelf; prospecting for minerals other than oil, gas, and sulphur; final rule. Fed Reg 53:25242–25260.

Department of the Interior (1989a) Outer continental shelf minerals and rights-of-way management; general leasing of minerals other than oil, gas, and sulphur in the outer continental shelf; final rule. Fed Reg 54:2042–2056.

Department of the Interior (1989b) Operations in the outer continental shelf for minerals other than oil, gas, and sulphur; postlease discovery, delineation, development and production; final rule. Fed Reg 54:2058–2080.

McMurray G (1986) The Gorda Ridge Technical Task Force: A cooperative federal-state approach to offshore mining issues. Marine Mining 5:467–475.

Zippin JP (1983) Draft environmental impact statement, proposed polymetallic sulfide lease offering, Gorda Ridge area, offshore Oregon and Northern California. Minerals Management Service.

Index

MIX
Papier aus verantwortungsvollen Quellen
Paper from responsible sources
FSC® C105338

If you have any concerns about our products,
you can contact us on
ProductSafety@springernature.com

In case Publisher is established outside the EU,
the EU authorized representative is:
Springer Nature Customer Service Center GmbH
Europaplatz 3, 69115 Heidelberg, Germany

Printed by Libri Plureos GmbH
in Hamburg, Germany